BARRETT'S ESOPHAGUS

BARRETT'S ESOPHAGUS

Emerging Evidence for Improved Clinical Practice

Edited by

DOUGLAS K. PLESKOW

Chief, Clinical Gastroenterology, Beth Israel Deaconess Medical Center
Associate Clinical Professor of Medicine, Harvard Medical School
Boston, MA, United States

TOLGA ERIM

Director of Endoscopy, Department of Gastroenterology, Digestive Disease Center, Cleveland Clinic Florida,
Weston, FL, United States

AMSTERDAM • BOSTON • HEIDELBERG • LONDON
NEW YORK • OXFORD • PARIS • SAN DIEGO
SAN FRANCISCO • SINGAPORE • SYDNEY • TOKYO

Academic Press is an imprint of Elsevier

Academic Press is an imprint of Elsevier
125, London Wall, London EC2Y 5AS, UK
525 B Street, Suite 1800, San Diego, CA 92101-4495, USA
50 Hampshire Street, 5th Floor, Cambridge, MA 02139, USA
The Boulevard, Langford Lane, Kidlington, Oxford OX5 1GB, UK

Notices
Knowledge and best practice in this field are constantly changing. As new research and experience broaden our understanding, changes in research methods, professional practices, or medical treatment may become necessary.

Practitioners and researchers may always rely on their own experience and knowledge in evaluating and using any information, methods, compounds, or experiments described herein. In using such information or methods they should be mindful of their own safety and the safety of others, including parties for whom they have a professional responsibility.

To the fullest extent of the law, neither the Publisher nor the authors, contributors, or editors, assume any liability for any injury and/or damage to persons or property as a matter of products liability, negligence or otherwise, or from any use or operation of any methods, products, instructions, or ideas contained in the material herein.

Library of Congress Cataloging-in-Publication Data
A catalog record for this book is available from the Library of Congress.

British Library Cataloguing-in-Publication Data
A catalogue record for this book is available from the British Library.

ISBN: 978-0-12-802511-6

For Information on all Academic Press publications
visit our website at https://www.elsevier.com/

Working together
to grow libraries in
developing countries

www.elsevier.com • www.bookaid.org

Publisher: Mica Haley
Acquisition Editor: Stacy Masucci
Editorial Project Manager: Samuel Young
Production Project Manager: Melissa Read
Designer: Matthew Limbert

Dedication

I would like to dedicate this to my family.
My daughters Sara, Heather, and Rebecca for their love and dedication.
My wife Randi Pleskow, MD, for her love and affection and always standing by my side.
—Douglas K. Pleskow

For Viviana, how lucky I am to share every day with you.
—Tolga Erim

Contents

List of Contributors

Kamar Belghazi Department of Gastroenterology and Hepatology, Academic Medical Center, Amsterdam, The Netherlands

Jacques J. Bergman Department of Gastroenterology and Hepatology, Academic Medical Center, Amsterdam, The Netherlands

Tyler M. Berzin Division of Gastroenterology, Center for Advanced Endoscopy, Department of Medicine, Beth Israel Deaconess Medical Center, Harvard Medical School, Boston, MA, United States

Kathryn Boom Department of Surgery, Houston Methodist Hospital, Houston, TX, United States

Amitabh Chak Advanced Technology & Innovation Center of Excellence, Division of Gastroenterology, Case Western Reserve University, Cleveland, OH, United States

Ram Chuttani Center for Advanced Endoscopy, Department of Medicine, Beth Israel Deaconess Medical Center, Harvard Medical School, Boston, MA, United States

Yaşar Çolak Department of Gastroenterology, Göztepe Education and Research Hospital, Istanbul Medeniyet University, Kadıköy, Istanbul, Turkey

John A. Dumot University Hospitals Digestive Health Institute, Case Western Reserve University, Cleveland, OH, United States; University Hospitals Ahuja Medical Center, Beachwood, OH, United States

Brian J. Dunkin Section of Endoscopic Surgery, Institute for Academic Medicine, Houston Methodist Hospital, Houston, TX, United States; Department of Surgery, Houston Methodist Hospital, Houston, TX, United States

Tolga Erim Department of Gastroenterology, Digestive Disease Center, Cleveland Clinic Florida, Weston, FL, United States

Rebecca C. Fitzgerald Medical Research Council Cancer Unit, University of Cambridge, Cambridge, United Kingdom

Alexander M. Frankell Medical Research Council Cancer Unit, University of Cambridge, Cambridge, United Kingdom

Martin Goetz Innere Medizin I, Universitätsklinikum Tübingen, Tübingen, Germany

Jeffrey D. Goldsmith Department of Pathology, Beth Israel Deaconess Medical Center, Harvard Medical School, Boston, MA, United States

Michalina J. Gora Wellman Center for Photomedicine, Massachusetts General Hospital, Boston, MA, United States; ICube Laboratory, Centre National de la Recherche Scientifique, University of Strasbourg, Strasbourg, France

Jennifer T. Higa Department of Medicine, University of Washington School of Medicine, Seattle, WA, United States

Joo Ha Hwang Gastroenterology Section, Harborview Medical Center, University of Washington School of Medicine, Seattle, WA, United States

Irving Itzkan Center for Advanced Biomedical Imaging and Photonics, Department of ObGyn and Reproductive Biology, Beth Israel Deaconess Medical Center, Harvard Medical School, Boston, MA, United States

Annalise C. Katz-Summercorn Medical Research Council Cancer Unit, University of Cambridge, Cambridge, United Kingdom

Umar Khan Center for Advanced Biomedical Imaging and Photonics, Department of ObGyn and Reproductive Biology, Beth Israel Deaconess Medical Center, Harvard Medical School, Boston, MA, United States

Gaurav Kistangari Department of Internal Medicine, Cleveland Clinic, Cleveland, OH, United States

Deepa T. Patil Cleveland Clinic Lerner College of Medicine, Cleveland, OH, United States; Department of Pathology, Robert J. Tomsich Pathology and Laboratory Medicine Institute, Cleveland Clinic, Cleveland, OH, United States

Lev T. Perelman Center for Advanced Biomedical Imaging and Photonics, Department of ObGyn and Reproductive Biology, Beth Israel Deaconess Medical Center, Harvard Medical School, Boston, MA, United States; Center for Advanced Endoscopy, Department of Medicine, Beth Israel Deaconess Medical Center, Harvard Medical School, Boston, MA, United States

Douglas K. Pleskow Clinical Gastroenterology, Beth Israel Deaconess Medical Center, Boston, MA, United States; Center for Advanced Endoscopy, Department of Medicine, Beth Israel Deaconess Medical Center, Harvard Medical School, Boston, MA, United States

Amareshwar Podugu Department of Gastroenterology, Digestive Disease Center, Cleveland Clinic Florida, Weston, FL, United States

Roos E. Pouw Department of Gastroenterology and Hepatology, Academic Medical Center, Amsterdam, The Netherlands

Le Qiu Center for Advanced Biomedical Imaging and Photonics, Department of ObGyn and Reproductive Biology, Beth Israel Deaconess Medical Center, Harvard Medical School, Boston, MA, United States

Nikhiel B. Rau Division of Gastroenterology, Beth Israel Deaconess Medical Center, Boston, MA, United States

Mandeep Sawhney Center for Advanced Endoscopy, Department of Medicine, Beth Israel Deaconess Medical Center, Harvard Medical School, Boston, MA, United States

Alison Schneider Department of Gastroenterology, Digestive Disease Center, Cleveland Clinic Florida, Weston, FL, United States

Ebubekir Şenateş Department of Gastroenterology, Göztepe Education and Research Hospital, Istanbul Medeniyet University, Istanbul, Turkey

Guillermo J. Tearney Wellman Center for Photomedicine, Massachusetts General Hospital, Boston, MA, United States; Department of Pathology, Massachusetts General Hospital, Boston, MA, United States; Harvard-MIT Division of Health Sciences and Technology, Boston, MA, United States

Prashanthi N. Thota Center of Excellence for Barrett's Esophagus, Department of Gastroenterology and Hepatology, Cleveland Clinic, Cleveland, OH, United States

George Triadafilopoulos Stanford University School of Medicine, Stanford, CA, United States

Vladimir Turzhitsky Center for Advanced Biomedical Imaging and Photonics, Department of ObGyn and Reproductive Biology, Beth Israel Deaconess Medical Center, Harvard Medical School, Boston, MA, United States

Edward Vitkin Center for Advanced Biomedical Imaging and Photonics, Department of ObGyn and Reproductive Biology, Beth Israel Deaconess Medical Center, Harvard Medical School, Boston, MA, United States

Fen Wang Center for Advanced Endoscopy, Department of Medicine, Beth Israel Deaconess Medical Center, Harvard Medical School, Boston, MA, United States

Eric U. Yee Department of Pathology, Beth Israel Deaconess Medical Center, Harvard Medical School, Boston, MA, United States

Lei Zhang Center for Advanced Biomedical Imaging and Photonics, Department of ObGyn and Reproductive Biology, Beth Israel Deaconess Medical Center, Harvard Medical School, Boston, MA, United States

Preface

In 1999, I started to treat Barrett's esophagus with photodynamic therapy. I was able to offer an endoscopic method to treat patients with high-grade dysplasia. The procedure was complicated but it offered our nonsurgical patients an endoscopic method to treat high-grade dysplasia. The patients were pleased that there was a way to ablate the disease without surgery. Unfortunately photodynamic therapy was not an optimal therapy. Minor and major complications were not infrequent. Buried Barrett's was also a frequent occurrence. Since that time we have made major advances in our understanding of Barrett's. Dr Erim and I identified a need to provide a reference for those interested in the treatment of patients with this disease.

Our goal in bringing together this book was to provide a framework for clinicians, clinical researchers, and basic scientists. We hoped to provide the practicing clinician, and the fellow in-training with an in-depth text which provides the basics and the state-of-the-art concepts in one place. At the end of each chapter, our experts provide where they believe the future research will be focused, providing invaluable insight into the minds of those who will shape the advances of the future.

This book brings the expertise of world leaders in the field of Barrett's Esophagus. Each author has provided a thorough review of their area of expertise. Within each chapter, there has been a special emphasis on emerging evidence with a focus on where the future research and clinical practice will be headed.

We would like to thank the authors for their commitment to providing their time and expertise. Each author is a recognized expert in their respective field. Special thanks to Dr. Helen Shields of Harvard Medical School for providing the photomicrograph on the front cover of this book. She is an outstanding resource for everything related to gastroenterology. In addition, we would like to thank the team at Elsevier for their skill in preparing this book. Lastly, I would like to thank Tolga Erim for his leadership, dedication, and hard work to this endeavor.

Douglas K. Pleskow
Chief, Clinical Gastroenterology,
Beth Israel Deaconess Medical Center,
Associate Clinical Professor of Medicine,
Boston, MA, United States

A Disease Entity Is Identified

Yaşar Çolak[1], Tolga Erim[2] and Douglas K. Pleskow[3]

[1]Department of Gastroenterology, Göztepe Education and Research Hospital, Istanbul Medeniyet University, Kadiköy, Istanbul, Turkey [2]Department of Gastroenterology, Digestive Disease Center, Cleveland Clinic Florida, Weston, FL, United States [3]Clinical Gastroenterology, Beth Israel Deaconess Medical Center, Boston, MA, United States

1.1 INTRODUCTION

Barrett's esophagus (BE) is the presence of metaplastic columnar epithelium in the lower portion of the esophagus, which is normally lined with stratified squamous epithelium. The main cause of the disease is theorized to be reflux esophagitis developed due to chronic acid exposure as a result of symptomatic or asymptomatic gastroesophageal reflux. In addition, the disease is clinically significant as a major risk factor for esophageal adenocarcinoma (EAC).

Barrett's esophagus is named after Norman Rupert Barrett, a highly regarded and successful thoracic surgeon of his period. Contrary to popular belief, however, Norman Barrett's contribution to the identification of the disease was quite limited. Barrett wrote of the presence of ulcers in the esophagus and the presence of columnar epithelium around ulcers in an article published in 1950 entitled "Chronic peptic ulcer of the oesophagus and 'oesophagitis'" [1]. However, there were number of inaccuracies in this article and the history of the disease dates back to much older times [2].

1.2 NORMAN RUPERT BARRETT (1903–1979)

Norman Rupert Barrett was born in North Adelaide, Australia, on May 16, 1903, the son of Alfred Barrett and Catherine Hill Connor [3]. His paternal grandfather was a wealthy malt manufacturer who moved to Australia from England in the 1880s. When Barrett was 10 years old, he moved from Australia to London together with his parents and a younger sister [4,5] where his brilliant academic career would start. Barrett would return to Australia after 50 years, as a visiting professor at Royal North Shore Hospital Sydney, in 1963.

Barrett received his education at Eton College (1917–1922), then continued in Trinity College, and graduated from Cambridge University in 1925. He completed his medical education at St Thomas Hospital (1925–1928). He continued

D. Pleskow & T. Erim (Eds): Barrett's Esophagus.
DOI: http://dx.doi.org/10.1016/B978-0-12-802511-6.00001-6

as resident assistant surgeon for the next 2 years at the same hospital and was elected to fellowship of the Royal College of Surgeons in 1930 and the postgraduate degree M. Chir in 1931. He married Annabel Elizabeth "Betty" Warington Smyth, his school friend, when he was 28. Then he began working at St Thomas as surgical staff, then as a consulting surgeon in 1935 and spent his entire professional career there. Barrett's first trip to America took place when he became entitled to participate in the Rockefeller Travelling Fellowship (1935–1936) program, a prestigious program at the Mayo Clinic. This program would also have a very important place in Barrett's career, since he would become interested in the emerging field of thoracic surgery, and he would continue the rest of his professional life as a thoracic surgeon. There was still not a thoracic surgery department when he returned to St Thomas. Therefore, he continued to work as both a consulting general surgeon and a consulting thoracic surgeon at the same time (Fig. 1.1).

He became a member of the British Thoracic Society, which was known as the Thoracic Society at that time. The Thoracic Society chose

FIGURE 1.1   Norman Rupert Barrett.

him as the first editor of Thorax journal and he served as the editor of the journal until 1971. The first article by Barrett in the literature was a report of two cases and an associated literature review that was titled "Surgical Emphysema During General Anesthesia" in 1944 [6]. He published an article about three cases with spontaneous esophageal perforation and a literature review, his second article, in the first issue of Thorax in 1946 [7]. He also pioneered many advances in the field of thoracic surgery. He successfully operated on a case with esophageal rupture, which had been previously considered fatal, and was the first to report it in the literature [8]. Barrett successfully operated on a case with esophageal diverticulum using a thoracic approach, again the first in the literature [9]. In addition, he wrote scientific articles on subjects such as removal of pulmonary cysts [10–12], surgical treatment of bronchial carcinoma [13], primary tumors of the rib [14], achalasia [15], mediastinal fibrosis [16], and congenital heart disease conditions [17,18]. His other important contribution to the medical literature was the successful detection of malignant cells in cytological examination of sputum with the "wet film" method in lung cancer patients [19]. However, it was the article titled "Chronic peptic ulcer of the oesophagus and 'oesophagitis'" in 1950 that made the name Norman Barrett famous in our day [1]. Barrett most likely could not estimate the great influence of his article at that time.

Barrett served as a president of the Thoracic Society as well as president of the Thoracic Surgeons of Great Britain and Ireland. He was a member of the Court of Examiners of the Royal College of Surgeons and he was an examiner at universities of Oxford, Cambridge, Birmingham, London, and Khartoum. He was awarded the Commander of the Most Excellent Order of the British Empire in 1969 [3]. He retired from St Thomas Hospital in 1970 after a long and productive career and passed away in London on January 8, 1979.

1.3 PHILIP ROWLAND ALLISON (1907–1974)

When we look back at the literature and sort through the evidence of how intestinal metaplasia (IM) of the esophagus was identified, we find a particular scientist other than Norman Barrett who played a key role. Dr Philip Rowland Allison (1907–1974) should arguably be more prominent than Barrett in receiving credit for the identification of the disease [20]. Allison was one of the leading cardiothoracic surgeons in England. He worked as a general surgeon and cardiothoracic surgeon for many years in the Leeds General Infirmary. One of his major achievements was the first successful cyanotic congenital heart disease surgery in 1948 in Leeds. In addition to cardiovascular surgery, he was a very successful surgeon in hernia surgery and published several scientific articles on this subject (Fig. 1.2).

Allison was the first person to use the "columnary lined esophagus" phrase, correctly identifying the histological change. Ironically, he was also the first person to use the expression "Barrett's ulcer" in the literature, when he

FIGURE 1.2   Philip Rowland Allison.

argued that Barrett had made a mistake in his article in 1950. Allison identified peptic ulcer of esophagus and used the expression "reflux esophagitis" first in the articles titled "Peptic ulcer of the oesophagus" in 1946 [21] and 1948 [22]. In addition, he described in detail and in an accurate manner the function of the cardia, that the esophageal epithelium is not resistant to gastric contents, and that gastric contents passing to the esophagus may cause esophagitis and ulceration in cases in which the cardia function was disabled due to reasons such as sliding hernia. Moreover, he mentioned radiological and histological findings of the disease and identified surgical treatments in a detailed manner in these articles. Allison, who was married and had three children, died in March 6, 1974.

1.4 HISTORY OF BARRETT'S ESOPHAGUS

Contrary to popular belief, the historical identification of the disease process started much earlier than in Barrett's lifetime. Boehm described gastroesophageal reflux first in the literature in 1722 as follows: "acute pain which reached down even to the stomach and which was accompanied by hiccup and a constant flow of serum from the mouth" [23]. Joanne Petro Frank [24] first used the expression "esophagitis" in 1792. Johann Friedrich Hermann Albers, a German physician and pathologist, was the first to propose the concept of esophageal ulcers in history in 1839. Quincke reported histopathological findings of esophageal ulcers with the presentation of three postmortem cases in 1879 [25,26]. The first scientific account of esophagitis was reported by Morell Mackenzie, a British laryngologist, in 1884 [27]. Mackenzie described acute esophagitis in the Disease of the Gullet section of his book as follows: "acute idiopathic inflammation of the mucous membranes of the esophagus, giving rise to extreme odynophagia, and often to aphagia."

The probable presence of columnar metaplasia in the esophagus was reported nearly two centuries ago. The presence of gastric mucosal islands in the esophagus was first described by Schmidt from Halle University in Germany in 1805 (about 150 years before Barrett) [2]. Schridde published in 1904 that aberrant columnar epithelial islands were present in the esophagus in autopsy cases [27]. Wilder Tileston, a Harvard pathologist, published the historically important article "Peptic ulcer of the oesophagus" in 1906 [25]. In this article, Tileston wrote that peptic ulcers may develop in the esophagus, that the lack of gastroesophageal junction is necessary for this development, and columnar epithelium, which is normally seen in the stomach, is present around the developed ulcer. In addition, this article is indeed the first in proposing the relationship between gastroesophageal reflux and esophageal ulcer disease, although it did not exactly use the expression gastroesophageal reflux disease (GERD) at that time. Moreover, Tileston described 12 different specific etiologies other than peptic ulcer in this article. A.L. Taylor from University of Leeds reported that macroscopic columnar mucosal islands may present in the distal esophagus and first suggested that this condition is a persistent process in 1927 [28].

Norman Barrett published the article titled "Chronic peptic ulcer of the oesophagus and 'oesophagitis'" which started the discussion in 1950 [1]. He proposed in his article as follows: "I believe that reflux oesophagitis is common and that it can give rise to ulceration of the oesophagus and stricture formation… I submit that most of these cases are in truth examples of congenital short esophagus, in which there is neither general inflammation nor stricture formation, but in which a part of the stomach extends upwards into the mediastinum, or even to the neck, and that in this stomach a typical chronic gastric ulcer can form." [1]. Barrett mentioned chronic peptic ulcer of the esophagus and presence of gastric-type epithelium in this

region. However, he believed that this region was not the esophagus but actually a part of the stomach that had slipped into the mediastinum due to congenital short esophagus and not due to hiatal hernia. Bosher and Taylor [29] published an article which was almost a response to this article and identified completely the histopathological features of BE for the first time in literature 1 year later. Authors mentioned aortic arch-level chronic peptic esophageal ulcer disease associated with columnar-lined epithelium and suggested that the epithelium in this region was similar to the gastric epithelium but without parietal cells and contained goblet cells and submucosal glands of the esophagus. Bosher and Taylor disproved Barrett's theory of the stomach sliding to the mediastinum due to congenital short esophagus with their article. This article is in fact the first article in the literature describing specialized (with goblet cells) IM. One year later Morson and Belcher [30] and 2 years later Allison and Johnstone [31] published articles confirming the presence of columnar epithelium with goblet cells.

The article published by Allison and Johnstone in 1953 has quite an important place in the history of BE. A series of 115 cases with esophageal ulcers and strictures was presented in this article titled "The oesophagus lined with gastric mucous membrane." They demonstrated the presence of columnar mucosa in the esophagus in 11 cases, presence of reflux esophagitis or hiatal hernia in 7 cases, and presence of EAC in 1 case. Authors pointed out that columnar-lined epithelium was not a rare condition, contrary to common belief. They emphasized that Norman Barrett made a mistake in 1950 and proposed that this condition may occur in the esophagus. The objection to Norman Barrett was published in Thorax journal, of which Barrett was still the editor. The article contained very important details. Authors expressed that: "Patients with the oesophagus lined by gastric mucous membrane are subject to gastric ulcers occurring in that part of the oesophagus lined by gastric mucosa, and it is suggested that, if these become chronic, they might be known as *Barrett's ulcers*. Such ulcers may occur alone or in association with reflux ulcers of the oesophagus." Ironically, although this article was published in objection to Barrett's article, the disease was referred for the first time in literature with his name and led to coining the term.

The precancerous potential creating the clinical significance of BE would also begin to be mentioned in those years. Before Allison and Johnstone, Carrie [32] published a presentation of 20 cases in 1950 and suggested first that EAC may develop from ectopic gastric epithelium. Subsequently, Morson and Belcher [30] published the first article revealing the relationship between the presence of columnar epithelium in the esophagus (still thought to be ectopic gastric mucosa) and EAC in 1952. They proposed IM with goblet cells to be a predisposing cause for EAC in this article. Allison and Johnstone in 1953 [31] and Thomas and Hay in 1954 [33] mentioned the relationship between EAC and columnar-lined epithelium. There would be many articles confirming this relationship in subsequent years [34–38].

Norman Barrett would renounce the idea of congenital short esophagus 7 years after his article but would continue to insist another inaccuracy; "it is probably the result of a failure of the embryonic lining of the gullet to achieve normal maturity" [39]. However, Barrett eventually gave up his insistence and offered in 1960 that "It would have been better if the term had never been introduced, because it has led to wrong thinking" [40]. As it can be seen, Norman Barrett had an important but nonetheless minor contribution on the etiology of the disease, much less so for the correct definition of its histology, and identification of the premalignant potential. It is one of the ironies of Medicine and History that the disease is still known with Norman Barrett's name to this day.

The eponym of "Barrett's syndrome" was used by Goldman and Beckman [41] in 1960;

and eventually eponym of *Barrett's esophagus* was used by Seaman and Wylie [42] in 1966 for the first time in the literature. The continued use of the eponym "Barrett's esophagus" has over the years led to objections. Bani-Hani mentioned the history of the disease in detail and objected to still referring to the disease with the name of Barrett in an article titled "Columnar-lined esophagus: Time to drop the eponym of 'Barrett': Historical review" [43]. They make a compelling argument, however, the disease has become more and more important in the recent decades and a name change at this point will be difficult.

There were many advances in identification of both etiology and histopathological features of the disease in the second half of the 20th century. The disease would be demonstrated to be associated with GERD and hiatal hernia and not to be congenital in publications by Moersch et al. [44], Hayward [45], and Adler [46]. Hayward stated that: "It is probably neither ectopic, nor congenital, nor permanent, nor in need of resection but metaplastic and reversible." Cohen et al. published the most detailed study ever conducted on the issue in 1963. They put an end to whether columnar-lined epithelium belongs to the stomach or the esophagus with this study. They examined this phenomenon with manometric, radiological, endoscopic, and histopathological methods. The region lined by columnar epithelium showed motor activity of the esophagus in manometric examination as well [47]. Naef et al. [48] demonstrated in 1972 that esophagitis findings decreased with Nissen fundoplication, but columnar epithelium did not regress. In subsequent years, it would also be shown that periodic acid suppression decreased esophagitis findings but only led to a partial regression of columnar epithelium; in addition, columnar epithelium did not fully regress even in cases after antireflux surgery, and adenocarcinoma could still develop as well [49–54].

Trier [55], Berenson et al. [56], and Paull et al. [57] published detailed studies on histopathological features of the disease in those years. Paull et al. [57] examined distal esophageal biopsies taken with manometric examination in detail in a series of 11 patients with BE in 1976. They showed the presence of three different types of columnar epithelium in their cases. The first one was intestinal-type epithelium, which was referred to as specialized (containing intestinal-type goblet cells but not parietal or chief cells) columnar epithelium, and it was present in squamous epithelium junction of the most proximal columnar-lined epithelium. The second one was junctional (cardia-type) epithelium that contained cells secreting mucus and presented in the middle parts. The third one was gastric fundic—type epithelium that contained parietal and chief cells and was located at the most distal end.

The disease was defined thoroughly in the 1980s, and there were many studies on the relationship of the disease with dysplastic changes and EAC [58—62]. Hayward proposed that the region until 2 cm from the distal of esophagus has cardia-type epithelium, and biopsies taken from this region by the endoscopist may mistakenly lead to a BE diagnosis, thus potentially causing overdiagnosis [45]. Therefore, there was a need for diagnostic criteria. Another problem was the misguided assumption that the disease was always associated with GERD symptoms and thus there was no need to conduct a biopsy in the asymptomatic [63]. Skinner et al. would argue for diagnostic criteria on the issue for the first time and recommended biopsy for asymptomatic patients as well in 1983 [64]. According to them, the 1—2 cm distal portion of the esophagus normally contained cardia-type columnar epithelium, and therefore columnar-lined epithelium was necessary to be ≥ 3 cm in the distal esophagus in order to be diagnosed as BE. In addition, they showed in this article that dysplasia may develop in asymptomatic patients as well [64]. The 3-m limit would survive for about 10 years, but the study by Spechler et al. [65]

showed that it was not very accurate in 1994. Squamocolumnar junction biopsies of 142 white and 114 non-white successive cases without BE on endoscopic examination were taken and examined. There was a presence of IM in 18% of the first group and 14% of the second group. In addition, they showed that endoscopic findings of GERD and GERD symptoms were not reliable diagnostic criteria. Following this study, a new concept came up: the presence of <3 cm IM on esophagus was called "short segment BE." There would be studies presenting the progressive potential of short segment BE and the relationship of short segment BE with dysplasia and adenocarcinoma in the following years [66—69].

Clinical guidelines began to be created in the 1990s. The American College of Gastroenterology guideline was first published in 1998 [70]. Then, different guidelines would be published and would be revised in subsequent years [71—77].

In the 2000s, diagnostic histopathological findings (such as the presence of the cardia-type metaplasia, and whether the goblet cells were present or not) as well as surveillance details would be the debated issues as will be described in detail in related sections.

Has Normal Barrett gotten too much credit for this disease? It may appear so when we examine the documented history. However, it does not appear that he himself asked for it and the discussion initiated by him in 1950 still continues today. This in itself deserves significant merit. We have certainly come a long way since the initial proposed argument. It is clear that IM of the esophagus leads to adenocarcinoma. We have identified several risk factors and intense research is being conducted on elucidating the genetics of the disease. In the future, we hope to be able to identify who is going to be at risk depending on genetic factors so that we can influence the lifestyle choices that lead to the disease. We have gotten better than ever at finding early disease and effective surveillance strategies are being evaluated to follow patients cost effectively. There have been several break-throughs in endoscopic and surgical treatment of BE. However, we are now finding that some of our assumed triumphs, like the duration of complete remission of intestinal metaplasia after ablative therapy, are not as long lasting as we had hoped. The next generation of therapies is sure to focus on noninvasive methods.

The following chapters will take us through the background of the different facets of the disease and provide emerging evidence that we hope will be used to improve clinical practice. At the end of each chapter, a short section has been provided for the authors to expand on where they believe the future research will focus, where the discussion will go.

References

[1] Barrett NR. Chronic peptic ulcer of the oesophagus and 'oesophagitis'. Br J Surg 1950;38(150):175–82.

[2] Schmidt FA. De Mammalian Oesophage Atque Ventriculo [dissertation]. Halle, Germany: University of Halle; 1805.

[3] Barrett, NR. (1903–1979). King's College London, College Archives. <http://www.aim25.ac.uk/cgi-bin/vcdf/detail?coll_id = 5587&inst_id = 6>. Published January 2004, accessed December 2015.

[4] Lord RV. Norman Barrett, "doyen of esophageal surgery". Ann Surg 1999;229(3):428–39.

[5] Barrett NR. The contribution of Australians to medical knowledge. Med History 1967;11;321–33.

[6] Barrett NR, Thomas D. Surgical emphysema during general anaesthesia. Br Med J 1944;2(4377):692–3.

[7] Barrett NR. Spontaneous perforation of the oesophagus: review of the literature and report of three new cases. Thorax 1946;1:48–70.

[8] Barrett NR. Report of a case of spontaneous perforation of the oesophagus successfully treated by operation. Br J Surg 1947;35(138):216–18.

[9] Barrett NR. Diverticula of the thoracic oesophagus: report of a case in which the diverticulum was successfully resected. Lancet 1933;1:1009–11.

[10] Barrett NR. The treatment of pulmonary hydatid disease. Thorax 1947;2(1):21–57.

[11] Barrett NR. Removal of simple univesicular pulmonary hydatid cyst. Lancet 1949;2(6571):234.

[12] Barrett NR. The anatomy and the pathology of multiple hydatid cysts in the thorax. Ann R Coll Surg Engl 1960;26:362–79.

[13] Barrett NR. The treatment of carcinoma of the bronchus. Med Press 1952;227(13):289–91.

[14] Barrett NR. Primary tumours of rib. Br J Surg 1955;43(178):113–32.

[15] Barrett NR. Achalasia: thoughts concerning the aetiology. Ann R Coll Surg Engl 1953;12(6):391–402.

[16] Barrett NR. Idiopathic mediastinal fibrosis. Br J Surg 1958;46(197):207–18.

[17] Barrett NR, Daley R. A method of increasing the lung blood supply in cyanotic congenital heart disease. Br Med J 1949;1(4607):699–702.

[18] Barrett NR, Hickie JB. Cor triatrium. Thorax 1957;12(1):24–7.

[19] Barrett NR. Examination of sputum for malignant cells and particles of malignant growth. J Thoracic Surg 1938;8:169–83.

[20] Lodge JP. Philip Rowland Allison BSc ChM FRCS 1907–1974. Ann R Coll Surg Engl 1988;70(4):189.

[21] Allison PR. Peptic ulcer of the esophagus. J Thorac Surg 1946;15:308–17.

[22] Allison PR. Peptic ulcer of the oesophagus. Thorax 1948;3:20–42.

[23] Mackenzie M. A manual of diseases of the throat and nose: including the pharynx. New York: W. Wood & Company; 1880.

[24] Frank JP. De curandis hominum morbis epitome praelectionibus academicis dicata. Viennae: Vilnius Universities; 1792.

[25] Albers JFH. Atlas der pathologischen Anatomie für praktische Ärzte; Erläuterungen zu dem Atlasse. Bonn: Henry & Cohen; 1832.

[26] Quincke H. Ulcus oesophagi ex digestione. Dtsch Archiv Klin Med 1879;24:72–9.

[27] Mackenzie M. Diseases of the throat and nose, including the pharynx, larynx, trachea, oesophagus, nasal cavities, and neck. Philadelphia, PA: P. Blakiston; 1880.

[28] Taylor AL. The epithelial heterotopias of the alimentary tract. J Pathol Bacteriol 1927;30:415–49.

[29] Bosher LH, Taylor FH. Heterotopic gastric mucosa in the esophagus with ulceration and stricture formation. J Thorac Surg 1951;21(3):306–12.

[30] Morson BC, Belcher JR. Adenocarcinoma of the oesophagus and ectopic gastric mucosa. Br J Cancer 1952;6:127–30.

[31] Allison PR, Johnstone AS. The oesophagus lined with gastric mucous membrane. Thorax 1953;8:87–101.

[32] Carrie A. Adenocarcinoma of the upper end of the oesophagus arising from ectopic gastric epithelium. Br J Surg 1950;37:474.

[33] Thomas JV, Hay LJ. Adenocarcinoma of the esophagus. Report of a case of glandular metaplasia of the esophageal mucosa. Surgery 1954;35:635–9.

[34] McCorkle RG, Blades B. Adenocarcinoma of the esophagus arising in aberrant gastric mucosa. Am Surg 1955;21(8):781–5.

[35] Armstrong RA, Carrera GM, Blalock JB. Adenocarcinoma of the middle third of the esophagus arising from ectopic gastric mucosa. J Thorac Surg 1959;37(3):398–403.

[36] Azzopardi JG, Menzies T. Primary oesophageal adenocarcinoma. Confirmation of its existence by the finding of mucous gland tumours. Br J Surg 1962;49:497–506.

[37] Dawson JL. Adenocarcinoma of the middle oesophagus arising in an oesophagus lined by gastric (parietal) epithelium. Br J Surg 1964;51:940–2.

[38] Raphael HA, Ellis Jr FH, Dockerty MB. Primary adenocarcinoma of the esophagus: 18-year review and review of literature. Ann Surg 1966;164(5):785–96.

[39] Barrett NR. The lower esophagus lined by columnar epithelium. Surgery 1957;41(6):881–94.

[40] Barrett NR. Hiatus hernia. Br Med J 1960;2:247–52.

[41] Goldman MC, Beckman RC. Barrett syndrome. Case report with discussion about concepts of pathogenesis. Gastroenterology 1960;39:104–10.

[42] Seaman WB, Wylie RH. Observations on the nature of the stricture in Barrett's esophagus (Allison and Johnstone's anomaly). Radiology 1966;87:30–2.

[43] Bani-Hani KE, Bani-Hani BK. Columnar-lined esophagus: time to drop the eponym of "Barrett": historical review. J Gastroenterol Hepatol 2008;23(5):707–15.

[44] Moersch RN, Ellis FH, McDonald JR. Pathologic changes occurring in severe reflux esophagitis. Surg Gynecol Obstet 1959;108:476–84.

[45] Hayward J. The lower end of the oesophagus. Thorax 1961;16:36–41.

[46] Adler RH. The lower esophagus lined by columnar epithelium. Its association with hiatal hernia, ulcer, stricture, and tumor. J Thorac Cardiovasc Surg 1963;45:13–34.

[47] Cohen BR, Wolf BS, Som M, Janowitz HD. Correlation of manometric, oesophagoscopic, and radiological findings in the columnar-lined gullet (Barrett syndrome). Gut 1963;4:406–12.

[48] Naef AP, Savary M, Jaques WA. Effectiveness of fundoplication in the treatment of reflux and peptic esophagitis (clinical aspects and endoscopy). Schweiz Med Wochenschr 1970;100(28):1228–9.

[49] Lanas A. Potent gastric acid inhibition in the management of Barrett's oesophagus. Drugs 2005;65 (1):75–82.

[50] Cooper BT, Chapman W, Neumann CS, Gearty JC. Continuous treatment of Barrett's oesophagus patients with proton pump inhibitors up to 13 years: observations on regression and cancer incidence. Aliment Pharmacol Ther 2006;23 (6):727–33.

[51] Csendes A, Burdiles P, Braghetto I, Korn O. Adenocarcinoma appearing very late after antireflux surgery for Barrett's esophagus: long-term follow-up, review of the literature, and addition of six patients. J Gastrointest Surg 2004;8(4):434–41.

[52] Hillman LC, Chiragakis L, Shadbolt B, Kaye GL, Clarke AC. Effect of proton pump inhibitors on markers of risk for high-grade dysplasia and oesophageal cancer in Barrett's oesophagus. Aliment Pharmacol Ther 2008;27(4):321–6.

[53] Csendes A, Burdiles P, Braghetto I, Smok G, Castro C, Korn O, et al. Dysplasia and adenocarcinoma after classic antireflux surgery in patients with Barrett's esophagus: the need for long-term subjective and objective follow-up. Ann. Surg 2002;235 (2):178–85.

[54] Csendes A, Braghetto I, Burdiles P, Puente G, Korn O, Díaz JC, et al. Long-term results of classic antireflux surgery in 152 patients with Barrett's esophagus: clinical, radiologic, endoscopic, manometric, and acid reflux test analysis before and late after operation. Surgery 1998;123(6):645–57.

[55] Trier JS. Morphology of the epithelium of the distal esophagus in patients with midesophageal peptic strictures. Gastroenterology 1970;58:444–61.

[56] Berenson MM, Herbst JJ, Freston JW. Enzyme and ultrastructural characteristics of esophageal columnar epithelium. Am J Dig Dis 1974;19:895–907.

[57] Paull A, Trier JS, Dalton MD, Camp RC, Loeb P, Goyal RK. The histologic spectrum of Barrett's esophagus. N Engl J Med 1976;295:476–80.

[58] Sjogren Jr RW, Johnson LF. Barrett's esophagus: a review. Am J Med 1983;74(2):313–21.

[59] Sarr MG, Hamilton SR, Marrone GC, Cameron JL. Barrett's esophagus: its prevalence and association with adenocarcinoma in patients with symptoms of gastroesophageal reflux. Am J Surg 1985;149(1): 187–93.

[60] Kalish RJ, Clancy PE, Orringer MB, Appelman HD. Clinical, epidemiologic, and morphologic comparison between adenocarcinomas arising in Barrett's esophageal mucosa and in the gastric cardia. Gastroenterology 1984;86(3):461–7.

[61] Zinner EN. Adenocarcinoma in Barrett's esophagus. N Engl J Med 1986;314(11):720.

[62] Cameron AJ, Ott BJ, Payne WS. The incidence of adenocarcinoma in columnar-lined (Barrett's) esophagus. N Engl J Med 1985;313(14):857–9.

[63] Skinner DB, Walther BC, Riddell RH, Schmidt H, Iascone C, DeMeester TR. Barrett's esophagus:

comparison of benign and malignant cases. Ann Surg 1983;198:554–65.

[64] Spechler SJ, Fitzgerald RC, Prasad GA, Wang KK. History, molecular mechanisms, and endoscopic treatment of Barrett's esophagus. Gastroenterology 2010;138(3):854–69.

[65] Spechler SJ, Zeroogian JM, Antonioli DA, Wang HH, Goyal RK. Prevalence of metaplasia at the gastro-oesophageal junction. Lancet 1994;344:1533–6.

[66] Hameeteman W, Tytgat GNJ, Houthoff HJ, van den Tweel JG. Barrett's esophagus: development of dysplasia and adenocarcinoma. Gastroenterology 1989; 96:1249–56.

[67] Schnell TG, Sontag SJ, Chejfec G. Adenocarcinomas arising in tongues or short segments of Barrett's esophagus. Dig Dis Sci 1992;37:137–43.

[68] Drewitz DJ, Sampliner RE, Garewal HS. The incidence of adenocarcinoma in Barrett's esophagus: A prospective study of 170 patients followed 4.8 years. Am J Gastroenterol 1997;92:212–15.

[69] Chak A, Faulx A, Eng C, Grady W, Kinnard M, Ochs-Balcom H, et al. Gastro-esophageal reflux symptoms in patients with adenocarcinoma of the esophagus or cardia. Cancer 2006;107:2160–6.

[70] Sampliner RE. Practice guidelines on the diagnosis, surveillance, and therapy of Barrett's esophagus. The Practice Parameters Committee of the American College of Gastroenterology. Am J Gastroenterol 1998; 93(7):1028–32.

[71] Wang KK, Sampliner RE. Practice Parameters Committee of the American College of Gastroenterology. Updated guidelines 2008 for the diagnosis, surveillance and therapy of Barrett's esophagus. Am J Gastroenterol 2008;103(3):788–97.

[72] Boyer J, Robaszkiewicz M. Guidelines of the French Society of Digestive Endoscopy: monitoring of Barrett's esophagus. The Council of the French Society of Digestive Endoscopy. Endoscopy 2000;32(6):498–9.

[73] Management of Barrett's esophagus. The Society for Surgery of the Alimentary Tract (SSAT), American Gastroenterological Association (AGA), American Society for Gastrointestinal Endoscopy (ASGE) Consensus Panel. J Gastrointest Surg 2000;4(2):115–16.

[74] American Gastroenterological Association, Spechler SJ, Sharma P, Souza RF, Inadomi JM, Shaheen NJ. American Gastroenterological Association medical position statement on the management of Barrett's esophagus. Gastroenterology 2011;140(3):1084–91.

[75] ASGE Standards of Practice Committee, Evans JA, Early DS, Fukami N, Ben-Menachem T, Chandrasekhara V, et al. The role of endoscopy in Barrett's esophagus and other premalignant conditions of the esophagus. Gastrointest Endosc 2012;76 (6):1087–94.

[76] Bennett C, Vakil N, Bergman J, Harrison R, Odze R, Vieth M, et al. Consensus statements for management of Barrett's dysplasia and early-stage esophageal adenocarcinoma, based on a Delphi process. Gastroenterology 2012;143(2):336–46.

[77] Fitzgerald RC, di Pietro M, Ragunath K, Ang Y, Kang JY, Watson P, et al. British Society of Gastroenterology. British Society of Gastroenterology guidelines on the diagnosis and management of Barrett's oesophagus. Gut 2014;63(1):7–42.

Fluctuating Risk Factors and Epidemiology

Gaurav Kistangari[1] and Prashanthi N. Thota[2]

[1]Department of Internal Medicine, Cleveland Clinic, Cleveland, OH, United States
[2]Center of Excellence for Barrett's Esophagus, Department of Gastroenterology and Hepatology, Cleveland Clinic, Cleveland, OH, United States

2.1 INTRODUCTION

The first description of Barrett's esophagus (BE) is attributed to Sir Norman Barrett in 1950 who reported ulcerations in the tubular segment of stomach that had been tethered within the chest by a congenitally short esophagus [1]. By the 1970s, it had been accepted that BE is an acquired condition associated with severe gastroesophageal reflux disease (GERD) and has a malignant predisposition [2,3]. The definition has evolved over the past several decades and the currently accepted definition is that of a condition in which any extent of metaplastic columnar epithelium that predisposes to cancer development and replaces the stratified squamous epithelium that normally lines the distal esophagus. In the United States, presence of specialized intestinal metaplasia is a prerequisite for diagnosis, whereas in United Kingdom and rest of Europe, gastric metaplasia alone is enough to make a diagnosis of BE.

2.2 PREVALENCE

The true prevalence of BE is difficult to estimate because most of the patients with BE are asymptomatic and the diagnosis of BE requires endoscopic evaluation and histologic confirmation. The available prevalence rates vary widely between 0.4% and over 25% based on the age, gender, ethnicity, and symptoms in the populations studied. Variation of BE prevalence among asymptomatic general population should be carefully interpreted based on age, sex, and prevalence of GERD symptoms.

2.2.1 Population-Based Studies/Routine Endoscopy

In population-based studies, the overall prevalence of BE in general population varied between 1.6% and 6.8% in western countries [4,5]. In a study by Gerson et al. [6], a 25% BE prevalence rate was noted in male veterans older than 50 years of age. In another study, both men and women above age 65 years undergoing screening colonoscopy were requested to undergo routine upper endoscopies after completing a detailed GERD questionnaire. Overall prevalence of BE was 16.7% and symptoms of heart burn were not significantly associated with BE [7].

2.2.2 Patients with Chronic Gastroesophageal Reflux Disease

Prevalence rates of BE in patients with symptomatic GERD are somewhat higher than in the general population with rates of 13–20% [7,8]. Long segment BE is found in 3–5% and short segment BE in 10–15% [9,10].

2.2.3 Autopsy Studies

Many individuals with BE remain undiagnosed. In one of the few autopsy-based studies, Cameron et al. [11] found a prevalence rate of 376 cases per 100,000 in Olmsted County, Minnesota, which was 21-fold higher than the clinically recognized cases in the county.

2.2.4 Geographic Variation

There is considerably more data on the prevalence of BE in western countries as compared to Asian countries. In general, the prevalence rates were higher in the western hemisphere compared to Asia. Two large prospective studies investigated the prevalence of BE in Japan and found an overall prevalence of BE of 0.9–1.2% [12]. The prevalence rate of BE in China from a pooled analysis of reports from 1989 to 2007 was 2.4% [13]. In South America, the prevalence rate of BE was 3.57% among patients with GERD symptoms [14].

2.3 INCIDENCE

Due to the rapidly increasing incidence of esophageal adenocarcinoma (EAC) in United States [15–17], there has been concern for rising incidence of BE, a precursor for EAC. Between 1980 and 1996, a steady increase in the incidence of BE at 0.08% per year has been reported with a median incidence of 1.17% [18]. It is not clear if this is a true increase or if it is due to increased recognition of BE from electronic health records and raising endoscopic volume. One of the largest studies from Northern California showed an increase in the incidence of BE from 14.5 to 22.9 per 100,000 person years between 1994 and 2007, even after adjusting for increase in endoscopy volume and age ($p < 0.01$), [19]. Similar increases in incidence have been reported in Europe as well [20].

2.4 RISK FACTORS

2.4.1 Gastroesophageal Reflux Disease

Though initially described as a congenital condition, BE had become widely accepted by the 1970s to be an acquired condition associated with severe GERD [3]. Since then, epidemiologic studies have confirmed that GERD is indeed the most important risk factor for BE. A recent meta-analysis showed that there is a threefold increased risk of BE with GERD (odds ratio (OR) 2.90, 95% confidence interval (CI), 1.86–4.54, $p = 0.0001$) and almost fivefold increased risk for long segment BE (OR 4.92, 95% CI 2.01–12.0, $p = 0.30$) [21]. BE patients are also more likely to have frequent reflux symptoms, nocturnal GERD, and longer duration of symptoms than controls [22]. It is not only the chronicity and severity of symptoms but also younger age of onset of symptoms that seem to increase the risk of BE in GERD [23].

Several physiologic mechanisms contribute to severe GERD in patients with BE (Table 2.1). Studies show a strong correlation between the abnormal and prolonged esophageal acid exposure and development of BE. In 1994, Neumann and Cooper [24] demonstrated that BE patients had longer duration of esophageal acid exposure and higher number of reflux episodes of greater than 5 min when compared with reflux esophagitis patients. In another study, compared to the patients with esophagitis, patients with BE had lower median lower esophageal sphincter (LES) pressure (10.5 vs 17.5 mmHg; $p = 0.013$) and higher median percentage of total time with pH less than 4 (48.2 vs 8.7 and 23.2 vs 5.2; $p = 0.0001$ for distal and proximal esophageal acid exposure, respectively) [25]. Duodenogastroesophageal reflux is a major contributing factor to the development of BE. Also, the pattern of acid and/or bile exposure seems to play a role in that only persistent exposure led to the development of columnar epithelium in contrast to short exposure [26].

The exact molecular pathways leading to the development of BE in GERD are not well elucidated. Chronic reflux of acid and bile can lead to BE by several mechanisms such as (1) oxidative DNA damage due to production of free radicals and nitric oxide [27], (2) inflammatory cytokine production by recruitment of immune cells such as naive T cells, macrophages, and dendritic cells into the esophagus, (3) deregulation of microRNAs which are short noncoding RNAs involved in a variety of cellular processes (eg, miRNA-145 was linked to the activation of BMP4 pathway which promotes squamous-to-columnar metaplasia), (4) express CDX2 (a gene known to play a key role in the development of intestinal epithelia) and MUC2 (a mucin normally found in intestinal goblet cells), and (5) decline in esophageal squamous progenitor cells as evidenced by reduction in p63 protein levels in esophageal cell cultures (a marker for squamous progenitor cells) [28].

Two major hypotheses have been proposed to explain the cellular origin of Barrett's metaplasia. As per the transdifferentiation theory, the cells in the esophagus that normally would differentiate into squamous cells instead differentiate into columnar cells triggered by surface epithelial damage from acid and bile [29]. In contrast, the stem cell theory proposes that columnar metaplasia results from proliferation of stem cells which may originate from the basal cell layer of the squamous epithelium, gastric cardia, submucosal glands, or from the bone marrow [30].

GERD is the most prevalent gastrointestinal disorder with an estimated prevalence of

TABLE 2.1 Physiologic Abnormalities in Barrett's Esophagus [90]

Abnormality	Contribution to GERD Severity
Gastric acid hypersecretion with or without duodenogastric reflux	Gastric contents available for reflux are highly caustic to the esophagus due to high concentrations of acid and, with duodenogastric reflux, bile
Defective lower esophageal sphincter	Impairment in antireflux barrier
Hiatal hernia	Impairment in antireflux barrier
Impaired esophageal peristalsis	Reduced ability to clear esophagus of refluxate and prolonged exposure
Diminished esophageal pain sensitivity	Reduced awareness of esophageal injury which can also decrease compliance with antireflux therapy
Delayed gastric emptying	Increased gastric volume leading to more reflux
Decreased salivary secretion of epidermal growth factor	Delayed healing of acid induced esophageal injury

18–28% in North America. However, less than 10% of patients with GERD develop BE. This suggests that BE is a multifactorial disease and several factors other than GERD may play a role in an individual's susceptibility for this condition. These contributing factors for BE are described in the following sections.

2.4.2 Age

Older age is a risk factor for BE although the age at which this risk increases is uncertain. Different population-based studies showed an increase in prevalence of BE from 30 to 70 years [31]. There is a 1.3-fold increase in risk of BE with each additional decade of age (95% CI 1.02–1.67, $p = 0.03$) [32]. More recent studies reveal an increased risk of BE with early age at onset of GERD symptoms. There seems to be a linear association of age at onset of GERD symptoms and risk of BE independent of the duration of GERD symptoms [23]; however, more studies are required to validate these results as there may be other risk factors such as childhood obesity, hiatus hernia, and genetic factors contributing to BE.

2.4.3 Gender

BE is thought to be a disease predominantly occurring in men. The risk of BE among men with GERD symptoms is 1.5- to 3-fold higher than that of women [19,33]. Several theories including differences in parietal cell mass, differences in LES function, and higher body mass index (BMI) among males have been proposed to explain the protective effect against the development of BE in women [34,35]. Women also seem to develop BE at a later age compared to men. This age shift leads to a lower prevalence of BE in females compared to the male population, with 2:1 male to female sex ratio [36]. This age lag seen in women explains to some extent the

marked gender-based differences that are also seen in EAC as men have BE earlier in their lives and have more time to develop dysplasia and EAC.

2.4.4 Race

BE is far more common in Caucasians compared to other ethnic groups. In an observational study from Kaiser Permanente in California, the prevalence of BE among non-Hispanic whites was twofold higher compared to Hispanic whites (247 vs 135/100,000 years, respectively, $p < 0.01$) and approximately fivefold higher than among blacks as shown in Fig. 2.1 (49/100,000 years, $p < 0.01$) [19]. Similarly, an endoscopy study from the United Kingdom showed an increased risk of BE among Caucasians compared to Asians (OR 6.03, CI 3.56–10.22) [37]. One plausible reason for higher prevalence of BE in Caucasians may be increased reflux of acid and a reduction in esophageal clearance leading to more severe GERD [38,39]. Studies assessing BE in the Hispanic population show no increased

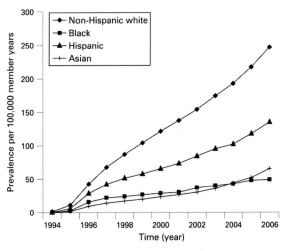

FIGURE 2.1   Prevalence of Barrett's esophagus by year of diagnosis and ethnicity, 1994–2006.

prevalence with 10% among Latinos and 12% among non-Latino whites, respectively [40].

A large database study of 280,075 endoscopic procedures showed that BE was found most often in whites (5% vs Hispanic 2.9% vs black 1.5%, Asian/Pacific islander 1.8%; $p < 0.0001$). Although the racial distribution of the database appeared to be similar to the Centers for Medicare and Medicaid Services (CMS) database, caution must be taken in generalizing the results of this study due to the inherent limitations of any large database such as risk of misclassification, as race group was entered by provider and not the patient, and the diagnosis of BE was made on endoscopic visualization without histologic confirmation [41].

2.4.5 Family History of Barrett's Esophagus or Esophageal Adenocarcinoma

Approximately 7% of the BE patients have a family history of BE or EAC [42,43]. Patients with familial BE were younger at onset of reflux symptoms and at EAC diagnosis compared to nonfamilial BE suggesting genetic factors contributing to the pathogenesis of BE [43]. Patients with BE are more likely to have first- or second-degree relatives with BE compared to normal population without BE (OR 12.2, CI 3.3–44.7) [44]. New susceptibility genes for BE have been identified as two large genome-wide analysis studies reported genetic variants in BE [45,46]. In the recent genome-wide study, Levine et al., after pooling data from 15 studies, identified genetic variants at three loci—CRTC1, FOXP1, and BARX1—that were significantly associated with BE and EAC. The inherited susceptibility to cancer appeared to be more related to early development of BE rather than progression of BE to EAC, emphasizing the need to understand the mechanisms/pathways that lead to the development of BE.

2.4.6 Visceral Obesity

Obesity is a well-known risk factor for GERD and EAC and possibly for BE. In a population-based case–control study, obese individuals with symptoms of acid reflux had markedly higher risk of BE (OR 34.4, 95% CI 6.3–188) than people with reflux alone (OR 9.3, 95% CI 1.4–62.2) or obesity alone (OR 0.7, 95% CI 0.2–2.4) [47]. However, data on whether BMI is an independent risk factor for BE are not robust. Earlier studies evaluating obesity in BE that used patients with normal upper endoscopy as controls suggested a significant association of BMI with BE [48,49]. However, subsequent large systemic reviews and meta-analyses did not find any significant risk of BE in patients with elevated BMI especially after adjusting for GERD; it was concluded that BMI was an indirect risk factor for BE through the precursor lesion caused by GERD [50–52].

Recent literature is supportive of visceral obesity, instead of BMI, as an important risk factor for BE. In a meta-analysis, abdominal obesity, measured by visceral adipose tissue area on abdominal CT, waist–hip ratio, or waist circumference, was significantly associated with BE independent of GERD (OR 1.98, CI 1.52–2.57 for greatest vs lowest category of abdominal obesity) [53]. Factors other than increased intra-abdominal pressure from the mechanical effects of obesity may play a role. Intra-abdominal fat is a metabolically active tissue producing adipokines such as adiponectin and leptin which have effects on cellular proliferation, apoptosis, angiogenesis, and inflammation, and are associated with increased risk of BE.

In contrast to abdominal obesity that is associated with BE, gluteofemoral obesity may have a protective effect in BE. In a study by Rubenstein et al. [54], each 5 cm increment in hip circumference was associated with 13% decreased risk of BE (OR 0.87, CI 0.76–0.99). Although several theories based on positive association of gluteofemoral obesity with insulin sensitivity and adiponectin levels have tried to explain the protective effect of gluteofemoral obesity, it is not clear how gluteofemoral obesity exerts a protective effect on BE. It is postulated that nonabdominal subcutaneous adipose tissue acts as a "metabolic sink" for nonesterified fatty acids thereby offering protection against their deleterious metabolic effects. This may explain the lower prevalence of BE in women in spite of their higher volumes of nonabdominal subcutaneous adipose tissue.

2.4.7 Smoking

The association between smoking and BE is unclear. Cigarette smoking is a strong risk factor for EAC and was long considered a traditional risk factor for BE. Being a potentially modifiable risk factor, several studies examined the association of smoking with BE. The strongest evidence of smoking as a risk factor for BE came from a pooled analysis from five case–control studies within the BEACON Consortium—an international Barrett's and Esophageal Adenocarcinoma Consortium—that found people with BE to be more likely to have ever-smoked compared to population-based controls (OR 1.67, CI 1.04–2.67) or compared to GERD controls (OR 1.61, CI 1.33–1.96). Moreover, the association was dose responsive with increased association seen up to 20 pack-years when it began to plateau [55]. This association of smoking with BE was further confirmed by a recent meta-analysis where having ever-smoked was associated with increased risk of BE compared to non-GERD controls (OR 1.44, 1.20–1.74) and GERD controls (OR 1.42, 1.15–1.76) [56]. On the contrary, in one large case–control study on BE in the United States where a systematic collection of detailed smoking data was obtained, Thrift et al. [57] demonstrated smoking as not being a strong risk factor for BE. Similar results were noted by other investigators and it is speculated that the increased susceptibility of smokers to *Helicobacter pylori* might explain a protective effect against BE [22,58,59].

2.4.8 Alcohol

Excessive alcohol consumption may increase esophageal reflux, cause inflammation, and lead to carcinogenesis. Several studies showed a significant association of heavy alcohol ingestion and esophageal squamous cell carcinoma [60]. However, the evidence on alcohol causing EAC and its precursor BE is limited. Recent studies have shown no association of alcohol intake and BE. A pooled analysis of individual patient data from multiple case–control studies, which included 1169 cases with BE, 1282 population-based controls, and 1418 GERD controls, found no association between BE and alcohol [57].

2.4.9 Diet

Diet is a modifiable risk factor for BE. However, only limited studies have evaluated the association of dietary intake and risk of BE, most of which are case–control studies. Although studies have shown inverse risk between BE and fruit and vegetable intake [61,62], this association is not consistent across all studies with a few studies reporting no association [63]. Two large population-based case–control studies in the United States have reported an inverse association [62,64]; in one study within the Kaiser Permanente northern California population where cases with BE ($n = 296$) were matched with GERD controls ($n = 308$) and population controls ($n = 309$), there was an inverse association of fruit and vegetable intake and BE. Subjects in the fourth quartile of fruit and vegetable consumption were 73% less likely to have BE when compared to first quartile [62]. However, this inverse association was not seen when compared to GERD controls suggesting that patients with GERD may not benefit in terms of risk of BE from increased intake of fruits and vegetables. In the second study, a similar inverse association was seen between fruit and vegetable intake and the risk of BE [64]. A study from Ireland reported similar findings with 40% reduction in BE risk among subjects with greater than 34 portions of fruits and vegetables per week, compared to those with less than 20 portions per week [22]. Authors hypothesize that fruits and vegetables contain antioxidants that are believed to suppress oxidative stress and decrease carcinogenesis [62]. Yet there is no strong evidence to support the protective effect of antioxidants intake and BE. Studies have failed to show any effect of antioxidant intake such as vitamin C, vitamin E, selenium, supplemental vitamins, or beta carotene level on the risk of BE [62–64].

Fiber intake has been inversely linked to the risk of BE [65,66]. Fiber intake is hypothesized to absorb carcinogens from food that pass through gastrointestinal tract and reduce risk of BE [67,68].

2.4.10 Diabetes Mellitus

There is sparse data studying diabetes mellitus (DM) as a risk factor for BE. A population-based study of over 8 million subjects found a significant association between BE and DM. Patients with BE (14,245 subjects) were matched with controls (70,361 subjects); and after adjusting for known risk factors such as obesity, smoking, and GERD, patients with BE were twofold more likely to have diabetes compared to controls (OR 2.03, CI 1.2–3.6) [69]. Being a case–control study, it is difficult to prove causation and more prospective studies are required to understand the link between BE and diabetes. The postulated mechanism of esophageal injury in diabetes involves activation of insulin-like

growth factor pathway that promotes tissue proliferation through activation of the phosphoinositol-3-kinase/AKT/mammalian target of rapamycin pathway and other similar pathways that lead to cell proliferation [70–72].

2.4.11 Obstructive Sleep Apnea

Recent studies have demonstrated obstructive sleep apnea (OSA) as a risk factor for BE independent of obesity and GERD. It has been hypothesized that a more negative intrathoracic pressure due to OSA can lead to increased nocturnal reflux and possibly weakening of the LES leading to the development of BE [73]. This is further supported by the fact that improvement in GERD is seen following the use of continuous positive airway pressure for OSA in several studies [74–76]. Furthermore, patients with OSA are more likely to be obese which predisposes them to a hiatal hernia, increasing the risk of acid reflux and prolonged exposure to gastric acid [77]. In a recent case–control study at the Mayo Clinic, patients with OSA had 80% increased risk of BE as compared to patients without OSA (OR 1.8, 95% CI 1.1–3.2), and this association was independent of other risk factors [78].

2.4.12 Erosive Esophagitis

Untreated erosive esophagitis is considered to be a risk factor for BE. In a recent prospective Swedish study, Ronkainen et al. [79] reported erosive esophagitis as a significant risk factor for BE (relative risk ratio (RR) 5.2, CI 1.2–22.9) and found 8 new cases of BE after following 90 cases of erosive esophagitis for 5 years.

2.4.13 Hiatal Hernia

Patients with BE have larger hiatal hernias than those without BE [80]. It is hypothesized that a hiatal hernia lowers LES pressure, increases the propensity to reflux, and therefore increases the risk of BE [81−83]. A recent meta-analysis of 4390 BE patients revealed a significant association of hiatal hernia and BE (OR 3.94, CI 3.02−5.13), and this association remained significant in subgroup analyses even after adjusting for clinically important confounders such as obesity and GERD [84].

2.4.14 *Helicobacter pylori*

One of the causes attributed to the increasing incidence of EAC and BE is the declining prevalence of *H. pylori* infection. *H. pylori* causes atrophic gastritis and the resulting reduction in parietal cell mass leads to less acid production. Therefore, GERD, erosive esophagitis, and BE are less likely to occur. However, the studies regarding the effect of *H. pylori* on BE are inconsistent. A recently published meta-analysis of 49 studies showed that *H. pylori* infection, especially with cag A strains, is protective against BE (RR, 0.46, 95% CI 0.35−0.60) [59].

2.5 FUTURE DIRECTIONS

American gastroenterology society guidelines recommend screening of high-risk populations for BE but do not specify threshold at which screening should be offered [85,86]. To address this issue, several risk prediction models have been reported. One of them is a prescreening panel for intestinal metaplasia ≥2 cm utilizing age, sex, heartburn, acid reflux, chest pain, abdominal pain, and medications for "stomach" symptoms with area under receiver−operator characteristic curves (AUC) of 0.81 (95% CI 0.76−0.86). This was validated in an independent cohort with AUC of 0.64 (95% CI 0.52−0.77) [87]. Another model, Michigan Barrett's Esophagus pREdiction Tool (M-BERET) has been proposed by Rubenstein et al. [88] incorporating age, GERD symptoms, waist−hip ratio, smoking history with area

under curve of 0.72. These risk prediction models are easy to use but far from perfect. Addition of biomarkers to these models may improve the accuracy of detection of BE. One of the biomarkers showing promise is the trefoil factor 3 (TFF3) identified in a gene expression microarray study to be expressed in high levels at the luminal surface of BE compared to absent expression in normal esophagus and gastric mucosa. In a case−control study using cytosponge TFF3, sensitivity for ≥3 cm BE was found to be 87.2% (95% CI 83.0−90.6) with specificity of 92.4% (95% CI 89.5−94.7) [89]. Other biomarkers which are being studied include microRNA panels to distinguish between GERD and BE and blood-based protein markers. On the other hand, genome-wide association studies have identified multiple susceptibility loci for BE at 6p21, 16q24, 3p13, 9q22, and 19p13.14 [45,46]. Though exciting, these findings are of uncertain clinical significance as these confer only slightly increased risk. Other risk factors for BE which continue to garner interest include increasing use of nitrate fertilizers, reduced sun exposure, and decreasing prevalence of *H. pylori*. Before these risk prediction models are incorporated in clinical practice, they need to be validated in large prospective cohort studies.

References

[1] Barrett NR. Chronic peptic ulcer of the oesophagus and 'oesophagitis'. Br J Surg 1950;38:175−82.
[2] Burgess JN, Payne WS, Andersen HA, Weiland LH, Carlson HC. Barrett esophagus: the columnar-epithelial-lined lower esophagus. Mayo Clinic Proc 1971;46:728−34.
[3] Naef AP, Savary M, Ozzello L. Columnar-lined lower esophagus: an acquired lesion with malignant predisposition. Report on 140 cases of Barrett's esophagus with 12 adenocarcinomas. J Thorac Cardiovasc Surg 1975;70:826−35.
[4] Ronkainen J, Aro P, Storskrubb T, Johansson SE, Lind T, Bolling-Sternevald E, et al. Prevalence of Barrett's esophagus in the general population: an endoscopic study. Gastroenterology 2005;129:1825−31.

[5] Rex DK, Cummings OW, Shaw M, Cumings MD, Wong RK, Vasudeva RS, et al. Screening for Barrett's esophagus in colonoscopy patients with and without heartburn. Gastroenterology 2003;125:1670−7.

[6] Gerson LB, Shetler K, Triadafilopoulos G. Prevalence of Barrett's esophagus in asymptomatic individuals. Gastroenterology 2002;123:461−7.

[7] Ward EM, Wolfsen HC, Achem SR, Loeb DS, Krishna M, Hemminger LL, et al. Barrett's esophagus is common in older men and women undergoing screening colonoscopy regardless of reflux symptoms. Am J Gastroenterol 2006;101:12−17.

[8] Westhoff B, Brotze S, Weston A, McElhinney C, Cherian R, Mayo MS, et al. The frequency of Barrett's esophagus in high-risk patients with chronic GERD. Gastrointest Endosc 2005;61:226−31.

[9] Winters Jr C, Spurling TJ, Chobanian SJ, Curtis DJ, Esposito RL, Hacker 3rd JF, et al. Barrett's esophagus. A prevalent, occult complication of gastroesophageal reflux disease. Gastroenterology 1987;92:118−24.

[10] Spechler SJ. Clinical practice. Barrett's esophagus. N Engl J Med 2002;346:836−42.

[11] Cameron AJ, Zinsmeister AR, Ballard DJ, Carney JA. Prevalence of columnar-lined (Barrett's) esophagus. Comparison of population-based clinical and autopsy findings. Gastroenterology 1990;99:918−22.

[12] Hongo M. Review article: Barrett's oesophagus and carcinoma in Japan. Aliment Pharmacol Ther 2004;20 (Suppl. 8):50−4.

[13] Chen X, Zhu LR, Hou XH. The characteristics of Barrett's esophagus: an analysis of 4120 cases in China. Dis Esophagus 2009;22:348−53.

[14] Andreollo NA, Michelino MU, Brandalise NA, Lopes LR, Trevisan MA, Leonardi LS. Incidence and epidemiology of Barrett's epithelium at the Gastrocentro-UNICAMP. Arq Gastroenterol 1997;34:22−6.

[15] Blot WJ, Devesa SS, Kneller RW, Fraumeni Jr. JF. Rising incidence of adenocarcinoma of the esophagus and gastric cardia. JAMA 1991;265:1287−9.

[16] Powell J, McConkey CC. Increasing incidence of adenocarcinoma of the gastric cardia and adjacent sites. Br J Cancer 1990;62:440−3.

[17] Skinner DB, Walther BC, Riddell RH, Schmidt H, Iascone C, DeMeester TR. Barrett's esophagus. Comparison of benign and malignant cases. Ann Surg 1983;198:554−65.

[18] Playford RJ. New British Society of Gastroenterology (BSG) guidelines for the diagnosis and management of Barrett's oesophagus. Gut 2006;55:442.

[19] Corley DA, Kubo A, Levin TR, Block G, Habel L, Rumore G, et al. Race, ethnicity, sex and temporal differences in Barrett's oesophagus diagnosis: a large community-based study, 1994-2006. Gut 2009;58:182−8.

[20] van Soest EM, Dieleman JP, Siersema PD, Sturkenboom MC, Kuipers EJ. Increasing incidence of Barrett's oesophagus in the general population. Gut 2005;54:1062−6.

[21] Taylor JB, Rubenstein JH. Meta-analyses of the effect of symptoms of gastroesophageal reflux on the risk of Barrett's esophagus. Am J Gastroenterol 2010;105 (1729):1730−7.

[22] Anderson LA, Watson RG, Murphy SJ, Johnston BT, Comber H, Mc Guigan J, et al. Risk factors for Barrett's oesophagus and oesophageal adenocarcinoma: results from the FINBAR study. World J Gastroenterol 2007;13:1585−94.

[23] Thrift AP, Kramer JR, Qureshi Z, Richardson PA, El-Serag HB. Age at onset of GERD symptoms predicts risk of Barrett's esophagus. Am J Gastroenterol 2013;108:915−22.

[24] Neumann CS, Cooper BT. 24 hour ambulatory oesophageal pH monitoring in uncomplicated Barrett's oesophagus. Gut 1994;35:1352−5.

[25] Singh P, Taylor RH, Colin-Jones DG. Esophageal motor dysfunction and acid exposure in reflux esophagitis are more severe if Barrett's metaplasia is present. Am J Gastroenterol 1994;89:349−56.

[26] Bajpai M, Liu J, Geng X, Souza RF, Amenta PS, Das KM. Repeated exposure to acid and bile selectively induces colonic phenotype expression in a heterogeneous Barrett's epithelial cell line. Lab Invest 2008;88:643−51.

[27] Dvorak K, Payne CM, Chavarria M, Ramsey L, Dvorakova B, Bernstein H, et al. Bile acids in combination with low pH induce oxidative stress and oxidative DNA damage: relevance to the pathogenesis of Barrett's oesophagus. Gut 2007;56:763−71.

[28] McQuaid KR, Laine L, Fennerty MB, Souza R, Spechler SJ. Systematic review: the role of bile acids in the pathogenesis of gastro-oesophageal reflux disease and related neoplasia. Aliment Pharmacol Ther 2011;34:146−65.

[29] Fitzgerald RC. Molecular basis of Barrett's oesophagus and oesophageal adenocarcinoma. Gut 2006;55:1810−20.

[30] Wang DH, Souza RF. Biology of Barrett's esophagus and oesophageal adenocarcinoma. Gastrointest Endosc Clin N Am 2011;21:25−38.

[31] Barrett's esophagus: epidemiological and clinical results of a multicentric survey. Gruppo Operativo per lo Studio delle Precancerosi dell'Esofago (GOSPE). Int J Cancer 1991;48:364−8.

[32] Eloubeidi MA, Provenzale D. Clinical and demographic predictors of Barrett's esophagus among patients with gastroesophageal reflux disease: a multivariable analysis in veterans. J Clin Gastroenterol 2001;33:306−9.

[33] Falk GW, Thota PN, Richter JE, Connor JT, Wachsberger DM. Barrett's esophagus in women: demographic features and progression to high-grade dysplasia and cancer. Clin Gastroenterol Hepatol 2005;3:1089−94.

[34] Adeniyi KO. Gastric acid secretion and parietal cell mass: effect of sex hormones. Gastroenterology 1991;101:66−9.

[35] Namiot Z, Sarosiek J, Rourk RM, Hetzel DP, McCallum RW. Human esophageal secretion: mucosal response to luminal acid and pepsin. Gastroenterology 1994;106:973−81.

[36] van Blankenstein M, Looman CW, Johnston BJ, Caygill CP. Age and sex distribution of the prevalence of Barrett's esophagus found in a primary referral endoscopy center. Am J Gastroenterol 2005;100:568−76.

[37] Ford AC, Forman D, Reynolds PD, Cooper BT, Moayyedi P. Ethnicity, gender, and socioeconomic status as risk factors for esophagitis and Barrett's esophagus. Am J Epidemiol 2005;162:454−60.

[38] Champion G, Richter JE, Vaezi MF, Singh S, Alexander R. Duodenogastroesophageal reflux: relationship to pH and importance in Barrett's esophagus. Gastroenterology 1994;107:747−54.

[39] Gillen P, Keeling P, Byrne PJ, Hennessy TP. Barrett's oesophagus: pH profile. Br J Surg 1987;74:774−6.

[40] Keyashian K, Hua V, Narsinh K, Kline M, Chandrasoma PT, Kim JJ. Barrett's esophagus in Latinos undergoing endoscopy for gastroesophageal reflux disease symptoms. Dis Esophagus 2013;26:44−9.

[41] Wang A, Mattek NC, Holub JL, Lieberman DA, Eisen GM. Prevalence of complicated gastroesophageal reflux disease and Barrett's esophagus among racial groups in a multi-center consortium. Dig Dis Sci 2009;54:964−71.

[42] Chak A, Ochs-Balcom H, Falk G, Grady WM, Kinnard M, Willis JE, et al. Familiarity in Barrett's esophagus, adenocarcinoma of the esophagus, and adenocarcinoma of the gastroesophageal junction. Cancer Epidemiol Biomarkers Prevention 2006;15:1668−73.

[43] Verbeek RE, Spittuler LF, Peute A, van Oijen MG, Ten Kate FJ, Vermeijden JR, et al. Familial clustering of Barrett's esophagus and esophageal adenocarcinoma in a European cohort. Clin Gastroenterol Hepatol 2014;12 1656−1663.e1

[44] Chak A, Lee T, Kinnard MF, Brock W, Faulx A, Willis J, et al. Familial aggregation of Barrett's oesophagus, oesophageal adenocarcinoma, and oesophagogastric junctional adenocarcinoma in Caucasian adults. Gut 2002;51:323−8.

[45] Su Z, Gay LJ, Strange A, Palles C, Band G, Whiteman DC, et al. Common variants at the MHC locus and at chromosome 16q24.1 predispose to Barrett's esophagus. Nat Genet 2012;44:1131−6.

[46] Levine DM, Ek WE, Zhang R, Liu X, Onstad L, Sather C, et al. A genome-wide association study identifies new susceptibility loci for esophageal adenocarcinoma and Barrett's esophagus. Nat Genet 2013;45:1487−93.

[47] Smith KJ, O'Brien SM, Smithers BM, Gotley DC, Webb PM, Green AC, et al. Interactions among smoking, obesity, and symptoms of acid reflux in Barrett's esophagus. Cancer Epidemiol Biomarkers Prevention 2005;14:2481−6.

[48] Stein DJ, El-Serag HB, Kuczynski J, Kramer JR, Sampliner RE. The association of body mass index with Barrett's oesophagus. Aliment Pharmacol Ther 2005;22:1005−10.

[49] Bu X, Ma Y, Der R, Demeester T, Bernstein L, Chandrasoma PT. Body mass index is associated with Barrett esophagus and cardiac mucosal metaplasia. Dig Dis Sci 2006;51:1589−94.

[50] Cook MB, Greenwood DC, Hardie LJ, Wild CP, Forman D. A systematic review and meta-analysis of the risk of increasing adiposity on Barrett's esophagus. Am J Gastroenterol 2008;103:292−300.

[51] Corley DA, Kubo A. Body mass index and gastroesophageal reflux disease: a systematic review and meta-analysis. Am J Gastroenterol 2006;101:2619−28.

[52] Hampel H, Abraham NS, El-Serag HB. Meta-analysis: obesity and the risk for gastroesophageal reflux disease and its complications. Ann Intern Med 2005;143:199−211.

[53] Singh S, Sharma AN, Murad MH, Buttar NS, El-Serag HB, Katzka DA, et al. Central adiposity is associated with increased risk of esophageal inflammation, metaplasia, and adenocarcinoma: a systematic review and meta-analysis. Clin Gastroenterol Hepatol 2013;11 1399−1412.e7

[54] Rubenstein JH, Morgenstern H, Chey WD, Murray J, Scheiman JM, Schoenfeld P, et al. Protective role of gluteofemoral obesity in erosive oesophagitis and Barrett's oesophagus. Gut 2014;63:230−5.

[55] Cook MB, Shaheen NJ, Anderson LA, Giffen C, Chow WH, Vaughan TL, et al. Cigarette smoking increases risk of Barrett's esophagus: an analysis of the Barrett's and Esophageal Adenocarcinoma Consortium. Gastroenterology 2012;142:744−53.

[56] Andrici J, Cox MR, Eslick GD. Cigarette smoking and the risk of Barrett's esophagus: a systematic review and meta-analysis. J Gastroenterol Hepatol 2013;28:1258−73.

[57] Thrift AP, Kramer JR, Richardson PA, El-Serag HB. No significant effects of smoking or alcohol

consumption on risk of Barrett's esophagus. Dig Dis Sci 2014;59:108−16.

[58] Steevens J, Schouten LJ, Goldbohm RA, van den Brandt PA. Alcohol consumption, cigarette smoking and risk of subtypes of oesophageal and gastric cancer: a prospective cohort study. Gut 2010;59:39−48.

[59] Fischbach LA, Nordenstedt H, Kramer JR, Gandhi S, Dick-Onuoha S, Lewis A, et al. The association between Barrett's esophagus and *Helicobacter pylori* infection: a meta-analysis. Helicobacter 2012;17 (3):163−75.

[60] Freedman ND, Murray LJ, Kamangar F, Abnet CC, Cook MB, Nyrén O, et al. Alcohol intake and risk of oesophageal adenocarcinoma: a pooled analysis from the BEACON Consortium. Gut 2011;60:1029−37.

[61] Veugelers PJ, Porter GA, Guernsey DL, Casson AG. Obesity and lifestyle risk factors for gastroesophageal reflux disease, Barrett esophagus and esophageal adenocarcinoma. Dis Esophagus 2006;19:321−8.

[62] Kubo A, Levin TR, Block G, Rumore GJ, Quesenberry Jr CP, Buffler P, et al. Dietary antioxidants, fruits, and vegetables and the risk of Barrett's esophagus. Am J Gastroenterol 2008;103:1614−23.

[63] Murphy SJ, Anderson LA, Ferguson HR, Johnston BT, Watson PR, McGuigan J, et al. Dietary antioxidant and mineral intake in humans is associated with reduced risk of esophageal adenocarcinoma but not reflux esophagitis or Barrett's esophagus. J Nutr 2010;140:1757−63.

[64] Thompson OM, Beresford SA, Kirk EA, Vaughan TL. Vegetable and fruit intakes and risk of Barrett's esophagus in men and women. Am J Clin Nutr 2009;89:890−6.

[65] Mulholland HG, Cantwell MM, Anderson LA, Johnston BT, Watson RG, et al. Glycemic index, carbohydrate and fiber intakes and risk of reflux esophagitis, Barrett's esophagus, and esophageal adenocarcinoma. Cancer Causes Control 2009;20:279−88.

[66] McFadden DW, Riggs DR, Jackson BJ, Cunningham C. Corn-derived carbohydrate inositol hexaphosphate inhibits Barrett's adenocarcinoma growth by pro-apoptotic mechanisms. Oncol Rep 2008;19:563−6.

[67] Burkitt DP. The protective properties of dietary fiber. N C Med J 1981;42:467−71.

[68] Brown LM, Swanson CA, Gridley G, Swanson GM, Schoenberg JB, Greenberg RS, et al. Adenocarcinoma of the esophagus: role of obesity and diet. J Natl Cancer Inst 1995;87:104−9.

[69] Iyer PG, Borah BJ, Heien HC, Das A, Cooper GS, Chak A. Association of Barrett's esophagus with type II diabetes mellitus: results from a large population-based case−control study. Clin Gastroenterol Hepatol 2013;11 1108−1114.e5

[70] Chen SC, Chou CK, Wong FH, Chang CM, Hu CP. Overexpression of epidermal growth factor and insulin-like growth factor-I receptors and autocrine stimulation in human esophageal carcinoma cells. Cancer Res 1991;51:1898−903.

[71] Liu YC, Leu CM, Wong FH, Fong WS, Chen SC, Chang C, et al. Autocrine stimulation by insulin-like growth factor I is involved in the growth, tumorigenicity and chemoresistance of human esophageal carcinoma cells. J Biomed Sci 2002;9:665−74.

[72] Takaoka M, Harada H, Andl CD, Oyama K, Naomoto Y, Dempsey KL, et al. Epidermal growth factor receptor regulates aberrant expression of insulin-like growth factor-binding protein 3. Cancer Res 2004;64:7711−23.

[73] Romero-Corral A, Caples SM, Lopez-Jimenez F, Somers VK. Interactions between obesity and obstructive sleep apnea: implications for treatment. Chest 2010;137:711−19.

[74] Kerr P, Shoenut JP, Millar T, Buckle P, Kryger MH. Nasal CPAP reduces gastroesophageal reflux in obstructive sleep apnea syndrome. Chest 1992;101:1539−44.

[75] Friedman M, Gurpinar B, Lin HC, et al. Impact of treatment of gastroesophageal reflux on obstructive sleep apnea-hypopnea syndrome. Ann Otol Rhinol Laryngol 2007;116:805−11.

[76] Tawk M, Goodrich S, Kinasewitz G, Orr W. The effect of 1 week of continuous positive airway pressure treatment in obstructive sleep apnea patients with concomitant gastroesophageal reflux. Chest 2006;130:1003−8.

[77] Siupsinskiene N, Adamonis K, Toohill RJ. Usefulness of assessment of voice capabilities in female patients with reflux-related dysphonia. Medicina (Kaunas, Lithuania) 2009;45:978−87.

[78] Leggett CL, Gorospe EC, Calvin AD, Harmsen WS, Zinsmeister AR, Caples S, et al. Obstructive sleep apnea is a risk factor for Barrett's esophagus. Clin Gastroenterol Hepatol 2014;12 583−588.e1

[79] Ronkainen J, Talley NJ, Storskrubb T, Johansson SE, Lind T, Vieth M, et al. Erosive esophagitis is a risk factor for Barrett's esophagus: a community-based endoscopic follow-up study. Am J Gastroenterol 2011;106:1946−52.

[80] Cameron AJ. Barrett's esophagus: prevalence and size of hiatal hernia. Am J Gastroenterol 1999;94:2054−9.

[81] Kahrilas PJ, Kim HC, Pandolfino JE. Approaches to the diagnosis and grading of hiatal hernia. Best Pract Res Clin Gastroenterol 2008;22:601−16.

[82] Buttar NS, Falk GW. Pathogenesis of gastroesophageal reflux and Barrett esophagus. Mayo Clinic Proc 2001;76:226—34.

[83] Gordon C, Kang JY, Neild PJ, Maxwell JD. The role of the hiatus hernia in gastro-oesophageal reflux disease. Aliment Pharmacol Ther 2004;20:719—32.

[84] Andrici J, Tio M, Cox MR, Eslick GD. Hiatal hernia and the risk of Barrett's esophagus. J Gastroenterol Hepatol 2013;28:415—31.

[85] Wang KK, Sampliner RE. Practice Parameters Committee of the American College of Gastroenterology. Updated guidelines 2008 for the diagnosis, surveillance and therapy of Barrett's esophagus. Am J Gastroenterol 2008;103:788—97.

[86] Spechler SJ, Sharma P, Souza RF, Inadomi JM, Shaheen NJ. American Gastroenterological Association medical position statement on the management of Barrett's esophagus. Gastroenterology 2011;140:1084—91.

[87] Liu X, Wong A, Kadri SR, Corovic A, O'Donovan M, Lao-Sirieix P, et al. Gastro-esophageal reflux disease symptoms and demographic factors as a pre-screening tool for Barrett's esophagus. PLoS One 2014;9:e94163.

[88] Rubenstein JH, Morgenstern H, Appelman H, Scheiman J, Schoenfeld P, McMahon Jr LF, et al. Prediction of Barrett's esophagus among men. Am J Gastroenterol 2013;108:353—62.

[89] Ross-Innes CS, Debiram-Beecham I, O'Donovan M, Walker E, Varghese S, Lao-Sirieix P, et al. Evaluation of a minimally invasive cell sampling device coupled with assessment of trefoil factor 3 expression for diagnosing Barrett's esophagus: a multi-center case-control study. PLoS Med 2015;12:e1001780.

[90] Spechler SJ. Barrett's esophagus. Semin Gastrointest Dis 1996;7:51.

Metaplasia and Dysplasia in Barrett's Esophagus

Deepa T. Patil[1,2]

[1]Cleveland Clinic Lerner College of Medicine, Cleveland, OH, United States
[2]Department of Pathology, Robert J. Tomsich Pathology and Laboratory Medicine Institute, Cleveland Clinic, Cleveland, OH, United States

3.1 INTRODUCTION

Barrett's esophagus (BE) occurs in the setting of chronic gastroesophageal reflux disease (GERD). It is characterized by conversion of the normal esophageal squamous epithelium into metaplastic columnar epithelium [1]. Since its original description by Tileston et al. [2] there have been several alterations of this definition. It is now widely accepted that BE is an alteration of the esophageal mucosa that is visible endoscopically and has a corresponding histologic abnormality. The endoscopic aspects about recognizing BE are reviewed in greater detail in Chapter 5: "Diagnosis of Barrett's Esophagus." The discussion here focuses on a brief review of the definition of BE, intestinal metaplasia at gastroesophageal junction (GEJ), and the gross and microscopic aspects of BE and BE-related dysplasia.

3.2 NORMAL ANATOMY AND HISTOLOGY

GEJ, as defined in North America, is the junction between the tubular esophagus and proximal stomach. The location of GEJ is approximated by the most proximal extent of the gastric folds [3]. In Asia, this is determined by locating the distal extent of palisade of longitudinal veins [4–6]. The squamocolumnar junction (SCJ), also known as the Z-line, is the junction of the squamous mucosa and columnar mucosa. It is normally somewhat irregular in appearance. The location of SCJ and GEJ may not coincide endoscopically. Furthermore, the precise anatomic localization remains difficult in many cases, particularly in the setting of a hiatal hernia [7]. Thus, the precise location of the biopsy is crucial in determining the presence of BE.

D. Pleskow & T. Erim (Eds): Barrett's Esophagus.
DOI: http://dx.doi.org/10.1016/B978-0-12-802511-6.00003-X

Traditionally, the narrow segment of mucus-secreting columnar mucosa distal to the squamous esophageal mucosa, but proximal to acid-secreting gastric oxyntic mucosa has been termed the gastric cardia. In recent years, the existence of the gastric cardia as a native structure has been called into question by some authors who believe that cardiac-type mucosa is always metaplastic, likely in response to gastroesophageal reflux [8,9]. While metaplastic cardiac-type mucosa undoubtedly is frequently identified in the distal esophagus, evidence from detailed studies of the anatomy and histology of the GEJ, including pediatric autopsy series, supports the notion that the gastric cardia is a native structure [10–12]. Thus, there is sufficient evidence to support the presence of a small zone of native cardiac mucosa in the most proximal stomach, and in many individuals, metaplastic cardiac-type mucosa of variable length in the distal esophagus.

3.3 HISTOLOGY OF BARRETT'S ESOPHAGUS

Although Tileston described the presence of reflux-induced esophageal ulcers in an autopsy series, it was not until nearly 50 years later that Dr Norman Barrett confirmed that this was indeed the lower esophagus that was lined by columnar epithelium [13]. However, it was not until later in 1976 that Paull et al. reported that columnar metaplasia of esophagus (or columnar-lined esophagus (CLE)) in fact is a mosaic of three different types of epithelia: (a) fundic type (with oxyntic glands), (b) junctional type (with cardiac-type glands), and (3) specialized type (with goblet cells) [14]. Multiple studies have documented that the risk of dysplasia and adenocarcinoma is significantly increased when the metaplastic epithelium harbors intestinal metaplasia. For this reason, in the United States, we currently diagnose BE based on the 2008 American College

of Gastroenterology Practice Parameters Committee Guidelines [15] and the recent Position Statement issued by the American Gastroenterological Association [16,17]. Per these guidelines, BE is defined as "change in the distal esophageal epithelium of any length that can be recognized as columnar-type mucosa at endoscopy and is confirmed to have intestinal metaplasia by biopsy of the tubular esophagus" [16]. Thus, the definition consists of a two-pronged approach—an endoscopically visible mucosal change/abnormality and a histologic correlate characterized by the identification of goblet cells. This definition, however, is not universally accepted. In fact, the British Society of Gastroenterology does not require the presence of intestinal metaplasia to diagnose BE [18]. Similarly, the Japanese require the documentation of CLE with or without goblet cells [19]. Additional details regarding evolving definition of Barrett's esophagus can be found in Chapter 5: "Diagnosis of Barrett's Esophagus." An international agreement on a definition of Barrett's was achieved in 2015 by the BOB CAT (Benign Barrett's and CAncer Taskforce) consensus group for the first time as "Barrett's is defined by the presence of columnar mucosa in the esophagus and it should be stated whether intestinal metaplasia is present above the gastroesophageal junction" [20].

3.3.1 Endoscopic Findings

BE is characterized by irregular tongues of salmon-colored mucosa within the esophagus with or without areas of erosion and ulceration. BE-related dysplasia can be subtle and difficult to recognize endoscopically. It may appear as erosions, ulcers, and slightly elevated mucosa to areas of nodularity, stricture, or polypoid growth. The current guidelines recommend the use of advanced imaging techniques that should be superimposed on high-resolution white light endoscopy using high-definition systems [21,22]. These techniques have the advantage of improving the detection rate of flat dysplastic lesions that are not readily visualized using standard endoscopic tools. Additional information regarding endoscopic techniques can be found in Chapter 6: "Screening and Surveillance of Barrett's Esophagus," Chapter 7: "In Vivo Optical Detection of Dysplasia in Barrett's Esophagus With Endoscopic Light Scattering Spectroscopy," Chapter 8: "Enhanced Imaging of the Esophagus: Optical Coherence Tomography," and Chapter 9: "Enhanced Imaging of the Esophagus: Confocal Laser Endomicroscopy."

3.3.2 Microscopic Findings

The surface and the crypt epithelium of Barrett's mucosa are composed of mucinous epithelium interspersed with goblet cells, enterocytes, and cells with features that are intermediate between intestinal and gastric foveolar cells. The surface epithelium usually shows neutrophilic inflammation, ulceration, or a villiform appearance in some cases. In addition to goblet cells, gastric foveolar-type cells (incomplete intestinal metaplasia), and intestinal absorptive-type cells (complete intestinal metaplasia), the metaplastic glands may also contain Paneth cells, neuroendocrine cells, and even pancreatic acinar cells. The lamina propria surrounding the glands contains variable numbers of inflammatory cells and fibroblasts.

Studies have shown that CLE is a heterogeneous mucosa that demonstrates considerable variability in the distribution of goblet cells, including a distinct gradient with the proximal segment containing a higher density of goblet cells compared to the distal segment [23]. The distal segment of BE has a higher density of oxyntic-type glands compared to the proximal segment [24]. Thus, the proportion of goblet cells is higher at the neo-squamocolumar junction and proximal esophagus. They may range from focal and few goblet cells per crypt to diffuse and numerous in other biopsy fragments. A retrospective study has shown that at least eight biopsy specimens are required to adequately assess intestinal metaplasia. In addition, the yield of intestinal metaplasia is lower in the presence of short-segment BE [25].

3.3.3 Goblet Cells versus Pseudogoblet Cells

Barrett's epithelium is typically composed of cells that resemble incomplete intestinal (type II or type III) metaplasia of stomach with chronic gastritis. Complete intestinal metaplasia (type I) is less common. Regardless, subtyping intestinal metaplasia has no clinical implication on the management of BE. Goblet cells are best identified by virtue of their shape and the chemical composition of intracytoplasmic mucin (Fig. 3.1). Because they are rich in acidic mucins (predominantly sialomucins admixed with lesser quantities of sulfated mucins), they acquire a basophilic cytoplasmic blush that is readily recognized on a routine hematoxylin and eosin-stained tissue section [26]. Histochemical stains for acidic mucins, such as Alcian blue at pH 2.5, show intense

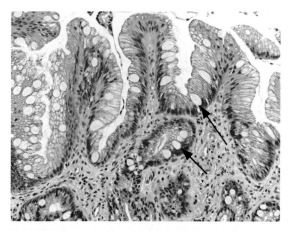

FIGURE 3.1 Barrett's esophagus. The biopsy shows esophageal squamous mucosa that is replaced by glandular mucosa containing goblet cells (arrows).

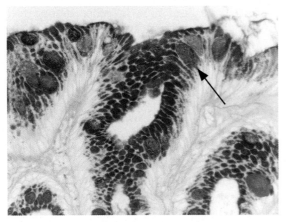

FIGURE 3.2   Periodic acid–Schiff (PAS)/Alcian blue at pH 2.5 demonstrates incomplete intestinal metaplasia. Goblet cells containing acid mucin stain intensely blue with Alcian blue (arrow), while the adjacent columnar cells containing neutral mucin stain with PAS (magenta staining cells within the epithelium).

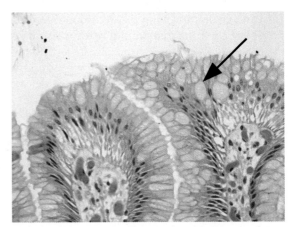

FIGURE 3.3   Pseudogoblet cells. These are markedly distended foveolar epithelial cells (arrow), which may mimic goblet cells. They are usually distributed in a continuous fashion compared to true goblet cells, which tend to be dispersed sporadically throughout the metaplastic epithelium.

dark-blue/magenta staining for this combination of sialomucins and sulfated mucins, which contrasts with the predominantly periodic acid–Schiff (PAS)-positive neutral mucins found within the adjacent gastric foveolar-type cells (Fig. 3.2). Routine Alcian blue staining is costly and time-consuming. Since goblet cells can be quite readily recognized on routine hematoxylin and eosin-stained slide, routine use of Alcian blue is not recommended for the diagnosis of BE.

On occasion, markedly distended foveolar epithelial cells may mimic goblet cells ("pseudogoblet cells") and cause diagnostic confusion. They are usually distributed in a continuous fashion compared to true goblet cells, which tend to be dispersed sporadically throughout the metaplastic epithelium (Fig. 3.3). Additionally, they stain pale eosinophilic on hematoxylin-eosin stain and contain neutral mucin that does not react with Alcian blue at pH 2.5. Caution must be exercised when interpreting histochemical stains for detecting goblet cells. The columnar cells located between the goblet cells contain small

quantities of acidic mucin and may show some Alcian blue positivity (so-called columnar blues). In the absence of goblet cell metaplasia, the identification of these cells does not fulfill the criteria for a definitive diagnosis of BE.

3.4 INTESTINAL METAPLASIA OF THE EGJ

The finding of intestinal metaplasia in biopsies obtained from the esophagogastric junction (EGJ) raises two diagnostic possibilities: ultrashort-segment BE or chronic carditis with intestinal metaplasia (CIM). Unfortunately, clinical, endoscopic, or pathologic findings do not allow one to accurately distinguish between these two entities. There is clear evidence that CIM carries a lower risk of neoplastic progression than either short- or long-segment BE [27–31] In a recent study, Srivastava et al. showed that in a mucosal biopsy from EGJ, the presence of the following features were significantly associated with a diagnosis of BE over CIM: (1) crypt disarray and atrophy, (2) incomplete and diffuse IM, (3) multilayered epithelium, (4) squamous epithelium overlying columnar crypts with IM, (5) hybrid glands, and (6) esophageal glands/ducts [32].

The expression of *CDX2*, a caudal homeobox gene expressed during development, is specific evidence of intestinal differentiation [33]. Several studies have shown that *CDX2* is expressed in BE-related intestinal metaplasia [34−36]. To date, no direct comparisons have been published with regard to potential expression differences between CIM and short-segment BE. Several other markers evaluated for this purpose include Das1, MUC1, MUC2, MUC5AC, MUC6, CK7/CK20 pattern, and CD10. These studies are hampered by differences in endoscopic biopsy protocols and study populations that contribute to apparent discrepancies in their results and lack of reproducibility. Thus, the clinical utility of evaluating intestinal metaplasia of the EGJ using these various biomarkers has not yet been established.

3.5 BARRETT'S ESOPHAGUS-RELATED DYSPLASIA

All patients with BE are at risk of developing esophageal adenocarcinoma [37]. The vast majority of adenocarcinomas arise through a metaplasia-dysplasia-carcinoma sequence. Mapping studies have documented epithelial dysplasia in mucosa adjacent to most adenocarcinomas in resection specimens, supporting a dysplasia-carcinoma sequence [38]. In addition, there are also studies that have reported patients progressing from dysplasia to adenocarcinoma in serial endoscopic biopsies [39,40]. Histologic evaluation of dysplasia in endoscopic biopsy samples remains the main method of risk assessment in patients with BE. Epithelial dysplasia, particularly high-grade dysplasia, is considered to be one of the most important risk factors for both synchronous

and metachronous esophageal adenocarcinoma [41−43]. Therefore, its identification is an integral part of cancer screening and surveillance programs as well as a trigger point for therapeutic intervention.

Dysplasia is defined as neoplastic change of the epithelium that remains confined within the basement membrane of the gland from which it arises (ie, intraepithelial neoplasia) [44]. Grossly, dysplastic epithelium may demonstrate a spectrum of mucosal changes that ranges from ulcers to flat or elevated/polypoid lesions. This morphologic spectrum forms the basis of the Paris classification used by many gastroenterologists [45]. On occasion, dysplastic mucosa may be indistinguishable from adjacent nondysplastic mucosa.

The most widely accepted histologic grading scheme for Barrett's-related dysplasia has been adapted from the classification system used for idiopathic inflammatory bowel disease-related dysplasia and is discussed below [44]. In most Western countries, including the United States, Barrett's-related dysplasia is classified as negative, indefinite, or positive (low grade or high grade). However, pathologists from Asia and Europe prefer the Vienna system of classification. The Vienna system of classifying dysplasia is very similar to the one described earlier, except that the term "noninvasive neoplasia" is used for dysplasia and "suspicious for invasive carcinoma" is used when the cytoarchitectural features are equivocal for tissue invasion. The categories in the Vienna system are: negative for neoplasia, indefinite for neoplasia, noninvasive low-grade neoplasia, noninvasive high-grade neoplasia (includes noninvasive carcinoma in situ and suspicious of invasive carcinoma), and invasive neoplasia (intramucosal and submucosal carcinoma).

3.5.1 Negative for Dysplasia

One of the unique features of metaplastic Barrett's mucosa is that there is a certain degree of "baseline atypia" which is most pronounced within the regenerative glandular compartment at the base of the mucosa. Importantly, these nuclear changes do not involve the surface epithelial cells (surface maturation) and as such, these biopsies are classified as negative for dysplasia.

Active inflammation, and its attendant neutrophil-mediated epithelial cell injury, is capable of producing profound cytologic alterations that overlap with those of Barrett's-related dysplasia. Distinguishing reactive cytologic atypia from dysplasia is frequently very difficult, if not impossible. The appearance from low magnification is critical in this evaluation, because truly dysplastic epithelium usually appears darker (hyperchromatic) than normal at this power. Confirmation of these changes is required at higher magnification and reveals nuclear enlargement, hyperchromasia, crowding, and irregular nuclear contours. In addition, inspection at higher power enables one to determine whether these changes extend onto the mucosal surface. Accurate assessment of the changes involving the mucosal surface is more difficult when faced with a tangentially sectioned biopsy specimen.

In contrast to dysplasia, reactive atypia has a more uniform appearance among the cells in question, whereas dysplastic nuclei are pleomorphic and thus vary considerably from one cell to the next. While cell size does not distinguish between reactive atypia and dysplasia, the nuclear to cytoplasmic (N:C) ratio is increased in the setting of dysplasia when compared with reactive cells. The chromatin distribution pattern is also helpful, as reactive nuclei have a more open chromatin pattern with prominent nucleoli, which contrasts with the more condensed chromatin pattern seen in dysplastic nuclei. In practice, one needs to weigh all of these features together when deciding whether or not the changes qualify as dysplasia.

3.5.2 Indefinite for Dysplasia

The diagnosis of indefinite for dysplasia should be reserved for cases where: (i) the cytologic and glandular architectural changes exceed the so-called baseline atypia of metaplastic specialized columnar epithelium, but fall short of low-grade dysplasia; (ii) when coexisting inflammation or ulceration is associated with striking cytologic atypia precluding a definitive distinction between regenerative atypia and dysplasia; or (iii) there is marked glandular distortion in the absence of surface nuclear changes which would be diagnostic of dysplasia.

3.5.3 Positive for Dysplasia

The most common form of Barrett's dysplasia is the intestinal ("adenomatous") type dysplasia. Two other forms of dysplasia that were recently characterized include the nonadenomatous type (gastric foveolar type) and basal crypt dysplasia. Intestinal-type dysplasia refers to the type of dysplasia that resembles a sporadic adenoma of the intestine. It is more commonly seen in its pure form, and on occasions, can be admixed with gastric foveolar type of dysplasia.

3.6 LOW-GRADE (INTESTINAL-TYPE) DYSPLASIA

The glandular architecture is mildly distorted in low-grade dysplasia, as the crypts remain parallel to one another, with minimal crypt branching or budding. The crypts are lined by cells with enlarged, hyperchromatic, and stratified nuclei with irregular nuclear membranes. These changes extend from the crypts to involve the mucosal surface (Fig. 3.4a). Goblet cells are often decreased in number and so-called dystrophic goblet cells, wherein the nucleus is located at the apical aspect of the cell, may also be present.

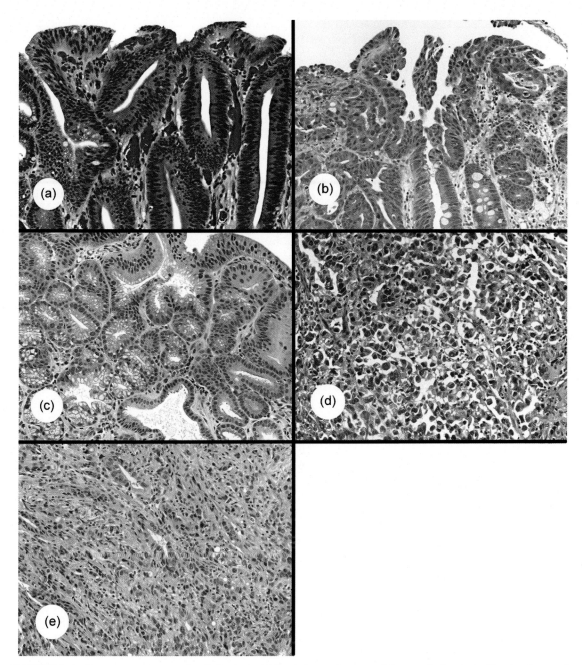

FIGURE 3.4   Barrett's esophagus-related dysplasia. (a) Barrett's esophagus with low-grade dysplasia. The dysplastic cells contain enlarged hyperchromatic nuclei with slightly irregular contours. The nuclear changes extend out from the base of the glands onto the mucosal surface where there is also significant overlapping and crowding. Note the small round nuclei of the nondysplastic glands beneath the dysplastic epithelium. (b) Barrett's esophagus with high-grade dysplasia. This focus of high-grade dysplasia is characterized by severe cytologic atypia, including markedly enlarged, irregular nuclei with coarse chromatin and small nucleoli. There is also an area of cribriform growth. (c) Barrett's gastric foveolar-type dysplasia, high-grade. H&E stained section of Barrett's esophageal biopsy showing full-thickness mucosal replacement by crowded glands and nonvilliform architecture. The cells demonstrate basally oriented monolayered and uniform nuclei with abundant pale eosinophilic to mucinous cytoplasm. (d) Intramucosal adenocarcinoma. The lamina propria is replaced by sheets of dysplastic glands with back-to-back arrangement of glands with very little intervening lamina propria. (e) Submucosal adenocarcinoma showing infiltrative, angulated dysplastic glands surrounded by stromal desmoplasia.

3.7 HIGH-GRADE (INTESTINAL-TYPE) DYSPLASIA

In high-grade dysplasia, both the cytologic atypia and architectural complexity are more pronounced. The crypts are crowded and show a "back-to-back" or cribriform arrangement. The nuclei show full-thickness stratification with marked hyperchromasia and prominent nucleoli. Loss of nuclear polarity, where the long axis of the nucleus is no longer perpendicular to the basement membrane, is a frequent finding (Fig. 3.4b).

3.8 GASTRIC FOVEOLAR-TYPE DYSPLASIA (NONADENOMATOUS DYSPLASIA)

More recently, a second type of Barrett's dysplasia, namely, the gastric foveolar type, was defined with respect to its prevalence, diagnostic criteria, and natural history [46—48]. Cumulative data shows that the overall incidence of this subtype of dysplasia ranges from 7% to 15% among patients with Barrett's-related dysplasia. It is more common in women compared to men and the patients are typically at least a decade older than those with intestinal-type dysplasia. Most cases of Barrett's gastric foveolar-type dysplasia are high grade and neoplastic progression occurs in up to 64% of patients [47]. Gastric foveolar dysplasia often shows immunoreactivity with MUC5AC but is negative for intestinal markers CDX2, villin, and MUC2.

In contrast to intestinal (adenomatous) dysplasia, Barrett's gastric foveolar-type dysplasia is characterized by a uniform monolayer of basally oriented nuclei with abundant apical cytoplasm (Fig. 3.4c). Architecturally, it is characterized by full-thickness replacement of the mucosa and in the great majority of cases, by a glandular rather than villiform growth pattern.

The grading system for gastric foveolar-type dysplasia is similar to intestinal (adenomatous) dysplasia (negative, indefinite, low grade, and high grade). However, the criteria for classifying the grade of dysplasia are somewhat different. Low-grade gastric foveolar-type dysplasia is characterized by slightly crowded glands lined by regular, nonstratified cells with nuclei that are 2—3 times the size of a small mature lymphocyte. There is mild nuclear pleomorphism. High-grade dysplasia is characterized by crowded glandular architecture and villiform growth pattern of the mucosa. The glands are lined by cells with nuclei that are 3—4 times the size of a small lymphocyte. The nuclei are still basally oriented and nucleoli are frequently present [47].

3.9 BASAL CRYPT DYSPLASIA

Molecular evidence suggests that BE-related dysplasia begins in the crypt bases and progressively involves the upper half of the crypts and the surface epithelium (so-called basal crypt dysplasia) [49]. Morphologically, basal crypt dysplasia has all of the features of traditional low-grade dysplasia, but is limited to the crypt bases. In their study, Lomo et al. found that basal crypt dysplasia was seen in biopsies devoid of acute inflammation, and in 47% of cases, there was evidence of full-thickness dysplasia elsewhere in the biopsies.

3.10 INTRAMUCOSAL ADENOCARCINOMA

Intramucosal adenocarcinoma (IMC) is defined by the presence of lamina propria or muscularis mucosae invasion. In addition to demonstrating invasion of single cells within the lamina propria, a diagnosis of IMC is also rendered based on architectural features. These features include the presence of sheets of dysplastic glands replacing the lamina propria (with very little intervening lamina propria; Fig. 3.4d), small, angulated, abortive glandular profiles infiltrating the lamina propria, and "never-ending" glandular pattern [50]. Although establishing a diagnosis of IMC may be relevant due to the small, but finite, risk of lymph node metastasis, as endoscopic mucosal resection is the therapeutic procedure of choice for both nodular HGD and IMC, its distinction from HGD is less important in biopsy specimens.

3.11 SUBMUCOSAL ADENOCARCINOMA

Submucosal adenocarcinoma is defined by the presence of dysplastic glands surrounded by desmoplastic stromal response (Fig. 3.4e). A diagnosis of submucosal adenocarcinoma can be challenging on biopsy specimens, in part due to the superficial nature of endoscopic biopsies that do not typically sample the submucosa. Additionally, Barrett's mucosa is associated with duplication of muscularis mucosae, which can pose diagnostic difficulty [51]. In this unique musculo-fibrous anomaly of Barrett's mucosa, there is development of a new inner layer of muscularis mucosae, which is separated from the deeper layer of true muscularis mucosae by loose fibrovascular stroma. Thus, on mucosal biopsy samples, invasion beyond this newly developed smooth muscle layer can be potentially misdiagnosed as submucosal invasion.

3.12 DIAGNOSTIC ISSUES IN BARRETT'S-RELATED DYSPLASIA

3.12.1 Sampling Error

Dysplasia may extend diffusely throughout a BE segment, or the changes may be focal and limited to a small area of one fragment in a patient with multiple biopsy specimens. When dysplasia is diffuse, there is a high likelihood that a rigorous biopsy protocol will detect foci of dysplasia at a high frequency; however, small foci may go unsampled. The need for a thorough biopsy sampling is further emphasized by the fact that high-grade dysplasia and even early adenocarcinoma may not be associated with an endoscopically visible lesion [41,52]. Harrison et al. showed that a minimum of eight biopsies was needed to maximize the yield of detecting dysplasia [25]. Four-quadrant, well-oriented biopsies obtained at an interval of 1−2 cm (Seattle protocol) showed a higher sensitivity of detecting dysplasia [53].

3.12.2 Observer Variation in Barrett's Esophagus-Related Dysplasia

Another problem facing the pathologist, gastroenterologist, thoracic surgeon, and ultimately, the patient is both the intra- and interobserver variation in the diagnosis of dysplasia. Given the spectrum of changes from baseline atypia to low-grade to high-grade dysplasia, it is not surprising that this variation exists. Reid et al. found this variation to be most significant in distinguishing negative for dysplasia from low-grade dysplasia or indefinite for dysplasia [54]. This study described overall agreement in terms of a percentage, which does not take into account agreement that may occur by chance alone. Two other studies by Montgomery et al. and Downs-Kelly et al., using kappa statistical analysis (which accounts for agreement occurring by chance alone), confirmed a high degree of intra- and interobserver variation among these same diagnostic categories, even among pathologists with a special interest in gastrointestinal pathology [55]. The study performed by Downs-Kelly and colleagues showed that it is difficult to reliably distinguish high-grade dysplasia from intramucosal and submucosal adenocarcinoma on biopsy specimens. This variation underscores the need to obtain multiple opinions in challenging cases and further supports the AGA requirement that a diagnosis of dysplasia should be confirmed by an expert gastrointestinal pathologist.

3.12.3 Squamous Overgrowth (Buried Barrett's Esophagus)

Buried Barrett's esophagus occurs in BE patients who have been on antireflux therapy or have undergone endoscopic ablation therapy. This phenomenon is recognized by the presence of islands of squamous mucosa within the Barrett's segment. In a recent study, buried Barrett's mucosa was found in 28% of BE patients without any history of ablation therapy [56]. The importance of recognizing this phenomenon is that it can underestimate the true length of BE at endoscopy, and rarely, patients may develop subsquamous neoplasia following ablation therapy [57].

3.13 SURROGATE BIOMARKERS FOR ASSESSING RISK OF ESOPHAGEAL ADENOCARCINOMA AND FUTURE DIRECTIONS

Given the limitations of light microscopy, several adjunctive techniques have been proposed as having a possible role in the screening or surveillance of patients with BE. For virtually every marker tested, there is an increased probability of finding an abnormality as one progresses along the dysplasia–carcinoma sequence. Certain markers are detectable early in the sequence, whereas others are found at later stages. The ideal marker would be detectable early in the metaplasia–dysplasia–carcinoma sequence, even before there is morphologic evidence of

dysplasia, and capable of distinguishing progressors from nonprogressors.

Numerous studies have evaluated p53 expression by immunohistochemistry, most of which attempt to correlate the degree of p53 expression with the grade of dysplasia or solely as a marker of increased risk of progressing to adenocarcinoma. p53 overexpression has been observed in 9–60% of cases with low-grade dysplasia and 55–100% of cases with high-grade dysplasia [58–61]. Although some have advocated the use of p53 immunohistochemistry to confirm a diagnosis of dysplasia and/or assist in grading of dysplasia, its use has not been widely accepted. There is some discrepancy between p53 expression as detected by immunohistochemistry and molecular alterations detectable at the gene level [62,63]. Also, the lack of a standardized immunohistochemical technique likely accounts for some of the discrepant data reported in the literature.

DNA content, as measured by flow cytometry, has also been evaluated in patients with BE, but the results are conflicting. A prospective study found that patients with negative, indefinite, or low-grade dysplasia histology and no evidence of aneuploidy or increased 4N fractions by flow cytometry had a cumulative 0% 5-year cancer risk, compared to a 28% risk for patients with either aneuploidy or increased 4N fractions [64]. Patients with baseline increased 4N, aneuploidy, and high-grade dysplasia had 5-year cancer rates of 56%, 43%, and 59%, respectively. In contrast to the results of Reid et al., Fennerty et al. found discordance between flow cytometric abnormalities and dysplasia in BE patients [65].

Although numerous others potential individual biomarkers of neoplastic progression in BE patients have also been evaluated with variable results (eg, Ki-67, *bcl*-2, cyclin D1, *p16*, EGFR, c-*erb*B-2), microarray-based technologies are well suited for surveying genomic abnormalities on a much broader scale. These methods allow for the rapid comparison of chromosomal copy numbers or relative expression of thousands of genes in a single assay, creating genomic profiles for the tissues tested. Not surprisingly, earlier studies [66–68] have identified a long list of chromosomal abnormalities and genes that are up or downregulated as one proceeds along the metaplasia–dysplasia–carcinoma sequence in BE. More recent studies have documented differential expression of miRNAs and protein glycosylation products in the Barrett's carcinogenesis pathway [69]. However, much work is needed to implement these assays in large-scale high-risk population screening to identify early preneoplastic changes and to determine whether or not they have a potential role in selecting those subset of patients who are at greatest risk of neoplastic progression.

References

[1] Haggitt RC. Barrett's esophagus, dysplasia, and adenocarcinoma. Hum Pathol 1994;25:982–93.

[2] Tileston W. Peptic ulcer of the oesophagus. Am J Med Sci 1906;132:240–65.

[3] Ofman JJ, Shaheen NJ, Desai AA, et al. The quality of care in Barrett's esophagus: endoscopist and pathologist practices. Am J Gastroenterol 2001;96:876–81.

[4] Choi DW, Oh SN, Baek SJ, et al. Endoscopically observed lower esophageal capillary patterns. Korean J Intern Med 2002;17:245–8.

[5] Takubo K, Aida J, Sawabe M, et al. The normal anatomy around the oesophagogastric junction: a histopathologic view and its correlation with endoscopy. Best Pract Res Clin Gastroenterol 2008;22:569–83.

[6] Takubo K, Vieth M, Aida J, et al. Differences in the definitions used for esophageal and gastric diseases in different countries: endoscopic definition of the esophagogastric junction, the precursor of Barrett's adenocarcinoma, the definition of Barrett's esophagus, and histologic criteria for mucosal adenocarcinoma or high-grade dysplasia. Digestion 2009;80:248–57.

[7] Spechler SJ, Goyal RK. The columnar-lined esophagus, intestinal metaplasia, and Norman Barrett. Gastroenterology 1996;110:614–21.

[8] Chandrasoma PT, Der R, Ma Y, Dalton P, Taira M. Histology of the gastroesophageal junction: an autopsy study. Am J Surg Pathol 2000;24:402–9.

[9] Chandrasoma PT, Lokuhetty DM, Demeester TR, et al. Definition of histopathologic changes in gastroesophageal reflux disease. Am J Surg Pathol 2000;24:344–51.

[10] Kilgore SP, Ormsby AH, Gramlich TL, et al. The gastric cardia: fact or fiction? Am J Gastroenterol 2000;95:921–4.

[11] Zhou H, Greco MA, Daum F, Kahn E. Origin of cardiac mucosa: ontogenic consideration. Pediatr Dev Pathol 2001;4:358–63.

[12] Derdoy JJ, Bergwerk A, Cohen H, et al. The gastric cardia: to be or not to be? Am J Surg Pathol 2003;27:499–504.

[13] Barrett N. The lower esophagus lined by columnar epithelium. Surgery 1957;41:881–94.

[14] Paull A, Trier JS, Dalton MD, et al. The histologic spectrum of Barrett's esophagus. The New England journal of medicine 1976;295:476–80.

[15] Sampliner RE. Updated guidelines for the diagnosis, surveillance, and therapy of Barrett's esophagus. Am J Gastroenterol 2002;97:1888–95.

[16] Wang KK, Sampliner RE. Gastroenterology. PPCotACo. Updated guidelines 2008 for the diagnosis, surveillance and therapy of Barrett's esophagus. Am J Gastroenterol 2008;103::788–97.

[17] Spechler SJ, Sharma P, Souza RF, Inadomi JM, Shaheen NJ. American Gastroenterological Association medical position statement on the management of Barrett's esophagus. Gastroenterology 2011;140:1084–91.

[18] Playford RJ. New British Society of Gastroenterology (BSG) guidelines for the diagnosis and management of Barrett's oesophagus. Gut 2006;55:442.

[19] Ogiya K, Kawano T, Ito E, et al. Lower esophageal palisade vessels and the definition of Barrett's esophagus. Dis Esophagus 2008;21:645–9.

[20] Bennett C, Moayyedi P, Corley DA, DeCaestecker J, Falck-Ytter Y, Falk G, et al. BOB CAT: A large-scale review and delphi consensus for management of Barrett's esophagus with no dysplasia, indefinite for, or low-grade dysplasia. Am J Gastroenterol May 2015;110:662–82.

[21] Bennett C, Vakil N, Bergman J, et al. Consensus statements for management of Barrett's dysplasia and early-stage esophageal adenocarcinoma, based on a Delphi process. Gastroenterology 2012;143:336–46.

[22] Sharma P, Savides TJ, Canto MI, et al. The American Society for Gastrointestinal Endoscopy PIVI (Preservation and Incorporation of Valuable Endoscopic Innovations) on imaging in Barrett's esophagus. Gastrointest Endosc 2012;76:252–4.

[23] Chandrasoma PT, Der R, Dalton P, et al. Distribution and significance of epithelial types in columnar-lined esophagus. Am J Surg Pathol 2001;25:1188–93.

[24] Antonioli DA, Wang HH. Morphology of Barrett's esophagus and Barrett's-associated dysplasia and adenocarcinoma. Gastroenterol Clin North Am 1997;26:495–506.

[25] Harrison R, Perry I, Haddadin W, et al. Detection of intestinal metaplasia in Barrett's esophagus: an observational comparator study suggests the need for a minimum of eight biopsies. Am J Gastroenterol 2007;102:1154–61.

[26] Haggitt RC, Reid BJ, Rabinovitch PS, Rubin CE. Barrett's esophagus. Correlation between mucin histochemistry, flow cytometry, and histologic diagnosis for predicting increased cancer risk. Am J Pathol 1988;131:53–61.

[27] Sharma P, Weston AP, Morales T, et al. Relative risk of dysplasia for patients with intestinal metaplasia in the distal oesophagus and in the gastric cardia. Gut 2000;46:9–13.

[28] Morales TG, Camargo E, Bhattacharyya A, Sampliner RE. Long-term follow-up of intestinal metaplasia of the gastric cardia. Am J Gastroenterol 2000;95:1677–80.

[29] Goldstein NS. Gastric cardia intestinal metaplasia: biopsy follow-up of 85 patients. Mod Pathol 2000;13:1072–9.

[30] Weston AP, Krmpotich PT, Cherian R, Dixon A, Topalovski M. Prospective evaluation of intestinal metaplasia and dysplasia within the cardia of patients with Barrett's esophagus. Dig Dis Sci 1997;42:597–602.

[31] Sharma P. Recent advances in Barrett's esophagus: short-segment Barrett's esophagus and cardia intestinal metaplasia. Semin Gastrointest Dis 1999;10:93–102.

[32] Srivastava A, Odze RD, Lauwers GY, et al. Morphologic features are useful in distinguishing Barrett esophagus from carditis with intestinal metaplasia. Am J Surg Pathol 2007;31:1733–41.

[33] Suh E, Traber PG. An intestine-specific homeobox gene regulates proliferation and differentiation. Mol Cell Biol 1996;16:619–25.

[34] Groisman GM, Amar M, Meir A. Expression of the intestinal marker Cdx2 in the columnar-lined esophagus with and without intestinal (Barrett's) metaplasia. Mod Pathol 2004;17:1282–8.

[35] Phillips RW, Frierson Jr. HF, Moskaluk CA. Cdx2 as a marker of epithelial intestinal differentiation in the esophagus. Am J Surg Pathol 2003;27:1442–7.

[36] Moons LM, Bax DA, Kuipers EJ, et al. The homeodomain protein CDX2 is an early marker of Barrett's oesophagus. J Clin Pathol 2004;57:1063–8.

[37] Haggitt RC, Tryzelaar J, Ellis FH, Colcher H. Adenocarcinoma complicating columnar epithelium-lined (Barrett's) esophagus. Am J Clin Pathol 1978;70:1–5.

[38] Spechler SJ, Goyal RK. Barrett's esophagus. The New England journal of medicine 1986;315:362–71.

[39] Reid BJ, Blount PL, Rubin CE, et al. Flow-cytometric and histological progression to malignancy in Barrett's esophagus: prospective endoscopic surveillance of a cohort. Gastroenterology 1992;102:1212–19.

[40] Hameeteman W, Tytgat GN, Houthoff HJ, van den Tweel JG. Barrett's esophagus: development of dysplasia and adenocarcinoma. Gastroenterology 1989;96:1249–56.

[41] Reid BJ, Weinstein WM, Lewin KJ, et al. Endoscopic biopsy can detect high-grade dysplasia or early adenocarcinoma in Barrett's esophagus without grossly recognizable neoplastic lesions. Gastroenterology 1988;94:81–90.

[42] Schmidt HG, Riddell RH, Walther B, Skinner DB, Riemann JF. Dysplasia in Barrett's esophagus. J Cancer Res Clin Oncol 1985;110:145–52.

[43] Smith RR, Hamilton SR, Boitnott JK, Rogers EL. The spectrum of carcinoma arising in Barrett's esophagus. A clinicopathologic study of 26 patients. Am J Surg Pathol 1984;8:563–73.

[44] Riddell RH, Goldman H, Ransohoff DF, et al. Dysplasia in inflammatory bowel disease: standardized classification with provisional clinical applications. Hum Pathol 1983;14:931–68.

[45] Update on the paris classification of superficial neoplastic lesions in the digestive tract. Endoscopy 2005;37:570-578.

[46] Rucker-Schmidt RL, Sanchez CA, Blount PL, et al. Nonadenomatous dysplasia in Barrett esophagus: a clinical, pathologic, and DNA content flow cytometric study. Am J Surg Pathol 2009;33:886–93.

[47] Mahajan D, Bennett AE, Liu X, Bena J, Bronner MP. Grading of gastric foveolar-type dysplasia in Barrett's esophagus. Mod Pathol 2010;23:1–11.

[48] Brown IS, Whiteman DC, Lauwers GY. Foveolar type dysplasia in Barrett esophagus. Mod Pathol 2010;23:834–43.

[49] Lomo LC, Blount PL, Sanchez CA, et al. Crypt dysplasia with surface maturation: a clinical, pathologic, and molecular study of a Barrett's esophagus cohort. Am J Surg Pathol 2006;30:423–35.

[50] Downs-Kelly E, Mendelin JE, Bennett AE, et al. Poor interobserver agreement in the distinction of high-grade dysplasia and adenocarcinoma in pretreatment Barrett's esophagus biopsies. Am J Gastroenterol 2008;103:2333–40 quiz 2341.

[51] Abraham SC, Krasinskas AM, Correa AM, et al. Duplication of the muscularis mucosae in Barrett esophagus: an underrecognized feature and its implication for staging of adenocarcinoma. Am J Surg Pathol 2007;31:1719–25.

[52] Falk GW, Rice TW, Goldblum JR, Richter JE. Jumbo biopsy forceps protocol still misses unsuspected

cancer in Barrett's esophagus with high-grade dysplasia. Gastrointest Endosc 1999;49:170−6.

[53] Reid BJ, Blount PL, Feng Z, Levine DS. Optimizing endoscopic biopsy detection of early cancers in Barrett's high-grade dysplasia. Am J Gastroenterol 2000;95:3089−96.

[54] Reid BJ, Haggitt RC, Rubin CE, et al. Observer variation in the diagnosis of dysplasia in Barrett's esophagus. Hum Pathol 1988;19:166−78.

[55] Montgomery E, Bronner MP, Goldblum JR, et al. Reproducibility of the diagnosis of dysplasia in Barrett esophagus: a reaffirmation. Hum Pathol 2001;32:368−78.

[56] Chennat J, Ross AS, Konda VJ, et al. Advanced pathology under squamous epithelium on initial EMR specimens in patients with Barrett's esophagus and high-grade dysplasia or intramucosal carcinoma: implications for surveillance and endotherapy management. Gastrointest Endosc 2009;70:417−21.

[57] Titi M, Overhiser A, Ulusarac O, et al. Development of subsquamous high-grade dysplasia and adenocarcinoma after successful radiofrequency ablation of Barrett's esophagus. Gastroenterology 2012;143 564−566 e561.

[58] Younes M, Lebovitz RM, Lechago LV, Lechago J. p53 protein accumulation in Barrett's metaplasia, dysplasia, and carcinoma: a follow-up study. Gastroenterology 1993;105:1637−42.

[59] Krishnadath KK, Tilanus HW, van Blankenstein M, Bosman FT, Mulder AH. Accumulation of p53 protein in normal, dysplastic, and neoplastic Barrett's oesophagus. J Pathol 1995;175:175−80.

[60] Jones DR, Davidson AG, Summers CL, Murray GF, Quinlan DC. Potential application of p53 as an intermediate biomarker in Barrett's esophagus. Ann Thorac Surg 1994;57:598−603.

[61] Ramel S, Reid BJ, Sanchez CA, et al. Evaluation of p53 protein expression in Barrett's esophagus by two-parameter flow cytometry. Gastroenterology 1992;102:1220−8.

[62] Hamelin R, Flejou JF, Muzeau F, et al. TP53 gene mutations and p53 protein immunoreactivity in malignant and premalignant Barrett's esophagus. Gastroenterology 1994;107:1012−18.

[63] Coggi G, Bosari S, Roncalli M, et al. p53 protein accumulation and p53 gene mutation in esophageal carcinoma. A molecular and immunohistochemical study with clinicopathologic correlations. Cancer 1997;79:425−32.

[64] Reid BJ, Levine DS, Longton G, Blount PL, Rabinovitch PS. Predictors of progression to cancer in Barrett's esophagus: baseline histology and flow cytometry identify low- and high-risk patient subsets. Am J Gastroenterol 2000;95:1669−76.

[65] Fennerty MB, Sampliner RE, Way D, et al. Discordance between flow cytometric abnormalities and dysplasia in Barrett's esophagus. Gastroenterology 1989;97 815−20.

[66] Xu Y, Selaru FM, Yin J, et al. Artificial neural networks and gene filtering distinguish between global gene expression profiles of Barrett's esophagus and esophageal cancer. Cancer Res 2002;62:3493−7.

[67] Brabender J, Marjoram P, Salonga D, et al. A multigene expression panel for the molecular diagnosis of Barrett's esophagus and Barrett's adenocarcinoma of the esophagus. Oncogene 2004;23:4780−8.

[68] Selaru FM, Zou T, Xu Y, et al. Global gene expression profiling in Barrett's esophagus and esophageal cancer: a comparative analysis using cDNA microarrays. Oncogene 2002;21:475−8.

[69] Shah AK, Saunders NA, Barbour AP, Hill MM. Early diagnostic biomarkers for esophageal adenocarcinoma—the current state of play. Cancer Epidemiol Biomarkers Prev 2013;22:1185−209.

Genetics and Biomarkers in Barrett's Esophagus and Esophageal Adenocarcinoma

Annalise C. Katz-Summercorn, Alexander M. Frankell and Rebecca C. Fitzgerald

Medical Research Council Cancer Unit, University of Cambridge, Cambridge, United Kingdom

4.1 INTRODUCTION

Barrett's esophagus (BE) is the premalignant lesion for esophageal adenocarcinoma (EAC): a malignancy with a very poor prognosis. The progression of BE from benign columnar-lined epithelium (CLE) to adenocarcinoma often occurs through a series of dysplastic stages termed low-grade dysplasia (LGD) and high-grade dysplasia (HGD). Recent evidence suggests a benefit for treating patients with dysplasia in order to prevent progression to adenocarcinoma. However, this strategy has several challenges. The causative molecular and cellular abnormalities predicting disease progression remain poorly understood. Moreover, there is a large proportion of patients with BE who remain undiagnosed within the population. Hence, in practice, there are problems of over and underdiagnosis, which hamper optimal clinical management.

Early detection and discovering better ways of predicting the course of the disease, particularly through understanding of the molecular genetics and developing biomarkers, is key to improving management of BE and thus survival from EAC.

4.2 GENETICS OF BARRETT'S ESOPHAGUS AND ESOPHAGEAL ADENOCARCINOMA

Research spanning the last 50 years has definitively shown that cancer is an acquired genetic disease whereby genomic instability within cells allows for an accumulation of advantageous genetic alterations leading to uncontrolled proliferation [1]. Initiation of BE appears to be caused by cellular damage from gastro-duodenal reflux components in the lower esophagus, causing cell death and as a

D. Pleskow & T. Erim (Eds): Barrett's Esophagus.
DOI: http://dx.doi.org/10.1016/B978-0-12-802511-6.00004-1

consequence cell proliferation to replenish the epithelium. This is accompanied by the acquisition of somatic mutations and epigenetic modifications, which lead to alterations in cell signaling. One of the key questions in the field is to precisely define the molecular and cellular alterations that drive the transition from BE to EAC, giving the cells the capacity to invade the underlying tissues and metastasize to other locations.

4.2.1 Genetic Susceptibility to Barrett's Esophagus and Esophageal Adenocarcinoma

Although the vast majority of genetic changes which contribute to cancer are acquired changes in the somatic tissue, heritable germ line gene variants are able to affect how subsequent somatic mutations cause cancer [2]. Recent data suggest that the development of both BE and EAC is associated with multiple low penetrance susceptibility loci and these may provide clues as to the pathogenesis of these conditions [3–5].

The first evidence that a proportion of BE and EAC cases may be heritable came from familial and twin association studies [6,7] but with the advent of relatively inexpensive genotyping on large cohorts it has now become possible to reveal the genomic variants responsible for such associations more easily. In the past 3 years several genome-wide association studies have, in, total, identified eight loci which contain single nucleotide polymorphisms (SNPs) associated with BE and/or EAC. Development of BE and EAC has been associated with loci in or adjacent to FOXF1, CRTC1, BARX1, FOXP1, ALDH1A2, and in the HLA region [4,5], and the development of BE, specifically, has revealed associations with loci in or adjacent to TBX5 and GDF7 [3]. Many of these genes (FOXF1, FOXP1, BARX1, and TBX1) are involved in embryonic

esophageal development [8–11]. Others have a variety of roles. CRTC1 itself has possible oncogenic roles [12] but SNPs in this locus may have oncogenic effects via regulation of PIK3R2 expression [13] and ALDH1A2 is required for the synthesis of retinoic acid, a developmental regulator [14]. GDF12, also known as BMP12, is a TGFβ-superfamily ligand in the BMP pathway, which is implicated in the development of BE [15] and the HLA region is a collection of genes vital for various functions of the immune system. Although we can speculate, it is difficult to determine precisely how these SNPs affect BE and EAC. Even the gene(s) which they affect are difficult to determine when they are in noncoding regions, as is common. Such SNPs may be directly affecting BE and EAC pathogenesis pathways or they may be affecting development of risk factors for BE and EAC such as gastro-esophageal reflux disease or obesity. Twenty-nine of the top forty genes in one study were linked with obesity for instance [4]. To evaluate and confirm their effects, functional studies will be required. It is also worth noting that germline genetic variations for BE and EAC susceptibility overlap significantly [16] suggesting that identified variants mediate their effect early in the pathogenesis sequence.

Pathway analysis of SNPs associated with BE has also been performed to detect enrichment of groups of genes with a similar function [3]. Such analysis is an important emerging aspect of cancer research as it has become apparent that genetic alterations may not necessarily target one gene, but whole sets of functionally related genes to achieve the same goal. The most significant pathways identified so far are type 1 diabetes mellitus, antigen processing and presentation and autoimmune thyroid disease [3]. This analysis suggests that the inflammatory component of BE and EAC may play an important role although the significance of this is still poorly understood [17,18].

4.2.2 Acquired Molecular Alterations in the Pathogenesis of Barrett's Esophagus

Molecular pathogenesis of BE involves dysregulation of a variety of signaling pathways. Such dysregulation partly has origins in genetic alterations but is also due to a complex series of events initiated by the reflux-damaged epithelium, involving inflammation and a wound response. Genomic mutations which alter these pathways can occur by a variety of mechanisms from changes in single nucleotides to whole chromosomes. These changes may also occur over a variety of time frames from gradual accumulation of mutations over decades to seemingly dramatic and sudden events which may be a vital part of the transition to EAC in some patients. Changes in the epigenome, altering the expression of a variety of genes, also contribute to dysregulation of cell signaling associated with BE carcinogenesis.

4.2.2.1 Altered Cell Signaling

Normal growth and division of cells are carefully controlled. Cells require a variety of growth factors, with their associated intracellular signaling machinery, to allow cell division and prevent apoptosis while keeping abnormal division under check. It is this fine balance of pro- and antiproliferative signals that cancer cells must alter to allow their uncontrolled proliferation [1].

BE cells have dysregulated this balance, leading to increased proliferation [19], however they do not generally contain the genetic alterations known to cause pro-proliferative growth factor signaling that are seen in some other precursor lesions, such as in the pancreas where such precancerous lesions appear to be initiated by KRAS mutations in 90% of cases [19]. However, there are direct effects of pulsatile pH and bile acids on cell cycle [20–22] and the pro-inflammatory environment may also play a role [23]. Inflammatory cytokines such as IL-8 are produced by the epithelial cells in response to reflux [24] and these cytokines act via pathways such as STAT3 signaling which lead to proliferation, intended to replenish the epithelium [25]. Other important inflammatory pathways are directly activated by reflux such as NF-κB [26]. Reflux also causes the production of Reactive Oxygen Species (ROS) via various pathways including inhibition of mitochondrial electron transport in epithelial cells and production by infiltrating immune cells [27]. These ROS species have several effects; they can induce DNA and protein damage, inducing the mutations required for the continued progression of BE [27], but can also affect signaling pathways which utilize endogenous ROS production to transmit signals. For instance in EGFR and PDGFR signaling, known to promote cell proliferation and carcinogenesis, the negative regulator PTEN can be specifically, reversibly inhibited by ROS causing increased proliferation and inhibition of apoptosis [28]. As well as causing an inflammatory response, reflux is thought to cause changes in cell and tissue morphology that are associated with BE. There is accumulating evidence that BE tissue has acquired a greater resistance to the reflux damage [29]. Hence it is presumed that death of cells due to reflux damage combined with inflammation associated proliferation provides an environment in which a subset of cells in the vicinity that are better able to survive reflux, either due to genomic alterations or a different differentiation program, come to predominate the tissue.

As well as the damage induced by inflammation and reflux, somatic alterations in BE appear to drive clonal expansion [30]. Genetic changes in BE are very common [31] but as usual in the development of cancer, most of these gene mutations are passengers rather than being causal in pathogenesis. Genes in which mutations appear to be selected above this background rate in BE include CDKN2A (p16) and TP53 (p53).

P16 is a small protein which binds to and inhibits cyclin dependent kinase 4 (CDK4) and CDK6, thereby inhibiting the phosphorylation of Rb, and preventing cell cycle progression and cell division [32]. It is activated by stimuli such as DNA damage and ROS [32], both caused by reflux, and hence it is likely that loss of this cell cycle inhibitor allows a greater rate of cell division in this environment. *CDKN2A* is commonly lost in BE either by mutation and loss of heterozygosity (LOH) or epigenetic silencing in approximately 15% and 60% of patients respectively [30,33] and these genetic changes are associated with expansions by spatial mapping of cell clones on the surface of BE [30,34]. P53 is a transcription factor which acts to inhibit cell proliferation and activate apoptosis via regulation of a variety of other genes. It is also activated in response to DNA damage and ROS. P53 function is lost in many different cancers at a high rate and hence may be a particularly vital node in these tumor-suppressive signaling pathways [35]. P53 mutation is relatively uncommon in Non-Dysplastic BE (NDBE) but prevalent in HGD occurring in approximately 86% of patients [31]. This occurs mostly via point mutations, is commonly accompanied by LOH, and is also associated with clonal expansion [30].

Evidence for specific gene mutations that demarcate the boundary between HGD and EAC, and hence may be important in this transition, is difficult to find. This was demonstrated in a recent study by Weaver and colleagues where Targeted Sequencing was used to compare single nucleotide variant (SNV) mutations in cases of NDBE, HGD, and EAC with a stable phenotype [31] and demonstrated dramatic heterogeneity across all three states. Weaver et al. identified only one gene mutation, *SMAD4*, which consistently associated with EAC rather than BE, and only at a low rate (13%). *SMAD4* is a central component for signaling via ligands of the transforming growth factor beta (TGFβ) superfamily [36]. *SMAD4* is commonly lost in many other cancers, for instance in pancreatic carcinoma where it is lost in 31% of cases [37] and carcinogenesis has been associated with a switch in TGFβ signaling from antiproliferative *SMAD4* dependent TGFβ signaling to invasion and migration of cells via *SMAD4* independent signaling [38]. The rate of point mutation in p53 is already high in HGD and did not increase with progression to EAC, but it was maintained [31]. *TP53* LOH and altered protein expression are associated with an increased risk of progression (see Table 4.3) and have pro-invasive and pro-migratory affects in vitro [39]. It is therefore thought that loss of *TP53* function is required for progression to EAC in the majority of patients. This is further evidenced by the mutation's effect on genomic stability [40], the importance of which in progression to EAC shall be discussed subsequently.

This genetic heterogeneity demonstrated by Weaver et al. has several possible explanations: it is possible that only a very few genetic changes are required for the transition to EAC, that other genomic changes, not detected in this study, such as large-scale chromosomal rearrangements-, copy number alterations or SNVs in genes not targeted in this study drive the process; or that, as discussed, whole gene networks are being targeted by such mutations rather than many individual genes. Hence pathway analysis of whole gene networks, using a wider variety of genomic alterations, in larger cohorts of BE and EAC could give a better indication of how this transition occurs, but may be limited by our incomplete understanding of these extremely complex networks. The genes in which genetic alterations occur have helped us little in understanding the molecular changes which underlie the development of EAC as yet, however more clues are perhaps to be found in the types of genetic alteration that occur.

4.2.2.2 *Mechanisms of Genetic Alteration*

Loss of genomic integrity is a vital constituent of the cancer phenotype [1]. Recent advances in next-generation sequencing technology have allowed the identification of a greater variety of genetic alterations and the increasing affordability has allowed larger patient cohorts to be investigated.

Millions of SNVs have been analyzed across thousands of patients in Many different cancers, and this has allowed statistical analyses to identify patterns termed mutational signatures [41]. These consist of biases in base conversions that occur in particular immediate sequence contexts, for instance it is common to find C-T conversions 5′ to a G nucleotide in many cancers due to deamination of Methyl-CpG dinucleotides. These signatures vary both between different cancers and between patients. In EAC an unusual signature of AA-CA conversions, with a preference for 5′ G nucleotides, has been identified [31,42], alongside other more common signatures. Some signatures appear to be associated with either specific types of damage, smoking for instance, or with mutations in particular DNA repair genes, such as BRCA2 [41]. Overall there is a high number of SNVs in EAC, comparable only with cancers driven by specific mutagens such as melanoma or lung cancer [41]. This caused some to suggest that this mutational signature could be due to reflux-associated oxidative stress in BE [41].

Deletions, insertions, inversions and translocations effecting large genomic regions are common in cancer, in particular EAC [43], and are collectively known as structural variants (SVs). With the resolution of new sequencing technologies it is becoming apparent that these changes account for a significant number of tumor suppressor inactivation events in EAC [43]. Deletions or amplifications occur in large sections of chromosomes, whole chromosomes and even the whole karyotype, amplifying and deleting oncogenes and tumor suppressors.

This is far easier to detect and has been known for many years [44]. A recent study identified such large-scale genomic variation as an important marker for the transition to EAC [45]. They identified changes in copy number across the genome in a longitudinal study of 248 BE patients. In patients who did not progress to EAC copy number alterations did not significantly alter over time—however, in the 79 patients who did progress, the mean number of copy number alterations increased rapidly beginning approximately 24 months before EAC was diagnosed. Importantly, these patients with high structural variability were still histologically diagnosed as BE during this 24-month period and hence this gain of large-scale genomic instability appears to be an important precursor step in the transition to EAC in many patients. As the degree of dysplasia in this study was not commented on we cannot be sure how this precursor step relates to the pathological state. Copy number changes lead to amplification of known oncogenes such as *MYC*, *ERBB2*, *EGFR*, and *KRAS* [44,45], however the only copy number change statistically associated with progressors rather than nonprogressors in this study was still deletion at the *SMAD4* locus. Whether this genomic instability, perhaps induced by *TP53* mutation which also occurs late in the progression of BE at the stage of HGD [31], is causally involved in progression or simply a consequence of changes which themselves lead to EAC development is unknown.

The spatial and temporal genomic distribution of mutations has long been presumed to be random, that is mutation events occur fairly evenly across the genome, even if they are then concentrated via selection, and in an independent manner over many cell generations. However local hypermutation of both SNVs and SVs has been identified in EAC, phenomena termed kataegis, and chromothripsis respectively [43,46] and there is evidence that chromothripsis may be due to single catastrophic events. Chromothripsis, Greek for "Chromosome Shattering," is thought to occur via a dramatic break event with currently unknown stimuli where the locus is broken into multiple pieces, and then stitched back together by the DNA repair machinery [47]. Variations in copy number state within these loci tend to be limited which has led many to the conclusion that the events occur at a single point in time [47]. Such events are not frequent in comparison to other types of mutation. Single chromothriptic-like events are only detected in 36% of tumors and 82% of tumors have fewer than 10 kataegic foci [43]. The importance of these events, relative to gradual accumulation of SVs and SNVs across the genome, is unknown.

Modification of gene function in cancer is not only achieved by alteration of base sequence, as has thus far been discussed, but also by other modifications both in DNA itself and the packaging proteins, histones. Such modifications are used in normal somatic cells to regulate gene expression, and are particularly important during development to allow differentiation [48]. These modifications consist of methylation directly onto DNA and a variety of chemical modifications that occur on specific residues of histone proteins [48]. These changes alter how DNA in these regions are packaged and so affects the availability of genes contained within these regions to RNA polymerases. Recent technological developments have allowed the DNA methylation profile of whole genomes to be assessed. Such assessment of histone modifications on a whole genome scale is much more difficult but is likely to be equally important in cancer development [49]. Hypermethylation of *CDKN2A* is common, as discussed, and clonally selected, and so possibly important in the pathogenesis of BE but such genome-wide approaches have identified multiple other genes which show similar patterns of hypermethylation such as APC, ESR1, REPRIMO, and many others [50–57]. Global

hypomethylation is also a feature of BE and EAC, as in many cancers, and results in upregulation of genes perhaps important in BE pathogenesis [58]. However many of these studies define aberrant methylation relative to squamous cells in the esophagus. These differences may therefore not all be important in the pathogenesis of BE but simply be fundamental differences in the epigenetic differentiation program between squamous and columnar tissues types [59]. Molecular analyses have provided insight into BE progression (Fig. 4.1). However, further work is required if we wish to predict and prevent progression in the clinic.

4.3 BIOMARKERS IN BARRETT'S ESOPHAGUS AND ESOPHAGEAL ADENOCARCINOMA

Alongside the rapidly increasing understanding of the genetic changes leading to the progression from BE to EAC is the search for diagnostic, prognostic, and predictive biomarkers. Special attention has been dedicated to looking for markers to diagnose BE in patients in the general population or with reflux symptoms who are not investigated; as well as to predict which 0.4% of cases of BE will progress to EAC. Vaughan and Fitzgerald have suggested a five-tier strategy, based on absolute risk, in order to target the population to the optimum prevention, screening, and treatment options [60]. These strategies are likely, in the long term, to lead to the biggest changes in the management of the disease, resulting in a massive reduction in morbidity and mortality.

Currently, British Society of Gastroenterology and American Gastroenterology Association guidelines do not recommend population screening until randomized controlled evidence is available [61,62], however they do suggest that endoscopic screening can be undertaken in higher risk groups such as males with increased BMI and persistent reflux symptoms.

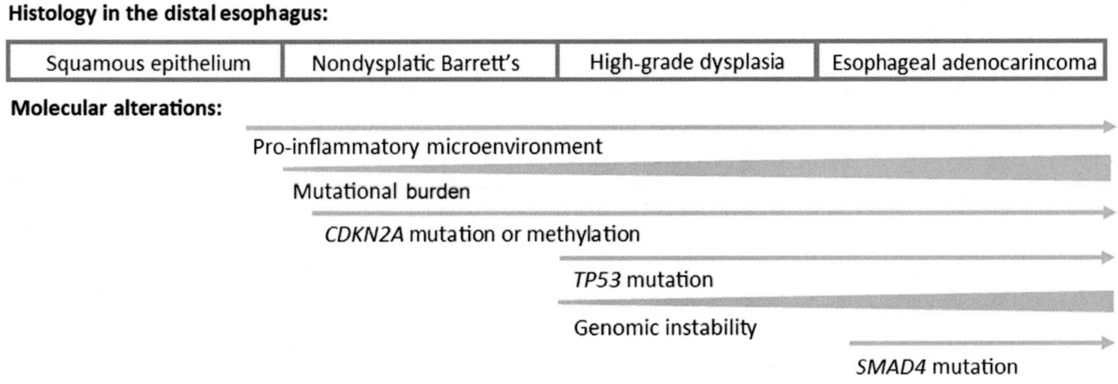

Histology in the distal esophagus:

Squamous epithelium	Nondysplatic Barrett's	High-grade dysplasia	Esophageal adenocarincoma

Molecular alterations:

Pro-inflammatory microenvironment

Mutational burden

CDKN2A mutation or methylation

TP53 mutation

Genomic instability

SMAD4 mutation

FIGURE 4.1   Molecular alterations that occur with progression of Barrett's esophagus (BE). From a histopathological perspective BE develops from the squamous esophagus in the context of chronic exposure to acid and bile reflux and then progresses in a minority of individuals through dysplastic stages to adenocarcinoma. At the molecular level changes are accompanied by an increased mutational burden, increasing copy number changes, and frequent loss of tumor suppressors CDKN2A (p16), TP53, and SMAD4 at early and late stages in this sequence respectively.

Once BE is diagnosed at endoscopy, four quadrant biopsies should be taken every 2 cm to look for the presence of dysplasia or EAC. Endoscopy is currently the only recommended method of monitoring or surveillance and there are a number of limitations with this method: it is expensive and time-consuming, unpleasant for patients, biopsies can miss focal areas of dysplasia or adenocarcinoma (sampling error), and the histological assessment is subjective.

For a biomarker to have the potential to be clinically useful it must have a number of characteristics:

- Easy to measure with inexpensive, widely available equipment for routine use
- Measured in an easily accessible biological sample: ideally blood, or from a nonendoscopic cell sampling device
- Sensitive and specific
- Facilitate early intervention

Routinely, the tissue collected at endoscopy is destined for paraffin embedding and histology and therefore potential biomarkers must be able to be used on these paraffin-embedded sections. However, increasingly fresh frozen biopsies are more routinely collected facilitating nucleic acid biomarkers. In addition, alternative methods of tissue collection are being developed on to apply biomarkers such as the Cytosponge. The Cytosponge is a swallowed capsule which dissolves in the stomach after 4—5 minutes releasing a sponge which collects cells as it is drawn back through the esophagus by the string attached to it [63]. The ample cells collected by this device can be tested for diagnostic and risk stratification biomarkers and is discussed subsequently. The ultimate aim would be to identify a circulating biomarker similar to that used in other cancers for example PSA in prostate cancer albeit ideally with a higher specificity. For EAC, although this concept is being considered, there is a long way to go before a serum biomarker will reach a clinical reality.

Traditionally, a potential biomarker was investigated based on knowledge of the disease process and its role as a potential candidate. The recent rapid developments in global screening and the "omics" revolution means that a huge number of genetic changes are being discovered as potential biomarkers in genes with known and unknown function.

TABLE 4.1   Early Detection Research Network Phases of Biomarker Development

Five Phases of Biomarker Development (Early Detection Research Network)	
Phase 1	Preclinical exploratory studies
Phase 2	Clinical assay development and validation
Phase 3	Retrospective longitudinal validation studies
Phase 4	Prospective screening validation studies
Phase 5	Population studies looking at impact of biomarker on disease burden and cancer control

Genome wide techniques generate huge amounts of data that require extensive validation. A number of biomarkers that are identified do not progress further in development because an accurate assay cannot be developed to measure them cheaply and effectively. The Early Detection Research Network (EDRN) has defined five stages for the development of biomarkers for clinical use [64] (Table 4.1).

Biomarkers have the potential for use in the following areas:

1. Screening of the population for BE
2. Risk stratification: identifying prevalent dysplasia in a more objective manner and/or predicting patients most likely to progress to dysplasia or adenocarcinoma in the future
3. Prognostic biomarkers for EAC
4. Defining treatment options for EAC (covered in Chapter 15: PostTreatment Surveillance, Risk for Recurrence of Barrett's Esophagus, and Adenocarcinoma After Treatment)

Overall, the aim is to produce biomarkers that will aid the clinical management of patients. This chapter focuses on overviewing the potential types of biomarker available and discussing those with real clinical potential in more detail.

4.3.1 Screening Biomarkers

As mentioned, one strategy for screening is a nonendoscopic cell collection device.

For a cytological cell collection device to be successful it is essential to couple the test with a biomarker since cytology alone is subjective and prone to inter- and intravariability. Trefoil Factor 3 (TFF3) has been identified as a strong candidate biomarker in the diagnosis of BE. Its expression is upregulated in Barrett's mucosa, yet absent in esophageal and gastric mucosa, and it is expressed on the luminal surface so can be easily sampled with brush cytology [65]. TFF3 is a member of the trefoil family, which is characterized by having at least one trefoil motif: a 40 amino acid domain containing three conserved disufhide bonds. It is a stable secretory protein expressed in the goblet cells of the gastrointestinal mucosa whose function is not defined, but may protect the mucosa from insults, stabilize the mucus layer, and affect healing of the epithelium [66].

The accuracy of immunostaining for TFF3 on cytological specimens, acquired using a nonendoscopic capsule sponge, Cytosponge, has been evaluated by the Fitzgerald laboratory in a multicenter BEST2 case-control study of over 1000 patients. It showed TFF3 testing to have a sensitivity of ~80% for short segments of 1 cm BE, increasing to 90% (95% CI 83.0−90.6) in longer segments or when swallowed for a second time with a specificity of 92.4% (95% CI 89.5−94.7) [67].

Further randomized controlled trial data in the primary care setting is required to establish this technique in routine clinical practice as a diagnostic triage prior to endoscopy in symptomatic patients or for use as a first-line diagnostic test.

4.3.2 Barrett's Dysplasia

Dysplasia is the only currently recognized marker of risk of progression for BE and it is

TABLE 4.2 Studies Assessing Low-Grade Dysplasia as a Marker of Progression [70−74]

Study	Finding	Sample Size	Type of Study
Bhat [71]	End-point HGD/EAC. HR 5.67, 95% CI 3.77−8.53, $p < 0.001$	8522: Northern Ireland BE database	Retrospective cohort
Sikkema [70]	Case = HGD/EAC. HR 3.6, 95% CI 1.6−8.1, $p = 0.002$	27 cases, 27 controls selected from cohort of 355 Erasmus MC University Medical Centre, Netherlands	Case-control
Kastelein [74]	Case = HGD/EAC. PPV 15% (can't access paper)	635 patients	Case-control
Bird-Lieberman [73]	Case = HGD/EAC. OR 11.78, 95% CI 4.31−32.18, $p < 0.001$	89 cases, 291 controls within Northern Ireland BE database	Nested case-control Phase 3
Hvid-Jensen [72]	End-point HGD/EAC. RR 5.1, 95% CI 3.4−7.6	11028: Danish pathology and cancer registries	Retrospective cohort

HGD, high-grade dysplasia; EAC, esophageal adenocarcinoma.

used clinically to direct management. Now that we routinely intervene with endoscopic treatment at the point of confirmed LGD and HGD, there is less information available from studies with EAC as an endpoint. The best recent data is from randomized controlled trials for radiofrequency ablation for dysplasia which have found up to a 19% (4/21) progression rate from HGD to EAC in 12 months [68] and an 8.8% (6/68) progression rate from LGD to EAC (median follow-up 36 months) [69].

There are ongoing studies analyzing the risk associated with LGD (Table 4.2). One thing to note in these studies is the variable way of expressing the data making comparison between studies difficult.

Although there is significant, quality evidence to support dysplasia as a risk factor, it does not excel as a biomarker candidate because its histopathological diagnosis is subjective and there is a propensity for overdiagnosis. In a recent large Dutch retrospective cohort study, two expert pathologists reviewed the slides of 293 patients diagnosed with LGD. There was agreement on first reading of 72% of cases. Seventy-three percent of cases were downgraded from LGD and none were upgraded to HGD. Interestingly, of those with a consensus diagnosis of LGD, the risk of progression to HGD/EAC was 9.1% per patient year versus 0.6% or 0.9% for those who were downgraded to NDBE or indeterminate dysplasia (IND) respectively [75]. In keeping with this, a recent meta-analysis found that studies with LGD/nondysplastic BE ratios (<0.15), which is indicative of a more stringent diagnostic criteria for dysplasia, reported a significantly higher annual incidence of cancer (0.76%, 95% CI 0.45−1.07) compared to studies with a ratio >0.15 (0.32%, 95% CI 0.07−0.58) [76].

Current British Society of Gastroenterology guidelines [61] recommend HGD to be the current best biomarker for the assessment of cancer risk, provided that it must be confirmed by two pathologists. However, the risk of progression of LGD has recently been acknowledged and it is now recommended that patients with LGD on more than one endoscopy, and confirmed by an expert GI pathologist, should also be offered radiofrequency ablation [77,78]. Ultimately, the development of other biomarkers, as discussed later in this chapter, that can be used either in conjunction with or as surrogate markers of dysplasia, would help further risk stratify this group and aid appropriate histopathological diagnosis.

4.3.3 P53

TP53 is the most well studied of candidate biomarkers in BE and is the only one to feature in the British Society of Gastroenterology guidelines: "The addition of p53 immunostaining to the histopathological assessment may improve the diagnostic reproducibility of a diagnosis of dysplasia in BE and should be considered as an adjunct to routine clinical diagnosis" Grade C recommendation. However, it is not yet recommended by the American Gastroenterological Association [79].

LOH for *TP53* was first shown in 2001 to be associated with a 16-fold increased risk of progression to EAC from nondysplastic Barrett's in a Phase 4 study of 325 patients [80]. As well as a single biomarker *TP53* LOH has also been studied as part of a biomarker panel [81] discussed later in the chapter. LOH detection in these studies required multiple technical steps including flow cytometry purification, DNA extraction, whole genome amplification, and then PCR with locus-specific primers. These methods are high cost and not easily applicable to routine clinical use.

Aside from LOH *TP53* immunostaining can also be used as a biomarker. *TP53* mutation in one of the alleles can result in an increased half-life of the protein by stabilizing it and preventing degradation. This protein accumulation, detected using immunohistochemistry (IHC), has been shown to predict progression from LGD to HGD/EAC [82], with a 63.6–100% sensitivity and 68–93% specificity [83–87]. Sikkema et al. found p53 overexpression to result in a fivefold increased risk of progression to HGD or EAC, independent of the presence of LGD (95% CI 2–14.5, $p = 0.004$) [70]. Since then, it has been realized that not all mutations stabilize the protein: they may truncate it or result in nonexpression, and so the absence of staining for p53 has been recognized to also have clinical utility. Kaye et al. looked back at the previous cohort to find that 7 of 53 (89%) had shown an abnormal p53 immunophenotype (either overexpression or loss of expression) [86].

P53 accumulation has been shown to precede development of HGD/EAC by several years, an important characteristic for a potential biomarker [87]. The two largest, most recent studies have quite differing findings and are worthy of note:

- Kastelein et al. performed a case-control study of 635 within a prospective cohort with BE. Over 12,000 specimens were analyzed independently by two histopathologists. During follow-up, 49 (8%) patients developed HGD or EAC. P53 overexpression was associated with an increased risk of neoplastic progression in patients with BE after adjusting for age, gender, Barrett length, and esophagitis (RR 5.6, 95% CI 3.1 to 10.3), but, interestingly, the risk was even higher with loss of p53 expression (RR 14.0, 95% CI 5.3 to 37.2). The positive predictive value for neoplastic progression increased from 15% with histological diagnosis of LGD to 33% with LGD and concurrent aberrant p53 expression [74].
- Bird-Leiberman et al. analyzed data from a nested case-control study performed using the Northern Ireland BE Register (1993–2005). P53 was not found to predict HGD/EAC progression in multivariate analysis. However, the presence of p53 showed a significant risk in EAC alone (OR 1.95, 95% CI 1.04–3.67) [73].

For both of these studies, the interpretation of the results is limited by the interlaboratory variability of p53 IHC and, again, the fact that not all p53 mutations lead to stabilization of the protein [73]. The following very recent study considers this by looking at the combination of p53 IHC and *TP53* LOH:

- Davelaar et al. performed a prospective cohort analysis of 100 patients with intestinal metaplasia (IM), IND, and LGD. They had first analyzed 116 patients with BE at baseline, finding that by using both p53 IHC and *TP53* LOH, *TP53* aberrancy could be detected in 100% of specimens, although only nine tested positive for both. Hundred patients were eligible for follow-up (median 71 months), with an endpoint of HGD/EAC. During this time, 7.5% (7/93) IM/IND progressed to BE/EAC, 42.9% (3/7), LGD progressed to HGD and 14.3% (1/7) LGD progressed to EAC. The combination of p53 IHC and *TP53* LOH positivity was associated with an OR 25.5 (95% CI 4.9−133, $p < 0.001$); 81.8% sensitivity and 85.0% specificity for predicting progression. They went on to break down the analysis to look at HGD and EAC separately, with smaller numbers but still significant results [87].

More recently the mutational status of *TP53* can be directly evaluated by next-generation sequencing techniques. *TP53* mutations were shown to occur in a stage-specific manner: 2.5% never-dysplastic BE ($n = 66$); 72% BE with HGD ($n = 43$) ($p < 0.0001$). More so, on brush cytology from a pilot set of samples obtained using the Cytosponge, *TP53* mutation analysis had a sensitivity of 86% and specificity of 100% for HGD in a pilot study [31].

This recent work on p53 mutational analysis is very promising as p53 mutation defines the disease state between NDBE and HGD, the key point of intervention, thereby making it the best candidate to date in risk stratification (see Table 4.3 for a summary).

4.3.4 Other Somatic Mutations

Knowledge of somatic mutations in EAC has, until recently, been limited to studies in small collections of tumors. Now, next-generation sequencing of large numbers of tumors, with comparison to BE and HGD has resulted in the discovery of a large number of mutated genes with knowledge of when they are mutated in the BE-LGD-HGD-EAC sequence. Some of the more recent findings are discussed below.

SMAD4 is a transcription factor and tumor suppressor gene in the TGF-β signaling pathway. *SMAD4* is mutated in EAC but not NDBE or HGD as discussed earlier. However, although this has provided a clear genetic distinction between HGD and EAC, it was only mutated at a low frequency (13%) making it a suboptimal candidate for biomarker development. Also, now that HGD is the main point of intervention for treatment, distinguishing it from NDBE is more important.

The other recent group of genes of interest are those involved in chromatin remodeling (*ARID1A* and *SMARCA4* which encode members of the SWI/SNF (Switch/sucrose nonfermentable) complex) have newly been found to be mutated in a fifth of EACs [44,88]. Weaver and colleagues found a low frequency of mutation of *ARID1A*, with similar frequencies in NDBE and HGD but analysis of its protein expression by IHC in a cohort of 298 EACs found absent or decreased expression in 41% of samples [31]. This suggests other mechanisms of downregulation. Interestingly, they observed that specimens with *ARID1A* loss had significantly less p53 accumulation ($p = 0.028$). Streppel et al. had similar findings: it was mutated in 15% of HGD and EAC. IHC revealed protein loss in 0% (0/76) NSE, 4.9% (2/40) BE, 14.3% (4/28) LGD, 16.0% (8/50) HGD, 12.2% (12/98) EAC [88]. Larger case-control studies of progressors versus nonprogressors are needed.

4.3.5 MicroRNAs as Biomarkers

There has been significant recent interest concerning the role of microRNAs (miRNAs)

TABLE 4.3   Summary of p53 as a Biomarker in Predicting Progression [70,73,74,80−84,86,87]

	Study	EDRN Stage	Sample Size	Finding
TP53 LOH using flow cytometry	Reid et al. [80]	Phase 3/4: prospectively collected samples, retrospective analysis	325	RR = 16, $p < 0.001$
	Galipeau et al. [81]	Phase 3/4: prospectively collected samples, retrospective analysis	243	RR 10.6 (95% CI 5.2−21.3, $p < 0.001$)
P53 positive on IHC	Younes et al. [83]	Phase 3 retrospective	5 progressors, 25 nonprogressors	Correlates with progression from LGD to HGD/EAC $p = 0.0108$; 100% sensitivity, 93% specificity in predicting progression
	Weston et al. [82]	Phase 4 prospective	5 progressors, 43 nonprogressors	Kaplan-Meier curves differed significantly between p53 positive and negative patients with progression from LGD
	Skacel et al. [84]	Phase 3 retrospective	8 progressors, 8 non-progressors	Correlates with progression from LGD to HGD/EAC $p = 0.017$; 88% sensitivity, 75% specificity in predicting progression
	Kaye et al. [86]	Phase 3 retrospective	154 progressors, 32 nonprogressors	80% sensitivity, 68% specificity in predicting progression
	Sikkema et al. [70]	Phase 4 prospective	27 progressors, 27 nonprogressors	HR 6.5 (95% CI 2.5−17.1)
	Kastelein et al. [74]	Phase 3/4 prospectively collected samples, retrospective analysis	X progressors, x nonprogressors	P53 overexpression (RR 5.6, 95% CI 3.1−10.3)
				Loss of p53 expression (RR 14.0, 95% CI 5.3−37.2)
	Bird-Lieberman et al. [73]	Phase 3/4 prospectively collected samples, retrospective analysis	Nested-case control	Risk of EAC alone (OR 1.95, 95% CI 1.04−3.67)
			89 progressors, 291 nonprogressors	P53 was not found to predict HGD/EAC progression in multivariate analysis
	Davelaar et al. [87]	Phase 4 prospective	116 patients	Progression to HGD/EAC 17 (95% CI 3.2−96, $p = 0.001$)
				Progression to HGD only (OR 30.8 95% CI 3.78−308, $p = 0.002$)
				IHC showed increased sensitivity (81.8%) but decreased specificity (85%) for progression to HGD when combined with FISH LOH

in cancer pathogenesis and their potential role as tumor suppressor genes or oncogenes. They are small, well-conserved, noncoding RNAs with normal biological roles in signaling, motility, apoptosis, proliferation, and angiogenesis. They regulate gene expression by either targeting mRNA for degradation or reducing its translation by binding the 3′ UTR. They have been shown to be differentially expressed in many cancers. Lu et al. published in *Nature* in 2005 demonstrating that tumors can be classified based on their miRNA expression profile [89]. Since then, their potential role in diagnosis, risk stratification, and likely response to treatment are being rapidly analyzed. They may also be affected by the same mutations as protein-coding genes such as chromosomal rearrangement, amplification, and deletion. Considering their role as potential biomarkers they are ideal candidates, they can be easily measured by simple PCR and more so, are present peripherally in the blood where they circulate in lipid or lipoprotein complexes [90].

Feber et al. were the first to look for miRNA biomarkers in BE in 2008, showing that differential expression could differentiate between the different tissue types [91]. Of the 328 human miRNA probes used, they found 2 miRNAs to show a 2−10-fold increase in expression in EAC tissue compared to normal esophageal tissue, mir-203 and mir-205. Since then, a number of different miRNAs have been found to correlate with the disease, but no dominant miRNA has emerged from the limited studies. The most consistently altered miRNAs across studies are: -192, -194, -203, -205, -215. See Table 4.4 for a summary.

In conclusion, knowledge in this field is rapidly increasing, however a number of different miRNA candidates have been identified in different studies, with no miRNA being common in all cases. Studies with larger patient numbers are needed to further analyze the statistical significance of the varying expression changes.

4.3.6 Epigenetic Biomarkers

Methylation-induced inactivation of p16 (*CDKN2A*) is one of the commonest changes observed in Barrett's metaplasia and it was the first gene found to be affected by methylation in BE [97]. Methylation of its promoter region is seen in 15% of BE tissue, yet it is unmethylated in normal tissue [98]. It seems to occur early in the process of progression and it is thought that the clonal expansion of *CDKN2A* -/- cells may form an environment conducive to the development of other genetic events, leading to EAC [97−99]. Overall, p16 would potentially be a good marker for the diagnosis of BE, however LOH or aneuploidy has been shown to be a better predictor of progression as p16 loss occurs early in the natural history prior to the development of dysplasia.

REPRIMO is a tumor suppressor gene which regulates p53-mediated cell cycle arrest and its methylation status is a potential screening biomarker for the development of BE. It is not methylated in normal esophageal tissue but it has been shown to have a significantly higher frequency of methylation in BE, 36% ($P = 0.001$), HGD, 64% ($P = 0.001$), and EAC, 63% ($P = 0.00003$). The study did not find a significant difference between BE and HGD/EAC.

Analysis of *REPRIMO* methylation in EAC versus normal esophagus (NE) revealed an area under the receiver-operator characteristic curve (AUROC) of 0.812 (95% CI 0.73−0.90, $p < 0.00001$) [100]. They also found that *REPRIMO* was methylated in significantly lower levels in chemoradiotherapy responders versus nonresponders.

The Meltzer laboratory in Baltimore have worked extensively on methylation, identifying genes whose hypermethylation is associated with progression from BE to EAC: *CDKN2A*, *RUNX3*, *HPP1* [101,102]. They went on to perform a large retrospective, multicenter validation study in 2009 of an eight-marker

TABLE 4.4   MicroRNA (miRNA) Biomarker Studies [92−96]

Study	EDRN Stage	Sample Size	Findings	
Fassan et al. [93]: *Screening*	Phase 2	326 human tumors: normal squamous vs Barrett's epithelium	Upregulation of mir-215, -192, and -194, downregulation of -203 and -205. Microarray	$p < 0.01$
Bansal et al. [95]: *Screening* to differentiate between GERD and BE	Phase 2	28 patients (19 with GERD, 11 with BE)	mi-192, -194, and -215 panel previously detected by next generation sequencing	91−100% sensitivity and 94% specificity
Skinner et al. [96]: *Prognosis*—prediction of response to neoadjuvant chemotherapy	Phase 2	65 patients in validation group	Profile of four miRNAs predicting a complete pathological response (mi-505, -99b, -451, -145. Microarray	AUROC 0.71 (95% CI 0.51−0.84)
Revilla-Nuin et al. [94]: *Risk of progression* to EAC	Phase 2	179 patients in validation group	4 miRNAs upregulated: mir-192, -194, -196a, -196b qRT-PCR and validated as a panel	71−85% sensitivity, 63−70% specificity
Mathe et al. [92]: *Survival*	Phase 2	62 with BE, 34 without	Mir-21, -223, -192, -194 expression elevated Microarray and qRT-PCR	Reduced mir-375 associated with poor prognosis (HR 0.31 95% CI 0.15−0.67) (independent of tumor stage or nodal status, cohort type, and chemoradiation therapy)

GERD, gastro esophageal reflux disease; AUROC, area under the receiver operator characteristic curve.

risk-of-progression panel combined with age. It was found that they were able to predict progression with a sensitivity of approaching 50% when specificity was set at 90% using AUROC curves [103]. The Fitzgerald group developed a different methylation panel [104]: *SLC22A18*, *PIGR*, *GJA12*, and *RIN2* which was able to distinguish between BE and HGD/EAC in a retrospective cohort (AUROC 0.988, 97% specificity and 94% sensitivity).

More recently, genome-wide methylation studies are shedding light on global patterns of methylation and "methylation signatures" [105] and have revealed large numbers of other genes showing altered methylation including *CXCL1*, *GATA6*, *DMBT1* [58]. Interestingly, it seems that in BE which progresses, global hypomethylation occurs early on, suggesting the importance of upregulation of growth-promoting genes, due to the inactivation of tumor suppressor genes [106]. Xu et al. showed that the overall methylation of CpG sites within the CpG islands was higher, but outside of the CpG islands was lower in BE and EAC tissues than in NE tissues. For discriminating BE from NE tissues, the AUROC was 0.965 (sensitivity: 94.81%, specificity: 91.49%), and for discriminating EAC from NE tissues, the AUROC was 0.973 (sensitivity: 94.87%, specificity: 93.62%), suggesting the excellent value of these differentially methylated CpG sites in discriminating BE or EAC tissues from NE tissues and a potential role in screening for Barrett's. However, there was no significant difference between BE and EAC, again suggesting that these methylation changes occur early [107].

Overall, gene methylation has been extensively studied. There have been many Phase 1 discovery studies highlighting a number of aberrantly methylated genes, however the studies generally have had few patients and have been retrospective. None of the potential markers have so far been subjected to rigorous prospective validation studies. Using gene methylation as a biomarker has its own technical consideration for widespread use: detection requires multiple steps including enzyme digestion and bisulfite treatment prior to probe hybridization. Hence it is not yet clear whether they could be used routinely in the clinical setting.

4.3.7 DNA Content Abnormalities

Chromosomal alterations, termed aneuploidy (an abnormal number of chromosomes) or tetraploidy (twice the normal number of chromosomes, 4N), have been known to correlate with the risk of progression of BE for over 20 years [108]. The Reid group have studied it extensively since then and have produced several Phase 4 studies, albeit with overlapping populations. In a large BE cohort, using flow cytometry, they found that the presence of both 4N fraction of >6% and aneuploid DNA content of >2.7N was highly predictive of cancer (RR 23, 95% CI 10−50), but not an aneuploidy of less than 2.7. They also found that this helped to risk stratify the group of NDBE and LGD with the presence of aneuploidy (>2.7N) and/or tetraploidy conferring a 25-fold increased risk of progression (RR 25, 95% CI 6.5−98) [109]. They also went on to show that having 17p (*TP53*) LOH at baseline resulted in increased 4N (RR 6.1, $p < 0.001$), and aneuploidy (RR 7.5, $p < 0.001$) [80] as well as the increased risk in progression, as discussed earlier. This is likely because of the increased genetic instability caused by *TP53* LOH and strengthens the concept of clonal expansion. The group published a prospective cohort study in 2004 showing that it was not just the presence of these abnormalities that increased the risk of progression to EAC, but also the clone size that was important: RR 1.31 x the length of the clone in cm (95% CI 1.07−1.60) for ploidy abnormalities [110]. This finding has helped increase our knowledge of BE transformation but has not yet been further considered as a clinical tool. Later smaller studies have not been able to reproduce this association with aneuploidy, finding it to be a late marker (occurring just prior to transformation to EAC [111] or no longer being predictive when in the presence of LGD, after multivariate analysis [70]. In another prospective cohort, Galipeau et al. demonstrated that the RR conferred by aneuploidy or tetraploidy for progression to EAC varied with time: for aneuploidy a RR 16.5 (95% CI 5.5−49.4) at 2 years versus RR 8.5 (95% CI 4.3−17) at 10 years [81].

Despite these promising, high-quality studies, there has been little progress in the use of aneuploidy or tetraploidy as a biomarker because of the need for flow cytometry and special media which are expensive and not easily applicable to the clinical setting. Fluorescence in situ hybridization (FISH) is an alternative method of detection of copy number abnormalities and has been shown in some studies to be more sensitive than flow cytometry in assessing ploidy status although it is quite labor intensive and expensive to perform [112]. Krishnadath and colleagues used a FISH panel, including probes to detect aneuploidy in a prospective multicenter trial. They found that there was a threefold risk of progression with p16 loss or aneuploidy after controlling for Barrett segment length and patient age (RR 3.23 95% CI 1.32−7.95). The absolute risk was 1.83% if the FISH panel was positive versus

0.58% if negative [113]. It has also been shown that a multicolor FISH panel may be able to predict survival in EAC independent of stage and grade [114].

More recently, alternative techniques are being explored, especially digital image cytometric DNA analysis, which has shown to be accurate in comparison to flow cytometry [115]. This technique was used in a population-based, nested case-control study showing that aneuploidy conferred more than three times the risk of progressing to HGD/EAC (RR 3.22 95% CI 1.73−6.00) although the overall sensitivity was low (44%) [73]. However, it may be useful as part of a biomarker panel. Finally, digital image cytometry has been used successfully on brush cytology specimens: which could result in a simple, easy way of identifying patients with LGD who are unlikely to progress [116].

4.3.8 Cell Cycle and Proliferation Markers

Proteins which drive the cell cycle and proliferation should make obvious potential candidates as biomarkers. Regarding cell-cycle control, genes coding proteins cyclin D and A have been shown to be upregulated but studies have disagreed on the link between this and progression to EAC [85,117−120]. The Fitzgerald laboratory have performed a prospective cohort study of 175 patients, considering the ability of a number of biomarkers to predict HGD/BE. They found cyclin A IHC to have a 79.5% sensitivity and an 85% specificity ($p < 0.01$). For detecting any dysplasia, the sensitivity dropped to 66.2%, however the specificity was higher at 88.8% [121].

Proliferation markers have also been implicated but again with some uncertainty: mini-chromosome maintenance proteins (mcm) 2 and 5 and Ki-67 [70,122−124]. Overexpression of Ki-67 may be a late change prior to HGD/

EAC development [111], which could detect those at high present risk of progression. The limitation is likely due to the confounding effect of inflammation, which also increases cell proliferation. Overall, there have not been any large prospective studies evaluating the clinical application of these markers.

4.3.9 Biomarker Panels

Ultimately, combinations of biomarkers are likely to predict progression with a higher sensitivity and specificity than singularly. As technology has improved, we can now analyze multiple molecular changes simultaneously. One of the earlier panels developed was the "Reid panel" of genetic instability biomarkers: 17p LOH, tetraploidy/aneuploidy, 9p LOH (locus for *CDKN2A*). In a large retrospective analysis of 243 patients it was able to predict risk of progression to EAC of 80% at 6 years (compared to patients with none of the abnormalities who had a 12% 10-year risk) with a relative risk of 38.7 [80]. More recently, di Pietro and colleagues developed a small, more practical panel of three biomarkers with the aim to detect dysplasia. P53 IHC, cyclin A IHC and aneuploidy had a sensitivity and a specificity of 95.8% (95% CI 76.9−99.8) and 88.6% (95% CI 79.7−94.1), respectively, for a diagnosis of HGD/EC, and 74.4% (95% CI 57.6−86.4) and 94.5% (95% CI 85.8−98.2), respectively, for a diagnosis of any grade of dysplasia [121]. Most other panels have either focused on this genetic instability or on methylation.

A Phase 2 study looking gene expression profiling of the stromal tissue developed an interesting set of biomarkers that has shown to discriminate between NDBE, HGD, and EAC ($p < 0.004$), and also predict outcome. The stromal compartment (extracellular matrix, mesenchymal, and nerve cells) is increasingly recognized to play a role in cancer and normal

TABLE 4.5 Panels as Biomarkers [73,81,102,103,121]

Study	Biomarker Panel	Type of Study	Sample Size	Finding
Galipeau et al. [81]	"Reid Panel": 17p LOH, tetraploidy/aneuploidy, 9p LOH	Phase 3/4 prospectively collected samples retrospective analysis	243 patients evaluated at baseline	RR of EAC 38.7; (95% CI 10.8−138.5, $p < 0.001$). 79% 10-year EAC incidence if all positive
Di Pietro et al. [121]	P53 IHC, cyclin A, aneuploidy combined with autofluorescence imaging	Cross-sectional prospective study	157 patients	Sensitivity 100%, specificity 85% for HGD/EAC. AUROC 0.97 (95% CI 0.95−0.99) for diagnosing HGD/EAC
Bird-Lieberman et al. [73]	Low-grade dysplasia, abnormal DNA ploidy, and *Aspergillus oryzae* lectin evaluated at baseline	Phase 3 retrospective nested case-control within Northern Ireland BE database	89 cases, 291 controls	For each additional positive biomarker, BE patients with LGD are almost at fourfold increased odds of progressing to develop EAC (OR, 3.90; 95% CI 2.39−6.37). TPR 24%, FPR 4%
Jin et al. [103]	8 methylation markers *CDKN2A, RUNX3, CDH13, TAC1, NELL1, AKAP12,* somatostatin plus age	Phase 3 retrospective case-control validation study	50 progressors, 145 nonprogressors	Sensitivity 80%, specificity 78%. AUROC ~0.84
Sato et al. [102]	Segment length, promoter methylation of *CDKN2A, RUNX3, HPP1,* baseline pathology (NDBE vs LGD) EAC	Phase 3 retrospective	35 progressors, 27 nonprogressors	Progression to HGD/EAC. 4-year prediction risk AUROC 0.79. 91% sensitivity, 52% specificity

epithelial cells depend on it to sustain their survival and proliferate. A five-gene overexpression signature in stromal tissue (*TMEPAI* and *TSP1* (TGFβ-related), *JMY* (potential modulator of invasion), *FAPα* (stromal activation marker), and *BCL6* (transcription factor)) was developed. Upregulation of *TMEPAI* and *JMY* was significantly associated with worse prognosis in EAC ($p < 0.005$) and upregulation of any one or more of these five targets had a worse prognosis than patients with no increase in expression of any of these stromal genes ($p = 0.022$) [18].

Table 4.5 summarizes some of the key Phase 3 and 4 studies concerning biomarker panels.

4.4 CONCLUSIONS

The molecular pathogenesis of BE, both concerning its development and its transition to EAC, is still poorly understood. Improvements to such knowledge will surely be vital to the segregation of BE patients into accurate, high-, and low-risk strata for progression, a goal which will hopefully both stem the flow of increasing EAC incidence in the Western world, and reduce the surveillance cost associated with the large number of BE patients presently with poorly defined risk.

Currently, management decisions for patients with BE are based on the histological diagnosis of dysplasia. LGD is, in itself, difficult to diagnose histologically and is often overcalled. The main problem that we face is understanding who with NDBE is at risk of progression to EAC. Having a biomarker panel that could risk stratify patients would allow those at high risk to be treated early while not subjecting those at low risk to multiple endoscopies. Accurately defining the dysplasia group is even more important with the advent of endoscopic ablation techniques.

P53 IHC is currently the only biomarker recognized to have a potential clinical role and it may be used to facilitate the diagnosis of dysplasia in cases of uncertainty. However, overall, large-scale prospective studies are still lacking.

A number of biomarkers other than dysplasia have been proposed to predict progression. The most promising of these remain *TP53* LOH, aneuploidy/tetraploidy, genomic diversity measures, and some of the combination biomarker panels.

4.5 FUTURE DIRECTIONS

Over the last 15 or so years, there have been leaps forward in our understanding of the genetic changes underlying BE and EAC. A huge number of somatic mutations and epigenetic modifications have been discovered but, on the whole, it is not known which of these are driver mutations in the transition from BE to EAC and which are passengers. New-generation whole-genome sequencing (WGS) technology is driving the field forward rapidly: facilitating comparisons of large numbers of tumors simultaneously with each other and their adjacent BE segments in order to identify the commonly mutated genes and the temporal acquisition of them [31,42,43]. This WGS is undoubtedly also going to lead to the discovery of large numbers of potential biomarkers which will require extensive validation and further investigation. These techniques are

rapidly decreasing in cost and clinically applicable assays are being developed apace. The Cancer Genome Atlas and International Cancer Genome Consortium are sequencing large numbers of tumors and will, when combined, provide a comprehensive dataset from enough cases to start to further unravel the statistically relevant molecular genetic changes. It is likely that, despite the heterogeneity of the disease, essential key molecular changes will be identified that have the potential to be biomarkers for risk stratification but perhaps it is unlikely that one biomarker alone will be found that it is sufficient to predict progression, and it would be a panel of markers that would ultimately be needed.

The analysis of larger chromosomal rearrangements lags behind point mutations but these may also be useful biomarkers. Early analysis of the genomes for structural rearrangements has highlighted a number of well-known cancer genes to be recurrently rearranged and future work to understand the temporal nature of this may lead to specific rearrangements being used to predict prognosis or understand which tumors are likely to progress.

It is also possible that loss of genomic integrity itself could itself act as a biomarker, with different mutational signatures conveying different prognostic features [43]. It has been suggested that rather than the presence or absence of individual copy number aberrations, it is the diversity of aberrations and mutations that is important and can be predictive for EAC development [125]. It will be important to determine how this can be quantified for a clinical assay.

Currently, there are many discovery and validation studies for new biomarkers and a number of retrospective studies. But to translate this to clinical applicability, prospective, well-powered studies of biomarkers are needed. More so, randomized-controlled trials based on biomarkers will be required.

In the long run, it may be that biomarkers superior to the current gold standard of HGD that better predict disease progression will be developed. Or perhaps they will be used as an adjunct to HGD. Ultimately, we need more accurate predictors of disease progression without the sampling-error and user-dependence of current techniques of biopsies to look for dysplasia.

References

[1] Hanahan D, Weinberg RA. Hallmarks of cancer: the next generation. Cell 2011;144(5):646−74.

[2] Lu Y, Ek WE, Whiteman D, et al. Most common "sporadic" cancers have a significant germline genetic component. Hum Mol Genet 2014;23(22):6112−18.

[3] Palles C, Chegwidden L, Li X, Findlay JM, Farnham G, Castro Giner F, et al. Polymorphisms near TBX5 and GDF7 are associated with increased risk for Barrett's esophagus. Gastroenterology 2015;148 (2):367−78.

[4] Su Z, Gay LJ, Strange A, Band G, Whiteman DC, Lescai F, et al. Common variants at the MHC locus and at chromosome 16q24.1 predispose to Barrett's esophagus. Nat Genet 2012;44(10):1131−6.

[5] Levine DM, Ek WE, Zhang R, Liu X, Onstad L, Sather C, et al. A genome-wide association study identifies new susceptibility loci for esophageal adenocarcinoma and Barrett's esophagus. Nat Genet 2013;45 (12):1487−93.

[6] Mohammed I, Cherkas LF, Riley SA, Spector TD, Trudgill NJ. Genetic influences in gastro-oesophageal reflux disease: a twin study. Gut 2003;52(8):1085−9.

[7] Chak A, Ochs-Balcom H, Falk G, Grady WM, Kinnard M, Willis JE, et al. Familiality in Barrett's esophagus, adenocarcinoma of the esophagus, and adenocarcinoma of the gastroesophageal junction. Cancer Epidemiol Biomarkers Prev 2006;15(9):1668−73.

[8] Shu W, Lu MM, Zhang Y, Tucker PW, Zhou D, Morrisey EE. Foxp2 and Foxp1 cooperatively regulate lung and esophagus development. Development 2007;134(10):1991−2000.

[9] Martin V, Shaw-Smith C. Review of genetic factors in intestinal malrotation. Pediatr Surg Int 2010;26 (8):769−81.

[10] Woo J, Miletich I, Kim BM, Sharpe PT, Shivdasani RA. Barx1-mediated inhibition of Wnt signaling in the mouse thoracic foregut controls tracheo-esophageal septation and epithelial differentiation. PLoS One 2011;6(7):e22493.

[11] Arora R, Metzger RJ, Papaioannou VE. Multiple roles and interactions of Tbx4 and Tbx5 in development of the respiratory system. PLoS Genet 2012;8(8):e1002866.

[12] Gu Y, Lin S, Li JL, Chen Z, Jin B, Tian L, et al. Altered LKB1/CREB-regulated transcription co-activator (CRTC) signaling axis promotes esophageal cancer cell migration and invasion. Oncogene 2012;31(4):469−79.

[13] Cheung LW, Hennessy BT, Li J, Yu S, Myers AP, Djordjevic B, et al. High frequency of PIK3R1 and PIK3R2 mutations in endometrial cancer elucidates a novel mechanism for regulation of PTEN protein stability. Cancer Discov 2011;1(2):170−85.

[14] Duester G. Retinoic acid synthesis and signaling during early organogenesis. Cell 2008;134(6):921−31.

[15] Castillo D, Puig S, Iglesias M, Seoane A, de Bolós C, Munitiz V, et al. Activation of the BMP4 pathway and early expression of CDX2 characterize non-specialized columnar metaplasia in a human model of Barrett's esophagus. J Gastrointest Surg 2012;16(2):227−37; discussion 237.

[16] Ek WE, Levine DM, D'Amato M, Pedersen NL, Magnusson PK, Bresso F, et al. Germline genetic contributions to risk for esophageal adenocarcinoma, Barrett's esophagus, and gastroesophageal reflux. J Natl Cancer Inst 2013;105(22):1711−18.

[17] O'Riordan JM, Abdel-latif MM, Ravi N, McNamara D, Byrne PJ, McDonald GS, et al. Proinflammatory cytokine and nuclear factor kappa-B expression along the inflammation-metaplasia-dysplasia-adenocarcinoma sequence in the esophagus. Am J Gastroenterol 2005;100(6):1257−64.

[18] Saadi A, Shannon NB, Lao-Sirieix P, O'Donovan M, Walker E, Clemons NJ, et al. Stromal genes discriminate preinvasive from invasive disease, predict outcome, and highlight inflammatory pathways in digestive cancers. Proc Natl Acad Sci U S A 2010;107 (5):2177−82.

[19] Hong MK, Laskin WB, Herman BE, Johnston MH, Vargo JJ, Steinberg SM, et al. Expansion of the Ki-67 proliferative compartment correlates with degree of dysplasia in Barrett's esophagus. Cancer 1995;75 (2):423−9.

[20] Fitzgerald RC, Omary MB, Triadafilopoulos G. Dynamic effects of acid on Barrett's esophagus. An ex vivo proliferation and differentiation model. J Clin Invest 1996;98(9):2120−8.

[21] Fitzgerald RC, Omary MB, Triadafilopoulos G. Acid modulation of HT29 cell growth and differentiation. An in vitro model for Barrett's esophagus. J Cell Sci 1997;110(Pt 5):663−71.

[22] Kaur BS, Ouatu-Lascar R, Omary MB, Triadafilopoulos G. Bile salts induce or blunt cell proliferation in Barrett's esophagus in an acid-dependent

fashion. Am J Physiol Gastrointest Liver Physiol 2000;278(6):G1000−9.

[23] Kavanagh ME, O'Sullivan KE, O'Hanlon C, O'Sullivan JN, Lysaght J, Reynolds JV. The esophagitis to adeno-carcinoma sequence; the role of inflammation. Cancer Lett 2014;345(2):182−9.

[24] Souza RF, Huo X, Mittal V, Schuler CM, Carmack SW, Zhang HY, et al. Gastroesophageal reflux might cause esophagitis through a cytokine-mediated mechanism rather than caustic acid injury. Gastroenterology 2009;137(5):1776−84.

[25] Jarnicki A, Putoczki T, Ernst M. Stat3: linking inflam-mation to epithelial cancer—more than a "gut" feel-ing? Cell Div 2010;5:14.

[26] Jenkins GJ, Mikhail J, Alhamdani A, Brown TH, Caplin S, Manson JM, et al. Immunohistochemical study of nuclear factor-kappaB activity and interleukin-8 abundance in oesophageal adenocarci-noma; a useful strategy for monitoring these biomar-kers. J Clin Pathol 2007;60(11):1232−7.

[27] Dvorak K, Payne CM, Chavarria M, Ramsey L, Dvorakova B, Bernstein H, et al. Bile acids in combina-tion with low pH induce oxidative stress and oxida-tive DNA damage: relevance to the pathogenesis of Barrett's oesophagus. Gut 2007;56(6):763−71.

[28] Lee SR, Yang KS, Kwon J, Lee C, Jeong W, Rhee SG. Reversible inactivation of the tumor suppressor PTEN by H2O2. J Biol Chem 2002;277(23):20336−42.

[29] Reid BJ, Li X, Galipeau PC, Vaughan TL. Barrett's oesophagus and oesophageal adenocarcinoma: time for a new synthesis. Nat Rev Cancer 2010;10(2):87−101.

[30] Leedham SJ, Preston SL, McDonald SA, Elia G, Bhandari P, Poller D, et al. Individual crypt genetic heterogeneity and the origin of metaplastic glandular epithelium in human Barrett's oesophagus. Gut 2008;57(8):1041−8.

[31] Weaver JM, Ross-Innes CS, Shannon N, Lynch AG, Forshew T, Barbera M, et al. Ordering of mutations in preinvasive disease stages of esophageal carcinogene-sis. Nat Genet 2014;46(8):837−43.

[32] Li J, Poi MJ, Tsai MD. Regulatory mechanisms of tumor suppressor P16(INK4A) and their relevance to cancer. Biochemistry 2011;50(25):5566−82.

[33] Wang JS, Guo M, Montgomery EA, Thompson RE, Cosby H, Hicks L, et al. DNA promoter hypermethyla-tion of p16 and APC predicts neoplastic progression in Barrett's esophagus. Am J Gastroenterol 2009;104 (9):2153−60.

[34] Wong DJ, Paulson TG, Prevo LJ, Galipeau PC, Longton G, Blount PL, et al. p16(INK4a) lesions are common, early abnormalities that undergo clonal expansion in Barrett's metaplastic epithelium. Cancer Res 2001;61(22):8284−9.

[35] Muller PA, Vousden KH. p53 mutations in cancer. Nat Cell Biol 2013;15(1):2−8.

[36] Massague J. TGFbeta in cancer. Cell 2008;134 (2):215−30.

[37] Waddell N, Pajic M, Patch AM, Chang DK, Kassahn KS, Bailey P, et al. Whole genomes redefine the muta-tional landscape of pancreatic cancer. Nature 2015;518 (7540):495−501.

[38] Voorneveld PW, Kodach LL, Jacobs RJ, Liv N, Zonnevylle AC, Hoogenboom JP, et al. Loss of SMAD4 alters BMP signaling to promote colorec-tal cancer cell metastasis via activation of Rho and ROCK. Gastroenterology 2014;147(1) 196−208 e113.

[39] Coffill CR, Muller PA, Oh HK, Neo SP, Hogue KA, Cheok CF, et al. Mutant p53 interactome identifies nardilysin as a p53R273H-specific binding partner that promotes invasion. EMBO Rep 2012;13(7):638−44.

[40] Gomez-Lazaro M, Fernandez-Gomez FJ, Jordan J. p53: twenty five years understanding the mechanism of genome protection. J Physiol Biochem 2004;60 (4):287−307.

[41] Alexandrov LB, Nik-Zainal S, Wedge DC, Aparicio SA, Behjati S, Biankin AV, et al. Signatures of muta-tional processes in human cancer. Nature 2013;500 (7463):415−21.

[42] Dulak AM, Stojanov P, Peng S, Lawrence MS, Fox C, Stewart C, et al. Exome and whole-genome sequencing of esophageal adenocarcinoma identifies recurrent driver events and mutational complexity. Nat Genet 2013;45(5):478−86.

[43] Nones K, Waddell N, Wayte N, Patch AM, Bailey P, Newell F, et al. Genomic catastrophes frequently arise in esophageal adenocarcinoma and drive tumorigene-sis. Nat Commun 2014;5:5224.

[44] Dulak AM, Schumacher SE, van Lieshout J, Imamura Y, Fox C, Shim B, et al. Gastrointestinal adenocarcino-mas of the esophagus, stomach, and colon exhibit dis-tinct patterns of genome instability and oncogenesis. Cancer Res 2012;72(17):4383−93.

[45] Li X, Galipeau PC, Paulson TG, Sanchez CA, Arnaudo J, Liu K, et al. Temporal and spatial evolution of somatic chromosomal alterations: a case-cohort study of Barrett's esophagus. Cancer Prev Res (Phila) 2014;7 (1):114−27.

[46] Zhang CZ, Leibowitz ML, Pellman D. Chromothripsis and beyond: rapid genome evolution from complex chromosomal rearrangements. Genes Dev 2013;27 (23):2513−30.

[47] Stephens PJ, Greenman CD, Fu B, Yang F, Bignell GR, Mudie LJ, et al. Massive genomic rearrangement acquired in a single catastrophic event during cancer development. Cell 2011;144(1):27−40.

[48] Goldberg AD, Allis CD, Bernstein E. Epigenetics: a landscape takes shape. Cell 2007;128(4):635−8.

[49] Esteller M. Cancer epigenomics: DNA methylomes and histone-modification maps. Nat Rev Genet 2007;8 (4):286−98.

[50] Eads CA, Lord RV, Kurumboor SK, Wickramasinghe K, Skinner ML, Long TI, et al. Fields of aberrant CpG island hypermethylation in Barrett's esophagus and associated adenocarcinoma. Cancer Res 2000;60 (18):5021−6.

[51] Jin Z, Mori Y, Yang J, Sato F, Ito T, Cheng Y, et al. Hypermethylation of the nel-like 1 gene is a common and early event and is associated with poor prognosis in early-stage esophageal adenocarcinoma. Oncogene 2007;26(43):6332−40.

[52] Jin Z, Olaru A, Yang J, Sato F, Cheng Y, Kan T, et al. Hypermethylation of tachykinin-1 is a potential biomarker in human esophageal cancer. Clin Cancer Res 2007;13(21):6293−300.

[53] Kuester D, Dar AA, Moskaluk CC, Krueger S, Meyer F, Hartig R, et al. Early involvement of death-associated protein kinase promoter hypermethylation in the carcinogenesis of Barrett's esophageal adenocarcinoma and its association with clinical progression. Neoplasia 2007;9(3):236−45.

[54] Jin Z, Cheng Y, Olaru A, Kan T, Yang J, Paun B, et al. Promoter hypermethylation of CDH13 is a common, early event in human esophageal adenocarcinogenesis and correlates with clinical risk factors. Int J Cancer 2008;123(10):2331−6.

[55] Jin Z, Hamilton JP, Yang J, Mori Y, Olaru A, Sato F, et al. Hypermethylation of the AKAP12 promoter is a biomarker of Barrett's-associated esophageal neoplastic progression. Cancer Epidemiol Biomarkers Prev 2008;17(1):111−17.

[56] Jin Z, Mori Y, Hamilton JP, Olaru A, Sato F, Yang J, et al. Hypermethylation of the somatostatin promoter is a common, early event in human esophageal carcinogenesis. Cancer 2008;112(1):43−9.

[57] Peng DF, Razvi M, Chen H, Washington K, Roessner A, Schneider-Stock R, et al. DNA hypermethylation regulates the expression of members of the Mu-class glutathione S-transferases and glutathione peroxidases in Barrett's adenocarcinoma. Gut 2009;58(1):5−15.

[58] Alvarez H, Opalinska J, Zhou L, Sohal D, Fazzari MJ, Yu Y, et al. Widespread hypomethylation occurs early and synergizes with gene amplification during esophageal carcinogenesis. PLoS Genet 2011;7(3):e1001356.

[59] di Pietro M, Lao-Sirieix P, Boyle S, Cassidy A, Castillo D, Saadi A, et al. Evidence for a functional role of epigenetically regulated midcluster HOXB genes in the development of Barrett esophagus. Proc Natl Acad Sci U S A 2012;109(23):9077−82.

[60] Vaughan TL, Fitzgerald RC. Precision prevention of oesophageal adenocarcinoma. Nat Rev Gastroenterol Hepatol 2015;12(4):243−8.

[61] Fitzgerald RC, di Pietro M, Ragunath K, Ang Y, Kang JY, Watson P, et al. British Society of Gastroenterology guidelines on the diagnosis and management of Barrett's oesophagus. Gut 2014;63(1):7−42.

[62] American Gastroenterological A, Spechler SJ, Sharma P, Souza RF, Inadomi JM, Shaheen NJ. American Gastroenterological Association medical position statement on the management of Barrett's esophagus. Gastroenterology 2011;140(3):1084−91.

[63] Kadri SR, Lao-Sirieix P, O'Donovan M, Debiram I, Das M, Blazeby JM, et al. Acceptability and accuracy of a non-endoscopic screening test for Barrett's oesophagus in primary care: cohort study. BMJ 2010;341:c4372.

[64] Pepe MS, Etzioni R, Feng Z, Potter JD, Thompson ML, Thornquist M, et al. Phases of biomarker development for early detection of cancer. J Natl Cancer Inst 2001;93 (14):1054−61.

[65] Lao-Sirieix P, Boussioutas A, Kadri SR, O'Donovan M, Debiram I, Das M, et al. Non-endoscopic screening biomarkers for Barrett's oesophagus: from microarray analysis to the clinic. Gut 2009;58(11):1451−9.

[66] TFF3 trefoil factor 3 (intestinal) [Homo sapiens (human)]. NCBI Gene profile. Available at: <http://www.ncbi. nlm.nih.gov/gene/7033>. [accessed 12.04.15].

[67] Ross-Innes CS, Debiram-Beecham I, O'Donovan M, Walker E, Varghese S, Lao-Sirieix P, et al. Evaluation of a minimally invasive cell sampling device coupled with assessment of trefoil factor 3 expression for diagnosing Barrett's esophagus: a multi-center case-control study. PLoS Med 2015;12(1):e1001780.

[68] Shaheen NJ, Sharma P, Overholt BF, Wolfsen HC, Sampliner RE, Wang KK, et al. Radiofrequency ablation in Barrett's esophagus with dysplasia. N Engl J Med 2009;360(22):2277−88.

[69] Phoa KN, van Vilsteren FG, Weusten BL, Bisschops R, Schoon EJ, Ragunath K, et al. Radiofrequency ablation vs endoscopic surveillance for patients with Barrett esophagus and low-grade dysplasia: a randomized clinical trial. JAMA 2014;311(12):1209−17.

[70] Sikkema M, Kerkhof M, Steyerberg EW, Kusters JG, van Strien PM, Looman CW, et al. Aneuploidy and overexpression of Ki67 and p53 as markers for neoplastic progression in Barrett's esophagus: a case-control study. Am J Gastroenterol 2009;104 (11):2673−80.

[71] Bhat S, Coleman HG, Yousef F, Johnston BT, McManus DT, Gavin AT, et al. Risk of malignant progression in Barrett's esophagus patients: results from a large population-based study. J Natl Cancer Inst 2011;103(13):1049−57.

[72] Hvid-Jensen F, Pedersen L, Drewes AM, Sorensen HT, Funch-Jensen P. Incidence of adenocarcinoma among patients with Barrett's esophagus. N Engl J Med 2011;365(15):1375—83.

[73] Bird-Lieberman EL, Dunn JM, Coleman HG, Lao-Sirieix P, Oukrif D, Moore CE, et al. Population-based study reveals new risk-stratification biomarker panel for Barrett's esophagus. Gastroenterology 2012;143(4) 927—935 e923

[74] Kastelein F, Biermann K, Steyerberg EW, Verheij J, Kalisvaart M, Looijenga LH, et al. Aberrant p53 protein expression is associated with an increased risk of neoplastic progression in patients with Barrett's oesophagus. Gut 2013;62(12):1676—83.

[75] Duits LC, Phoa KN, Curvers WL, Ten Kate FJ, Meijer GA, Seldenrijk CA, et al. Barrett's oesophagus patients with low-grade dysplasia can be accurately risk-stratified after histological review by an expert pathology panel. Gut 2015;64(5):700—6.

[76] Singh S, Manickam P, Amin AV, Samala N, Schouten LJ, Iyer PG, et al. Incidence of esophageal adenocarcinoma in Barrett's esophagus with low-grade dysplasia: a systematic review and meta-analysis. Gastrointest Endosc 2014;79(6):897—909.

[77] Tham T. Guidelines on the Diagnosis and Management of Barrett's Oesophagus—An Update. British Society of Gastroenterology 2015;. Available at: <http://www.bsg.org.uk/clinical-guidelines/oesophageal/guide-lines-on-the-diagnosis-and-management-of-barrett-s-oesophagus.html>. [accessed 1.05.15].

[78] Epithelial radiofrequency ablation for Barrett's oesophagus. National Institute for Health and Care Excellence; 2010.

[79] Spechler SJ, Sharma P, Souza RF, Inadomi JM, Shaheen NJ. American Gastroenterological Association. American Gastroenterological Association technical review on the management of Barrett's esophagus. Gastroenterology 2011;140(3):e18—52.

[80] Reid BJ, Prevo LJ, Galipeau PC, Sanchez CA, Longton G, Levine DS, et al. Predictors of progression in Barrett's esophagus II: baseline 17p (p53) loss of heterozygosity identifies a patient subset at increased risk for neoplastic progression. Am J Gastroenterol 2001;96 (10):2839—48.

[81] Galipeau PC, Li X, Blount PL, Maley CC, Sanchez CA, Odze RD, et al. NSAIDs modulate CDKN2A, TP53, and DNA content risk for progression to esophageal adenocarcinoma. PLoS Med 2007;4(2):e67.

[82] Weston AP, Banerjee SK, Sharma P, Tran TM, Richards R, Cherian R. p53 protein overexpression in low grade dysplasia (LGD) in Barrett's esophagus: immunohistochemical marker predictive of progression. Am J Gastroenterol 2001;96(5):1355—62.

[83] Younes M, Ertan A, Lechago LV, Somoano JR, Lechago J. p53 Protein accumulation is a specific marker of malignant potential in Barrett's metaplasia. Dig Dis Sci 1997;42(4):697—701.

[84] Skacel M, Petras RE, Rybicki LA, Gramlich TL, Richter JE, Falk GW, et al. p53 expression in low grade dysplasia in Barrett's esophagus: correlation with interobserver agreement and disease progression. Am J Gastroenterol 2002;97(10):2508—13.

[85] Murray L, Sedo A, Scott M, McManus D, Sloan JM, Hardie LJ, et al. TP53 and progression from Barrett's metaplasia to oesophageal adenocarcinoma in a UK population cohort. Gut 2006;55(10):1390—7.

[86] Kaye PV, Haider SA, Ilyas M, James PD, Soomro I, Faisal W, et al. Barrett's dysplasia and the Vienna classification: reproducibility, prediction of progression and impact of consensus reporting and p53 immunohistochemistry. Histopathology 2009;54(6):699—712.

[87] Davelaar AL, Calpe S, Lau L, Timmer MR, Visser M, Ten Kate FJ, et al. Aberrant TP53 detected by combining immunohistochemistry and DNA-FISH improves Barrett's esophagus progression prediction: a prospective follow-up study. Genes Chromosomes Cancer 2015;54(2):82—90.

[88] Streppel MM, Lata S, DelaBastide M, Montgomery EA, Wang JS, Canto MI, et al. Next-generation sequencing of endoscopic biopsies identifies ARID1A as a tumor-suppressor gene in Barrett's esophagus. Oncogene 2014;33(3):347—57.

[89] Lu J, Getz G, Miska EA, Alvarez-Saavedra E, Lamb J, Peck D, et al. MicroRNA expression profiles classify human cancers. Nature 2005;435(7043):834—8.

[90] Calin GA, Croce CM. MicroRNA-cancer connection: the beginning of a new tale. Cancer Res 2006;66 (15):7390—4.

[91] Feber A, Xi L, Luketich JD, Pennathur A, Landreneau RJ, Wu M, et al. MicroRNA expression profiles of esophageal cancer. J Thorac Cardiovasc Surg 2008;135 (2):255—60.

[92] Mathe EA, Nguyen GH, Bowman ED, Zhao Y, Budhu A, Schetter AJ, et al. MicroRNA expression in squamous cell carcinoma and adenocarcinoma of the esophagus: associations with survival. Clin Cancer Res 2009;15(19):6192—200.

[93] Fassan M, Volinia S, Palatini J, Pizzi M, Fernandez-Cymering C, Balistreri M, et al. MicroRNA expression profiling in the histological subtypes of Barrett's metaplasia. Clin Transl Gastroenterol 2013;4:e34.

[94] Revilla-Nuin B, Parrilla P, Lozano JJ, de Haro LF, Ortiz A, Martínez C, et al. Predictive value of MicroRNAs in the progression of Barrett esophagus to adenocarcinoma in a long-term follow-up study. Ann Surg 2013;257(5):886—93.

[95] Bansal A, Hong X, Lee IH, Krishnadath KK, Mathur SC, Gunewardena S, et al. MicroRNA expression can be a promising strategy for the detection of Barrett's esophagus: a pilot study. Clin Transl Gastroenterol 2014;5:e65.

[96] Skinner HD, Lee JH, Bhutani MS, Weston B, Hofstetter W, Komaki R, et al. A validated miRNA profile predicts response to therapy in esophageal adenocarcinoma. Cancer 2014;120(23):3635–41.

[97] Wong DJ, Barrett MT, Stoger R, Emond MJ, Reid BJ. p16INK4a promoter is hypermethylated at a high frequency in esophageal adenocarcinomas. Cancer Res 1997;57(13):2619–22.

[98] Eads CA, Lord RV, Wickramasinghe K, Long TI, Kurumboor SK, Bernstein L, et al. Epigenetic patterns in the progression of esophageal adenocarcinoma. Cancer Res 2001;61(8):3410–18.

[99] Maley CC, Galipeau PC, Li X, Sanchez CA, Paulson TG, Reid BJ. Selectively advantageous mutations and hitchhikers in neoplasms: p16 lesions are selected in Barrett's esophagus. Cancer Res 2004;64(10):3414–27.

[100] Hamilton JP, Sato F, Jin Z, Greenwald BD, Ito T, Mori Y, et al. Reprimo methylation is a potential biomarker of Barrett's-associated esophageal neoplastic progression. Clin Cancer Res 2006;12(22):6637–42.

[101] Schulmann K, Sterian A, Berki A, Yin J, Sato F, Xu Y, et al. Inactivation of p16, RUNX3, and HPP1 occurs early in Barrett's-associated neoplastic progression and predicts progression risk. Oncogene 2005;24 (25):4138–48.

[102] Sato F, Jin Z, Schulmann K, Wang J, Greenwald BD, Ito T, et al. Three-tiered risk stratification model to predict progression in Barrett's esophagus using epigenetic and clinical features. PLoS One 2008;3(4): e1890.

[103] Jin Z, Cheng Y, Gu W, Zheng Y, Sato F, Mori Y, et al. A multicenter, double-blinded validation study of methylation biomarkers for progression prediction in Barrett's esophagus. Cancer Res 2009;69 (10):4112–15.

[104] Alvi MA, Liu X, O'Donovan M, Newton R, Wernisch L, Shannon NB, et al. DNA methylation as an adjunct to histopathology to detect prevalent, inconspicuous dysplasia and early-stage neoplasia in Barrett's esophagus. Clin Cancer Res 2013;19(4):878–88.

[105] Kaz AM, Wong CJ, Luo Y, Virgin JB, Washington MK, Willis JE, et al. DNA methylation profiling in Barrett's esophagus and esophageal adenocarcinoma reveals unique methylation signatures and molecular subclasses. Epigenetics 2011;6(12):1403–12.

[106] Agarwal R, Jin Z, Yang J, Mori Y, Song JH, Kumar S, et al. Epigenomic program of Barrett's-associated neoplastic progression reveals possible involvement of insulin signaling pathways. Endocr Relat Cancer 2012;19(1):L5–9.

[107] Xu E, Gu J, Hawk ET, Wang KK, Lai M, Huang M, et al. Genome-wide methylation analysis shows similar patterns in Barrett's esophagus and esophageal adenocarcinoma. Carcinogenesis 2013;34(12):2750–6.

[108] Reid BJ, Blount PL, Rubin CE, Levine DS, Haggitt RC, Rabinovitch PS. Flow-cytometric and histological progression to malignancy in Barrett's esophagus: prospective endoscopic surveillance of a cohort. Gastroenterology 1992;102(4 Pt 1):1212–19.

[109] Rabinovitch PS, Longton G, Blount PL, Levine DS, Reid BJ. Predictors of progression in Barrett's esophagus III: baseline flow cytometric variables. Am J Gastroenterol 2001;96(11):3071–83.

[110] Maley CC, Galipeau PC, Li X, Sanchez CA, Paulson TG, Blount PL, et al. The combination of genetic instability and clonal expansion predicts progression to esophageal adenocarcinoma. Cancer Res 2004;64 (20):7629–33.

[111] Kerkhof M, Steyerberg EW, Kusters JG, van Dekken H, van Vuuren AJ, Kuipers EJ, et al. Aneuploidy and high expression of p53 and Ki67 is associated with neoplastic progression in Barrett esophagus. Cancer Biomark 2008;4(1):1–10.

[112] Rygiel AM, Milano F, Ten Kate FJ, de Groot JG, Peppelenbosch MP, Bergman JJ, et al. Assessment of chromosomal gains as compared to DNA content changes is more useful to detect dysplasia in Barrett's esophagus brush cytology specimens. Genes Chromosomes Cancer 2008;47(5):396–404.

[113] Timmer MR, Brankley SM, Gorospe EC, Sun G, Lutzke LS, Iyer PG, et al. Prediction of response to endoscopic therapy of Barrett's dysplasia by using genetic biomarkers. Gastrointest Endosc 2014;80(6):984–91.

[114] Geppert CI, Rummele P, Sarbia M, Langer R, Feith M, Morrison L, et al. Multi-colour FISH in oesophageal adenocarcinoma-predictors of prognosis independent of stage and grade. Br J Cancer 2014;110 (12):2985–95.

[115] Dunn JM, Mackenzie GD, Oukrif D, Mosse CA, Banks MR, Thorpe S, et al. Image cytometry accurately detects DNA ploidy abnormalities and predicts late relapse to high-grade dysplasia and adenocarcinoma in Barrett's oesophagus following photodynamic therapy. Br J Cancer 2010;102(11):1608–17.

[116] Vogt N, Schonegg R, Gschossmann JM, Borovicka J. Benefit of baseline cytometry for surveillance of patients with Barrett's esophagus. Surg Endosc 2010;24(5):1144–50.

[117] Bani-Hani K, Martin IG, Hardie LJ, Mapstone N, Briggs JA, Forman D, et al. Prospective study of cyclin D1 overexpression in Barrett's esophagus:

association with increased risk of adenocarcinoma. J Natl Cancer Inst 2000;92(16):1316−21.

[118] Lao-Sirieix P, Lovat L, Fitzgerald RC. Cyclin A immunocytology as a risk stratification tool for Barrett's esophagus surveillance. Clin Cancer Res 2007;13(2 Pt 1):659−65.

[119] Shi XY, Bhagwandeen B, Leong AS. p16, cyclin D1, Ki-67, and AMACR as markers for dysplasia in Barrett esophagus. Appl Immunohistochem Mol Morphol 2008;16(5):447−52.

[120] van Dekken H, Hop WC, Tilanus HW, Haringsma J, van der Valk H, Wink JC, et al. Immunohistochemical evaluation of a panel of tumor cell markers during malignant progression in Barrett esophagus. Am J Clin Pathol 2008;130(5):745−53.

[121] di Pietro M, Boerwinkel DF, Shariff MK, Liu X, Telakis E, Lao-Sirieix P, et al. The combination of autofluorescence endoscopy and molecular biomarkers is a novel diagnostic tool for dysplasia in Barrett's oesophagus. Gut 2015;64 (1):49−56.

[122] Going JJ, Keith WN, Neilson L, Stoeber K, Stuart RC, Williams GH. Aberrant expression of minichromosome maintenance proteins 2 and 5, and Ki-67 in dysplastic squamous oesophageal epithelium and Barrett's mucosa. Gut 2002;50(3):373−7.

[123] Sirieix PS, O'Donovan M, Brown J, Save V, Coleman N, Fitzgerald RC. Surface expression of minichromosome maintenance proteins provides a novel method for detecting patients at risk for developing adenocarcinoma in Barrett's esophagus. Clin Cancer Res 2003;9(7):2560−6.

[124] Chao DL, Sanchez CA, Galipeau PC, Blount PL, Paulson TG, Cowan DS, et al. Cell proliferation, cell cycle abnormalities, and cancer outcome in patients with Barrett's esophagus: a long-term prospective study. Clin Cancer Res 2008;14(21):6988−95.

[125] Li D, Wang C, Li N, Zhang L. Propofol selectively inhibits nuclear factor-kappaB activity by suppressing p38 mitogen-activated protein kinase signaling in human EA.hy926 endothelial cells during intermittent hypoxia/reoxygenation. Mol Med Rep 2014;9(4):1460−6.

Diagnosis of Barrett's Esophagus

Alison Schneider[1], Amitabh Chak[2] and Amareshwar Podugu[1]

[1]Department of Gastroenterology, Digestive Disease Center, Cleveland Clinic Florida, Weston, FL, United States [2]Advanced Technology & Innovation Center of Excellence, Division of Gastroenterology, Case Western Reserve University, Cleveland, OH, United States

5.1 INTRODUCTION

In 1950, Norman Barrett wrote the paper entitled "Chronic peptic ulcer of the oesophagus and 'oesophagitis'" where he described a case in which a portion of the stomach was noted to be within the chest [1]. It was later determined by Allison and Johnstone in a detailed description of seven cases that what Barrett had described was actually a columnar-lined esophageal segment. They also noted an association with esophageal reflux disease [2]. Since that time there have been dramatic changes in the understanding of this condition, which it seems will forever bear Norman Barrett's eponym.

Barrett's esophagus is an acquired condition present in 10–15% of patients with chronic gastroesophageal reflux disease (GERD). It is currently defined as a condition in which normal stratified squamous epithelium is replaced with columnar epithelium with intestinal type mucosa in the distal esophagus [3]. While Barrett's esophagus is asymptomatic, its clinical importance is that it is one of the only known risk factors for esophageal adenocarcinoma. Esophageal adenocarcinoma is the cancer with the most rapidly increasing incidence in the Western world [4]. Barrett's is thought to progress in stages from intestinal metaplasia to low-grade dysplasia (LGD), high-grade dysplasia (HGD), and then finally to invasive adenocarcinoma.

In the United States, the diagnosis of Barrett's esophagus first begins with the endoscopic visualization of proximally displaced squamous epithelial lining of the distal esophagus followed by biopsy specimens obtained from the true tubular esophagus showing specialized intestinal metaplasia. Biopsies are obtained not only to enable a diagnosis of Barrett's but also to identify dysplasia and early cancers at a stage when curative treatment may be possible. A clear diagnosis is very important for patients as, once diagnosed, they will likely be placed into routine endoscopic surveillance based on current society guidelines and there will be personal fears and concerns about their risk for esophageal cancer in the future. A diagnosis of Barrett's may also have implications for health and life insurance in the United States [5].

D. Pleskow & T. Erim (Eds): Barrett's Esophagus.
DOI: http://dx.doi.org/10.1016/B978-0-12-802511-6.00005-3

5.2 HISTOPATHOLOGIC DIAGNOSIS OF BARRETT'S

Barrett's esophagus (BE) is well established as a complication of GERD [6,7]. Reflux of various substances such as acid and bile from the stomach, along with other factors such as increased transient lower esophageal sphincter relaxations (TRLES) and decreased esophageal motility allow for a prolonged exposure of the esophagus to refluxate from the stomach. This causes chronic inflammation and subsequent repair of the esophageal mucosa. The repair process in some individuals results in the replacement of normal esophageal epithelium with intestinal metaplasia. These changes may be the result of a number of genetic alterations and changes in cellular molecular pathways. It has been proposed that loss of p63 expression in squamous mucosa is involved [8,9]. Other studies have suggested overexpression of CDX2 and BMP4 to play a role in the pathogenesis of Barrett's [10–12]. Loss of TGFβ signaling as seen in knockout mice may lead to altered cell differentiation and intestinal metaplasia [13].

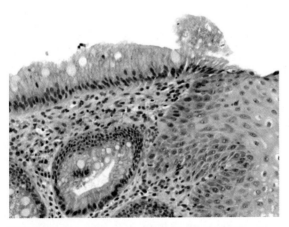

Barrett's esophagus. The columnar type mucosa shows intestinal metaplasia (goblet cells) with no dysplastic changes. Squamous mucosa is on the right (H&E, 200×).

For many years, there has been disagreement on what defines Barrett's esophagus histologically. In the United States, the definition of Barrett's esophagus as per the *American Gastroenterological Association* (AGA) is "the condition in which any extent of metaplastic columnar epithelium that predisposes to cancer development replaces the stratified squamous epithelium that normally lines the distal esophagus" [3]. Thus, the presence of goblet cells is required for a diagnosis of Barrett's esophagus in the United States and many other countries (Fig. 5.1). The *British Society of Gastroenterology* requires the presence of endoscopically visible features of Barrett's esophagus with metaplastic columnar epithelium on esophageal biopsies [4]. It does not necessarily require the detection of intestinal metaplasia in the affected segment. The Japanese do not require the presence of intestinal type epithelium either, but do require the presence of endoscopically evident palisading vessels to delineate the gastroesophageal junction [14–16].

There is much debate about whether intestinal metaplasia must be present for the diagnosis of BE or if only the presence of columnar mucosa is enough. Although most esophageal adenocarcinomas occur in the presence of intestinal metaplasia, esophageal adenocarcinomas may also occur in the presence of other types of mucosa such as cardia type in the absence of intestinal metaplasia [16,17]. Some studies have suggested that columnar cell epithelium may have analogous immunohistochemical profiles to intestinal metaplasia based on similar intestinal markers such as CDX-1, cytokeratins, mucin, and similar chromosomal instability [18–21]. But much of what is known about cancer risk in Barrett's esophagus is based on an association with intestinal metaplasia either primarily or exclusively. For the past several decades, most studies on Barrett's esophagus have used specialized intestinal metaplasia as an entry criterion.

More recently in 2015, experts conducted an international, multidisciplinary, systematic search and evidence-based review of BE (BOB CAT) and established an international agreement for a definition of BE. The definition states "BE is defined by the presence of columnar mucosa in the esophagus and it should be stated whether intestinal metaplasia (IM) is present above the gastroesophageal junction" [22]. This definition works to combine endoscopic and pathological diagnosis of various worldwide GI societies. It recognizes that intestinal metaplasia may not always be sampled on biopsies and there are cases where EA is not always preceded by intestinal metaplasia (see sections later). There are likely to be future modifications of this working definition.

5.2.1 Definition of Intestinal Metaplasia

Barrett's esophagus is composed of columnar epithelium and may contain goblet cells, enterocytes, Paneth cells, and some cells with combined gastric and intestinal features. There may also be multilayered epithelium, which has characteristics of both squamous and columnar features [23]. Goblet cells are specialized columnar epithelial cells normally found in the intestinal lining that function to produce gel-forming mucins. When these cells are identified in the stomach or esophagus, they represent intestinal metaplasia [24]. Goblet cells are identified both with hematoxylin—eosin stain and ancillary stains like Alcian blue/periodic acid Schiff stain. On hematoxylin—eosin stains, Goblet cells are dispersed on a background of nongoblet neutral mucin-containing cells similar to those of the gastric mucosa. Goblet cells contain acid mucin that imparts blue discoloration to the mucin vacuole, which compresses the nucleus and laterally displaces the membranes of adjacent cells. The Alcian blue/periodic acid Schiff stain facilitates identification of goblet cells. This stain colors the neutral mucin

of gastric foveolar epithelium red, while the acidic mucin of goblet cells is blue (Fig. 5.1). However, a variety of goblet cell mimics may also show Alcian blue positivity, thereby representing diagnostic pit falls when these stains are utilized. For example, injured, hyperplastic foveolar type epithelial cells, columnar cells that contain blue-tinged mucin, columnar cells lining ducts that drain mucosal and submucosal glands, submucosal mucinous glands, and multilayered epithelium cells may also stain positive for Alcian blue stains (Alcian blue-positive goblet cell mimics). None of these goblet cell mimics have been shown to confer as a high risk of malignant progression and should not be considered to represent specialized epithelium [25].

Columnar metaplasia of the distal esophagus is difficult to distinguish from the proximal gastric cardia histologically. Cardiac epithelium, which has traditionally been considered the normal lining of the most proximal portion of the stomach (the gastric cardia), is almost exclusively composed of mucus secreting gastric foveolar type cells without goblet cells. Observations suggest that cardiac epithelium found in the esophagus might not be a normal type of epithelium but rather an abnormal, metaplastic epithelium acquired as a consequence of GERD [26,27]. Although lacking the goblet cells of specialized intestinal metaplasia, cardiac epithelium nevertheless expresses molecular markers of intestinal differentiation (eg, villin and CDX2) [28]. Cardiac epithelium can be considered "intestinalized" even without goblet cells. Furthermore, cardiac epithelium exhibits genetic abnormalities similar to those found in specialized intestinal metaplasia. Thus, cardiac epithelium might be predisposed to malignancy despite its lack of goblet cells [29]. In one series of 141 patients with small esophageal adenocarcinoma removed primarily by endoscopic mucosa resection, more than 70% of the lesions were adjacent to cardiac/fundic-type mucosa rather than intestinal metaplasia. Intestinal metaplasia was

actually not found in 56.6% of biopsy specimens [13]. Indeed, even in specialized intestinal metaplasia, the highly differentiated goblet
cell seems an unlikely candidate to be the
malignant cell of origin. Examples of low-
grade, high-grade and indefinite for dysplasia
are shown in (Figs. 5.2–5.4).

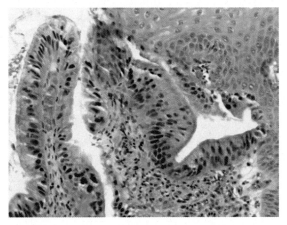

FIGURE 5.2 Low-grade dysplasia (LGD). Squamocolumnar mucosa showing LGD with crowding and nuclear
enlargement cells of the glandular component. The nuclei
do not reach the apical portions of the cells and mitotic
activity is absent (H&E, original magnification, 200×).

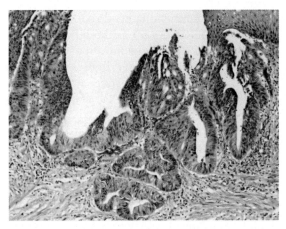

FIGURE 5.3 Barrett's esophagus with high-grade dysplasia (HGD). There is nuclear hyperchromasia with loss of
nuclear polarity accompanied by glandular architectural
distortion with crowding and back-to-back glands (H&E,
original magnification, 100×).

5.3 ENDOSCOPIC DIAGNOSIS OF BARRETT'S

Although the diagnosis of Barrett's esophagus requires histologic confirmation by biopsy,
it will first be suspected in those patients
where the junction of the squamous esophageal
epithelium and columnar epithelium of the
stomach, or Z-line, has been displaced proximal to the gastroesophageal junction (GEJ).
Thus it is critical to accurately identify the GEJ.
The definition of the GEJ is controversial and
varies depending on geographical location. For
example, in the United States, this area is
defined as the most proximal location of gastric
folds. Alternately in Japan, it is defined as the
lowest extent of the esophageal palisade blood
vessels [14,15] (Fig. 5.4).

Traditionally, these changes in the Z-line
location, or tongues of Barrett's esophagus,
were classified as short segment (<3 cm) or
long segment Barrett's esophagus (≥3 cm) but
there was often significant inter- and intraobserver variation especially in relation to shorter

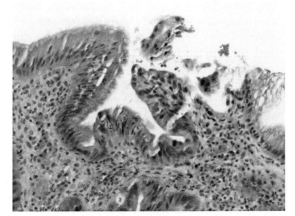

FIGURE 5.4 Indefinite for dysplasia. Although there is
hyperchromasia and stratification of the nuclei, these
changes are associated with the presence of acute inflammatory cells in the columnar epithelium, thus precluding a
definite diagnosis of dysplasia as the changes may be reactive in nature and attributed to the inflammation (H&E,
original magnification, 200×).

segments of Barrett's esophagus [30–32]. Due to controversies over the endoscopic definition of Barrett's esophagus and to help improve the definition, the Prague Classification of Barrett's esophagus was developed in 2006 by Sharma et al. [33]. This established a standardized criterion for the endoscopic diagnosis of circumferential (C) and maximal length (M) of endoscopically visualized Barrett's mucosa. A main component of this grading system looks at the landmarks of the squamocolumnar junction (SCJ), the gastroesophageal junction (GEJ), the extent of circumferential columnar lining (C), and the most proximal extension of the columnar mucosa (M) of Barrett's esophagus. (Fig. 5.5). The depth of these landmarks is determined at the point just before such findings are seen in full view on scope withdrawal, as the scope is straighter in this position. The SCJ is also termed the Z-line and is the most distal portion of the squamous epithelium and appears as a serrated line [34]. The anatomical location of the GEJ is determined at the point where the distal end of the tubular esophagus meets the upper gastric folds. Sometimes the distal end of the palisade vessels, which enter the submucosa at the GEJ, may be visible. Normally, the SCJ coincides with the GEJ. In patients without Barrett's esophagus and in the absence of a hiatal hernia, these junctions will be located just proximal to the diaphragmatic hiatus. It is important to recognize any hiatal hernia that is present by distinguishing the diaphragmatic hiatal impression, where the caliber of the lumen becomes narrow, or sphincter "pinch," from the GEJ. This can be difficult to see at times during endoscopy due to movements that occur from peristalsis and with respirations [35]. Additionally, in the setting of large hiatal hernias this may be hard to identify and desufflating the stomach may assist. The location of these landmarks, diaphragmatic pinch, GEJ, and SCJ, should be measured in centimeters from the incisors and documented in the endoscopy report (Figs. 5.5 and 5.6).

In the setting of Barrett's esophagus, the SCJ will be displaced proximal to the GEJ. Suspected Barrett's esophagus may be in the form of circumferential segments, tongues, or islands of columnar metaplasia. The difference in the depth of the circumferential and maximum extent from the depth of the endoscope insertion at the GEJ is then reported as C & M, respectively. It should be noted that islands of Barrett's esophagus are not included in this criteria and short segments <1 cm are not included in the Prague criteria [36].

The Prague criteria were developed by the International Working Group for Classification of Oesophagitis (IWGCO) based on video endoscopic recordings of 29 patients. The criteria were then validated by 29 external expert endoscopists who reviewed and scored these recordings. The reliability coefficients (RC) for determining the GEJ was 0.88 and for the location of a hiatus hernia 0.85. For the criteria evaluating the extent of Barrett's C and M, RC was 0.95 and 0.94, respectively. The overall RC for the endoscopic recognition of Barrett's esophagus ≥1 cm was 0.72 but decreased to a coefficient of 0.22 for Barrett's less than 1 cm in length [37]. Thus these criteria have a higher degree of validity for Barrett's esophagus segments when they are >1 cm in length (Fig. 5.6).

A later study compared interobserver agreement of the Prague C & M classification between expert endoscopists and community hospital endoscopists. A total of 187 patients with Barrett's esophagus underwent two upper endoscopies by each group of endoscopists and esophageal landmarks were documented. Absolute agreement was 74% (95% CI 68–80) for circumferential Barrett's length, 68% (95% CI 62–75) for maximum length, and 63% (95% CI 56–70) for hiatal hernia length. The relative agreement between the groups was 91%. There was less agreement for shorter Barrett's segments compared to longer segments. Overall, the authors concluded that the overall agreement between the groups was good [38]. The

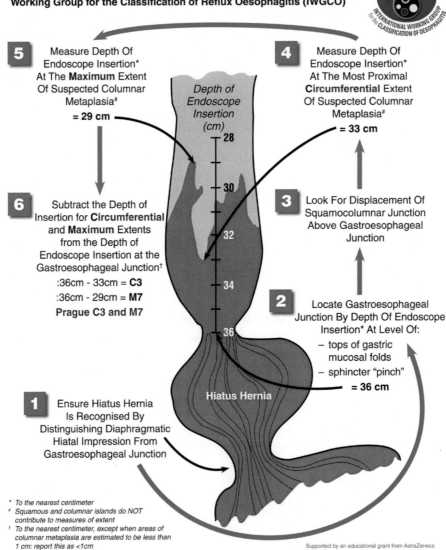

FIGURE 5.5   The Prague Criteria.

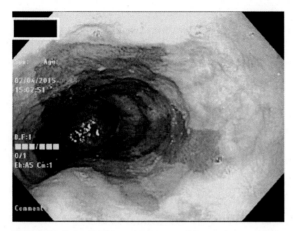

FIGURE 5.6 Endoscopic image of Barrett's esophagus showing an area of C: circumferential metaplasia and M: maximal extent of the metaplasia.

performance of the Prague criteria has also been studied among gastroenterology trainees and showed a high interobserver reliability that was similar at all levels of training [39].

Issues arise with short segments and an "irregular" Z-line. DeNarid and Ridell's description was that "the Z-line consists of small projections of red gastric epithelium, up to 5 mm long and 3 mm wide, extending upward into the pink-white squamous epithelium" [40]. The Z-line can be well demarcated and circular where there is no suggestion of tongues or islands of columnar epithelium. When the Z-line is irregular, there are noticeable linear extensions of columnar-appearing epithelium [34]. Debate has existed as to whether biopsies are needed for small variations in the Z-line. It can sometimes be difficult to find intestinal metaplasia in these very short segments [41]. In one retrospective analysis of 2000 upper endoscopies with no previous diagnosis of Barrett's, 166 (8.3%) of patients were identified with an irregular Z-line and 43% of these patients were found to have specialized intestinal metaplasia on biopsy. Risk factors included male sex and findings of a hiatal hernia [42]. The ProGERD study looked at data of

patients with specialized intestinal metaplasia at the cardia without endoscopic evidence of Barrett's esophagus and found that 25.8% of patients had progression to Barrett's esophagus in 2–5 years [43]. It is not fully known if very short segments of intestinal metaplasia actually have an increased risk of esophageal malignancy. However, one could argue to these being short segment Barrett's and it is known that dysplasia has been described in short segments and short segments of intestinal metaplasia have been found on surgical resections for adenocarcinoma [44,45].

One criterion that has been proposed to better identify short segment Barrett's is the Japanese criteria, which defines the GEJ as the lowest limit of the palisade vessels in the lamina propria of the esophagus. Kinjo et al. prospectively compared the Japanese criteria with the Prague C & M criteria in a group of 110 Japanese patients. A greater number of patients were diagnosed with endoscopic Barrett's esophagus using the Japanese criteria (39%) compared with the C & M criteria (26%). Identification of the GEJ was also higher with the Japanese criteria versus the C & M criteria (95% vs 86%). The authors concluded that the Japanese criteria might be better suited for their population. It is unknown if the results are applicable outside the Japanese population [14]. An important limitation of this study was that only two endoscopists interpreted the findings and the study was not designed to assess interobserver correlation for both criteria. More studies are needed to assess these possible differences.

It is worth considering the fact that the importance of Barrett's esophagus is that this condition increases the risk for development of cancer. Surveying patients with Barrett's esophagus may lead to prevention or early detection. Broadening the definition of Barrett's esophagus to include intestinal metaplasia in the presence of an irregular Z-line will identify more people at risk for progressing to adenocarcinoma but at the same time, given the large number of people with irregular Z-lines, a broader definition will greatly decrease the risk of progression to cancer. The lower the risk of progression to cancer the lower the benefit of surveillance.

5.3.1 Tissue Sampling

To make a diagnosis of Barrett's esophagus and to detect intestinal metaplasia and goblet cells, it has been suggested that a minimum of eight biopsies be obtained within long segment of Barrett's. In a study of 1646 biopsy samples from 125 patients with a mean Barrett's length of 4 cm, obtaining 8 biopsies led to a diagnosis of Barrett's esophagus with the presence of intestinal metaplasia in 67.9% of endoscopies. When only four biopsies were obtained, the yield dropped to 34.7%. Additional Alcian blue periodic acid Schiff staining only increased the diagnostic yield by 5.4% [46].

There have been a few publications on the technique for obtaining esophageal biopsies to improve the diagnostic yield of Barrett's esophagus. One way for acquiring better tissue sampling is with a "turn-and-suction" endoscopic biopsy technique. The biopsy forceps is drawn back to the endoscope tip in an open position. The endoscope is then turned toward the esophageal wall followed by suctioning and then slight advancement with simultaneous closure of the biopsy forceps to obtain tissue for sampling. Using this technique, biopsy samples are taken in a perpendicular orientation to the esophageal wall. Biopsies taken in this manner have been reported to be over 50% longer [47]. Care must be taken that biopsies are taken from the true esophagus and not the gastric cardia when biopsying at the EGJ.

Another way to improve biopsy sampling is with large capacity or jumbo biopsy forceps, which can sample larger dimensions of tissue with greater mean width and depth [48−51]. Interestingly, one study noted no difference in

the findings of esophageal cancer in patients with Barrett's esophagus and high-grade dysplasia with the use of jumbo biopsy forceps compared with standard biopsy forceps [52]. Additionally, a study of 65 patients undergoing surveillance for Barret's who were randomized to standard, large capacity, or jumbo forceps, adequate biopsy samples were not significantly different between the three groups. Jumbo forceps also had the lowest percentage of well-oriented biopsies (p = 0.001) [53].

Currently, the most commonly utilized protocol for performing esophageal biopsies for surveillance is systematic four-quadrant biopsies obtained every 2 cm in the Barrett's segment [54]. Nodules need to be identified based on the fact that they are raised or if there is a change in vascularity noted in the Barrett's mucosa. Targeted biopsies should first be taken of any nodules or visible lesions before other biopsies are taken. Also there should not be active esophageal inflammation, such as reflux esophagitis, at the time of biopsy [55]. Biopsies with this protocol have been shown to be more effective for finding LGD, HGD, and intramucosal cancers than older methods with random biopsy sampling. In one study comparing the detection of Barrett's dysplasia and adenocarcinoma in two cohorts (random biopsy vs systematic four-quadrant biopsy protocols), the prevalence of LGD per patient was 18.9% in the group with systematic biopsy sampling versus 1.6% ($p < 0.001$) with random biopsies. The prevalence of HGD was 2.8% versus 0% ($p = 0.03$) in the systematic and random groups, respectively. No patients who had the systematic biopsy protocol developed advanced cancer whereas three patients in the other group died of esophageal adenocarcinoma [56]. Similarly, in another study from the United Kingdom, the yield of findings of HGD and early esophageal cancers increased from 1% to 4.6% with the use of a systematic multiple biopsy protocol [57].

The Seattle biopsy protocol technique, targeted and four-quadrant jumbo biopsies every

1 cm, has been adapted by the American College of Gastroenterology and the British Society of Gastroenterology [17,58,59]. It should be noted that the Seattle protocol still has potential disadvantages, while it may decrease the risk of missed dysplasia, it still does not eliminate the issue completely. Additionally, adherence to biopsy protocol has been shown to be low. Abrams et al. showed that in 2245 Barrett's esophagus surveillance biopsy cases, only 51.2% adhered to guidelines. Longer segments of Barrett's were more likely to have reduced adherence and this was associated with decreased detection of dysplasia (OR 0.53, 95% CI 0.35–0.82) [60].

5.4 ADVANCES IN IMAGING AND DIAGNOSIS OF BARRET'S

Biopsy methods for Barrett's still remain a random sampling of Barrett's mucosa. Even with biopsy protocols it is possible to miss dysplasia and if dysplasia is found, it is still possible to miss invasive cancers. Novel endoscopic imaging techniques are being developed, which may improve on the endoscopic diagnosis of Barrett's dysplasia and early esophageal cancers. These techniques include chromoendoscopy, narrow band imaging (NBI), optical coherence tomography (OCT), and confocal laser endoscopy. "Red flag" techniques such as chromoendoscopy try to highlight potential patches of intestinal type mucosa that can subsequently be targeted for biopsy, whereas high magnification techniques such as confocal laser endoscopy attempt to make a histologic diagnosis of Barrett's esophagus and dysplasia during endoscopy.

5.4.1 Magnification Techniques and High-Resolution Endoscopy

Most standard endoscopes have the ability to magnify images and create very good

mucosal detail of Barrett's esophagus. There are specialized magnifying endoscopes that can magnify images up to 200 fold. These work best to focus on specific lesions rather than a large area when on magnification. Studies show lower interobserver agreement [61]. Additionally, prospective data have not found this technique to be superior to targeted biopsies for detection of intestinal metaplasia [62].

Most endoscopes now are able to produce high-definition resolution images and may have a higher sensitivity to detect Barrett's esophagus [63]. The current high-resolution endoscopes have the ability to generate million pixel images as opposed to 300,000 pixel images generated by traditional white light endoscopes. This generates images with fine details enabling the endoscopists to identify early target lesions with greater sensitivity. These images are best viewed with high-definition televisions in the endoscopy suite. Kara et al. illustrated that even in the hands of an expert endoscopists, targeted biopsy using high-resolution endoscopy only identified 79% of dysplasia cases and there was difficulty in characterizing LGD [63,64].

5.4.2 Chromoendoscopy

Chromoendoscopy is a technique in which different colored dyes are sprayed over the mucosa to highlight mucosal detail. Agents that have been studied include methylene blue, toluidine blue, acetic acid, and indigo carmine [65–69]. The mechanism by which each of the agents work is either as a contrast stain or through an absorptive stain process.

For example, methylene blue is an absorptive stain and can be absorbed by goblet cells to identify intestinal type epithelium. A mucolytic agent must first be applied to the esophagus to enhance the staining process. Methylene blue at a concentration of 0.5% can then be applied to an area of suspected Barrett's esophagus using an appropriate "spray catheter" to topically apply a fine mist of blue color. After washing off excess dye, biopsies can be obtained from segments that stain intensely blue to improve the diagnostic yield. This technique is operator dependent with variations in sensitivity and specificity [67,68]. Because of this learning curve, the fact that the technique can be untidy, and some theoretical concerns that methylene blue might be a carcinogen, this technique has not found favor in clinical practice [66].

Another strategy for highlighting mucosal detail in columnar mucosa is to spray dilute acetic acid on the surface of columnar epithelium. This technique has been borrowed from examination of the cervix during colposcopy. Acetic acid temporarily denatures surface proteins whitening the squamous epithelium and giving the Barrett's epithelium a reddish hue. Circular or round pits are seen in non-Barrett's sections while a villous or ridged pattern corresponds to intestinal type epithelium can be identified for biopsy [70]. A study by Longcroft-Wheaton et al. of high-resolution acetic acid chromoendoscopy showed an excellent correlation between predicted histology on chromoendoscopy (normal, HGD, and invasive neoplasia) to actual histology ($r = 0.98$) [71]. Again, although this is a simple technique and is easy to perform, it has not had clinical uptake because it requires high magnification and no standardized classification systems have been developed.

One advantage of chromoendoscopy is that the stains are relatively inexpensive (such as acetic acid) and can be used with any endoscope.

5.4.3 Virtual Chromoendoscopy and Other Novel Imaging

Virtual chromoendoscopy is a technology by which spectral features are modified with filters to narrow the band width of transmittance. The filters are adjusted to a blue light, which is similar to the absorption of hemoglobin. Restricting the spectrum of light can thus emphasize mucosal detail by enhancing the visualization of vascular structures in the mucosa [72,73]. As vascular changes are important in premalignant conditions, this technology may allow early detection of dysplasia and cancer [74]. Virtual chromoendoscopy can be performed by using narrow band filters (red, green, and blue bands) incorporated into the light source termed narrow band imaging. Postprocessing algorithms incorporated into the electronic processors can also obtain the same effect. NBI (Olympus, Inc.), I-Scan (Pentax Medical), and FICE (Fujifilm Global) are virtual chromoendoscopic technologies that are incorporated into commercial endoscopes. The same circular pattern or ridged and villous pattern characteristic of cardia or intestinal type epithelium, respectively, that can be identified by chromoendoscopy can easily be visualized with the press of a button using virtual chromoendoscopy [63]. The endoscopists can switch between standard and NBI modes as needed during a procedure.

There have been proposed classification systems for NBI and Barrett's mucosa [75]. The mucosal patterns in nondysplastic Barrett's have been described as a villous/gyrus pits with regular microvasculature forming pattern (Fig. 5.7a). Metaplastic Barrett's epithelium also displays normal long branching blood vessels. Irregular, mucosal, and vascular patterns have a high sensitivity for the diagnosis of HGD and cancer [76,77] (Fig. 5.7b).

Virtual chromoendoscopy can help detect Barrett's esophagus with fewer targeted biopsies compared to the random four-quadrant biopsy technique. It may also detect more areas of dysplasia. There have been several randomized studies comparing NBI with HD-WLE. Sharma et al. performed the first, randomized, cross-over trial comparing HD-white light versus NBI in 123 patients with Barrett's [78]. The NBI detected more areas of dysplasia versus HD-WLE (30% vs 21%, $p = 0.01$). They noted

(a)   (b)

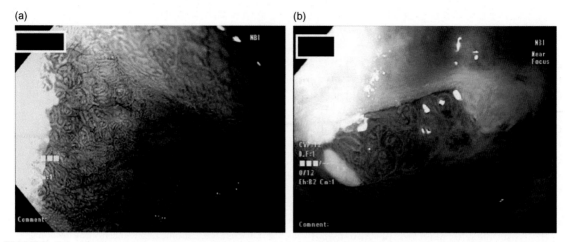

FIGURE 5.7   High-resolution endoscopic NBI image of (a) nondysplastic Barrett's with regular mucosal villous/ridge pits and vascular pattern and (b) high-grade dysplasia showing irregularity of the mucosal and vascular patterns.

that areas of HGD had irregular mucosal and vascular patterns with NBI views. None of the areas with a regular pattern had dysplasia. Thus it would seem that one could avoid biopsies in those areas, improving proficiency by performing fewer biopsies. One meta-analysis of 446 patients found overall sensitivity for detecting intestinal metaplasia using NBI of 95% and specificity of 65%. The sensitivity and specificity for detection of HGD with NBI was 96% and 94%, respectively [79]. The results of another recent meta-analysis of CE and VC found a 34% increase in the diagnostic yield of dysplasia or cancer for Barrett's [80]. The benefit of VC and CE was similar. Study heterogeneity was noted, and random effects models were used in their calculations. The authors concluded their findings and of other studies may be clinically large enough to consider a change to current Barrett's surveillance guidelines. The findings were in agreement to another meta-analysis by Ngamreungphong, which reported an 11% increase in the diagnostic yield [67]. It should be noted that many of these studies were from medical centers with high volume and expertise. Virtual techniques such as NBI are likely to be used first as they are easier to

do without adding the extra material costs of dyes and does not add significant time to endoscopic procedures. In addition, VC does not have the concerns of risks such as oxidative damage like the use of methylene blue has. Many questions still need to be answered regarding these new imaging techniques; one is if there is a high enough degree of sensitivity and specificity. It is not clear if virtual chromoendoscopy will increase the proportion of patients detected with Barrett's esophagus as compared to the technique of randomly obtaining four-quadrant biopsies every 2 cm although virtual chromoendoscopy may detect more areas with dysplasia [76].

Other newer imaging techniques include autofluorescence, confocal laser endomicroscopy, spectroscopy, and optical coherence tomography. These applications are discussed more fully in other chapters. Many of these techniques are still in trials and are not used routinely at this time in clinical practice. Random 4 quadrant biopsies with standard white light endoscopy are still considered the best approach until other techniques are proven to have high diagnostic yield and are widely available in clinical practice.

5.5 FUTURE DIRECTIONS

Advanced imaging techniques for Barrett's esophagus have been a very exciting area of research and continue to grow. One of the major challenges in preventing esophageal adenocarcinoma is that over 90% of cancers occur in previously undiagnosed Barrett's esophagus. Ongoing research will assess advanced imaging technologies and biomarkers as potential improvements to a refined diagnosis.

In addition, alternative methods that are more accessible and less expensive than upper endoscopy clearly need to be developed. A novel strategy for Barrett's esophagus screening may be with a swallowed capsule tethered to a string. The capsule dissolves in the stomach releasing a sponge, which can then be withdrawn. Cytological material from the esophagus adherent to the sponge is then stained for *TFF-3*, a biomarker that is highly specific for Barrett's esophagus [81]. Aberrant methylation of Vimentin is another molecular biomarker that has high sensitivity and specificity for Barrett's esophagus. This DNA marker can be detected in brushings from the esophagus and could enable a DNA test using nonendoscopic esophageal sampling devices [82]. Other targeted molecular markers are fluorescently labeled peptides or lectins which may bind to specific targets in Barrett's neoplasia. The concept is to visualize an altered expression and activity of neoplastic cells [83]. Another approach for diagnosing Barrett's esophagus is based on advancements in optical frequency domain imaging. A tethered capsule that incorporates this technology has been reported to provide in vivo histologic images of Barrett's esophagus [84] and these novel technologies will need to be tested in large-scale screening studies. Finding a low cost technique that is acceptable to patients and physicians will enable the identification of a greater proportion of Barrett's esophagus. Only then will it be possible to improve the prognosis of esophageal adenocarcinoma.

Acknowledgments

We would like to thank Dr Pablo Bejarano, Department of Pathology, Cleveland Clinic Florida, for his assistance with the histopathology images for this chapter.

References

[1] Barrett NR. Chronic peptic ulcer of the oesophagus and "oesophagitis". Br J Surg 1950;38(150):175–82.

[2] Allison PR, Johnstone AS. The oesophagus lined with gastric mucous membrane. Thorax 1953;8(2):87–101.

[3] Spechler SJ, Sharma P, Souza RF, Inadomi JM, Shaheen NJ, American Gastroenterological Association. American Gastroenterological Association technical review on the management of Barrett's esophagus. Gastroenterology 2011;140(3):e18–52 quiz e13

[4] Fitzgerald RC, di Pietro M, Ragunath K, Ang Y, Kang JY, Watson P, et al. British Society of Gastroenterology guidelines on the diagnosis and management of Barrett's oesophagus. Gut 2014;63(1):7–42.

[5] Shaheen NJ, Dulai GS, Ascher B, Mitchell KL, Schmitz SM. Effect of a new diagnosis of Barrett's esophagus on insurance status. Am J Gastroenterol 2005;100 (3):577–80.

[6] Azuma N, Endo T, Arimura Y, Motoya S, Itoh F, Hinoda Y, et al. Prevalence of Barrett's esophagus and expression of mucin antigens detected by a panel of monoclonal antibodies in Barrett's esophagus and esophageal adenocarcinoma in Japan. J Gastroenterol 2000;35(8):583–92.

[7] Conio M, Cameron AJ, Romero Y, Branch CD, Schleck CD, Burgart LJ, et al. Secular trends in the epidemiology and outcome of Barrett's oesophagus in Olmsted County, Minnesota. Gut 2001;48(3):304–9.

[8] Daniely Y, Liao G, Dixon D, Linnoila RI, Lori A, Randell SH, et al. Critical role of p63 in the development of a normal esophageal and tracheobronchial epithelium. Am J Physiol Cell Physiol 2004;287(1): C171–81.

[9] Wang X, Ouyang H, Yamamoto Y, Kumar PA, Wei TS, Dagher R, et al. Residual embryonic cells as precursors of a Barrett's-like metaplasia. Cell 2011;145 (7):1023–35.

[10] Buttar NS, Wang KK, Leontovich O, Westcott JY, Pacifico RJ, Anderson MA, et al. Chemoprevention of esophageal adenocarcinoma by COX-2 inhibitors in an animal model of Barrett's esophagus. Gastroenterology 2002;122(4):1101–12.

[11] Liu T, Zhang X, So CK, Wang S, Wang P, Yan L, et al. Regulation of Cdx2 expression by promoter methylation, and effects of Cdx2 transfection on morphology and gene expression of human esophageal epithelial cells. Carcinogenesis 2007;28(2):488–96.

[12] Kazumori H, Ishihara S, Kinoshita Y. Roles of caudal-related homeobox gene Cdx1 in oesophageal epithelial cells in Barrett's epithelium development. Gut 2009;58 (5):620–8.

[13] Crawford SE, Stellmach V, Murphy-Ullrich JE, Ribeiro SM, Lawler J, Hynes RO, et al. Thrombospondin-1 is a major activator of TGF-β1 in vivo. Cell 1998;93 (7):1159–70.

[14] Kinjo T, Kusano C, Oda I, Gotoda T. Prague C&M and Japanese criteria: shades of Barrett's esophagus endoscopic diagnosis. J Gastroenterol 2010;45(10):1039–44.

[15] Ogiya K, Kawano T, Ito E, Nakajima Y, Kawada K, Nishikage T, et al. Lower esophageal palisade vessels and the definition of Barrett's esophagus. Dis Esophagus 2008;21(7):645–9.

[16] Takubo K, Aida J, Naomoto Y, Sawabe M, Arai T, Shiraishi H, et al. Cardiac rather than intestinal-type background in endoscopic resection specimens of minute Barrett adenocarcinoma. Hum Pathol 2009;40 (1):65–74.

[17] Bhat S, Coleman HG, Yousef F, Johnston BT, McManus DT, Gavin AT, et al. Risk of malignant progression in Barrett's esophagus patients: results from a large population-based study. J Natl Cancer Inst 2011;103(13):1049–57.

[18] DeMeester SR, Wickramasinghe KS, Lord RV, Friedman A, Balaji NS, Chandrasoma PT, et al. Cytokeratin and DAS-1 immunostaining reveal similarities among cardiac mucosa, CIM, and Barrett's esophagus. Am J Gastroenterol 2002;97(10):2514–23.

[19] Phillips RW, Frierson HF, Moskaluk CA. Cdx2 as a marker of epithelial intestinal differentiation in the esophagus. Am J Surg Pathol 2003;27(11):1442–7.

[20] Glickman JN, Chen Y-Y, Wang HH, Antonioli DA, Odze RD. Phenotypic characteristics of a distinctive multilayered epithelium suggests that it is a precursor in the development of Barrett's esophagus. Am J Surg Pathol 2001;25(5):569–78.

[21] Chaves P, Crespo M, Ribeiro C, Laranjeira C, Pereira AD, Suspiro A, et al. Chromosomal analysis of Barrett's cells: demonstration of instability and detection of the metaplastic lineage involved. Modern Pathol 2007;20(7):788–96.

[22] Bennett C, Moayyedi P, Corley DA, DeCaestecker J, Falck-Ytter Y, Falk G, et al. BOB CAT: a large-scale review and Delphi consensus for management of Barrett's esophagus with no dysplasia, indefinite for, or low-grade dysplasia. Am J Gastroenterol 2015;110 (5):662–82.

[23] Odze RD. Pathology of the gastroesophageal junction. Semin Diagn Pathol 2005;22(4):256–65.

[24] Spechler SJ. Barrett's esophagus: is the goblet half-empty? Clin Gastroenterol Hepatol 2012;10(11):1237.

[25] Younes M, Ertan A, Ergun G, Verm R, Bridges M, Woods K, et al. Goblet cell mimickers in esophageal biopsies are not associated with an increased risk for dysplasia. Arch Pathol Lab Med 2007;131(4):571–5.

[26] Chandrasoma P. Pathophysiology of Barrett's esophagus. Semin Thorac Cardiovasc Surg 1997;9(3):270–8.

[27] Chandrasoma PT. Histologic definition of gastroesophageal reflux disease. Curr Opin Gastroenterol 2013;29(4):460–7.

[28] Hahn HP, Blount PL, Ayub K, Das KM, Souza R, Spechler S, et al. Intestinal differentiation in metaplastic, non-goblet columnar epithelium in the esophagus. Am J Surg Pathol 2009;33(7):1006.

[29] Liu W, Hahn H, Odze RD, Goyal RK. Metaplastic esophageal columnar epithelium without goblet cells shows DNA content abnormalities similar to goblet cell-containing epithelium. Am J Gastroenterol 2009;104(4):816–24.

[30] Sharma P, Morales TG, Sampliner RE. Short segment Barrett's esophagus—the need for standardization of the definition and of endoscopic criteria. Am J Gastroenterol 1998;93(7):1033–6.

[31] Kim SL, Waring JP, Spechler SJ, Sampliner RE, Doos WG, Krol WF, et al. Diagnostic inconsistencies in Barrett's esophagus. Department of Veterans Affairs Gastroesophageal Reflux Study Group. Gastroenterology 1994;107(4):945–9.

[32] Dekel R, Wakelin DE, Wendel C, Green C, Sampliner RE, Garewal HS, et al. Progression or regression of Barrett's esophagus—is it all in the eye of the beholder? Am J Gastroenterol 2003;98(12):2612–15.

[33] Sharma P, Dent J, Armstrong D, Bergman JJ, Gossner L, Hoshihara Y, et al. The development and validation of an endoscopic grading system for Barrett's esophagus: the Prague C & M criteria. Gastroenterology 2006;131(5):1392–9.

[34] Wallner B, Sylvan A, Janunger KG. Endoscopic assessment of the "Z-line" (squamocolumnar junction) appearance: reproducibility of the ZAP classification among endoscopists. Gastrointest Endosc 2002;55 (1):65–9.

[35] McClave SA, Boyce Jr. HW, Gottfried MR. Early diagnosis of columnar-lined esophagus: a new endoscopic

diagnostic criterion. Gastrointest Endosc 1987;33 (6):413—16.

[36] Wang KK, Sampliner RE, Practice Parameters Committee of the American College of Gastroenterology. Updated guidelines 2008 for the diagnosis, surveillance and therapy of Barrett's esophagus. Am J Gastroenterol 2008;103(3):788—97.

[37] Zhang X, Huang Q, Goyal RK, Odze RD. DNA ploidy abnormalities in basal and superficial regions of the crypts in Barrett's esophagus and associated neoplastic lesions. Am J Surg Pathol 2008;32(9):1327—35.

[38] Alvarez Herrero L, Curvers WL, van Vilsteren FG, Wolfsen H, Ragunath K, Wong Kee Song LM, et al. Validation of the Prague C&M classification of Barrett's esophagus in clinical practice. Endoscopy 2013;45(11):876—82.

[39] Vahabzadeh B, Seetharam AB, Cook MB, Wani S, Rastogi A, Bansal A, et al. Validation of the Prague C & M criteria for the endoscopic grading of Barrett's esophagus by gastroenterology trainees: a multicenter study. Gastrointest Endosc 2012;75(2):236—41.

[40] DeNardi FG, Riddell RH. The normal esophagus. Am J Surg Pathol 1991;15(3):296—309.

[41] Jones TF, Sharma P, Daaboul B, Cherian R, Mayo M, Topalovski M, et al. Yield of intestinal metaplasia in patients with suspected short-segment Barrett's esophagus (SSBE) on repeat endoscopy. Dig Dis Sci 2002;47 (9):2108—11.

[42] Dickman R, Levi Z, Vilkin A, Zvidi I, Niv Y. Predictors of specialized intestinal metaplasia in patients with an incidental irregular Z line. Eur J Gastroenterol Hepatol 2010;22(2):135—8.

[43] Leodolter A, Nocon M, Vieth M, Lind T, Jaspersen D, Richter K, et al. Progression of specialized intestinal metaplasia at the cardia to macroscopically evident Barrett's esophagus: an entity of concern in the ProGERD study. Scand J Gastroenterol 2012;47 (12):1429—35.

[44] Spechler SJ, Zeroogian JM, Antonioli DA, Wang HH, Goyal R. Prevalence of metaplasia at the gastro-oesophageal junction. Lancet 1994;344(8936):1533—6.

[45] Clark GW, Ireland AP, Peters JH, Chandrasoma P, DeMeester TR, Bremner CG. Short-segment Barrett's esophagus: a prevalent complication of gastroesophageal reflux disease with malignant potential. J Gastrointest Surg 1997;1(2):113—22.

[46] Harrison R, Perry I, Haddadin W, McDonald S, Bryan R, Abrams K, et al. Detection of intestinal metaplasia in Barrett's esophagus: an observational comparator study suggests the need for a minimum of eight biopsies. Am J Gastroenterol 2007;102 (6):1154—61.

[47] Levine DS, Reid BJ. Endoscopic biopsy technique for acquiring larger mucosal samples. Gastrointest Endosc 1991;37(3):332—7.

[48] Faigel DO, Eisen GM, Baron TH, Dominitz JA, Goldstein JL, Hirota WK, et al. Tissue sampling and analysis. Gastrointest Endosc 2003;57(7):811—16.

[49] Bernstein DE, Barkin JS, Reiner DK, Lubin J, Phillips RS, Grauer L. Standard biopsy forceps versus large-capacity forceps with and without needle. Gastrointest Endosc 1995;41(6):573—6.

[50] Komanduri S, Swanson G, Keefer L, Jakate S. Use of a new jumbo forceps improves tissue acquisition of Barrett's esophagus surveillance biopsies. Gastrointest Endosc 2009;70(6):1072—8. e1.

[51] Elmunzer BJ, Higgins PD, Kwon YM, Golembeski C, Greenson JK, Korsnes SJ, et al. Jumbo forceps are superior to standard large-capacity forceps in obtaining diagnostically adequate inflammatory bowel disease surveillance biopsy specimens. Gastrointest Endosc 2008;68(2):273—8 quiz 334, 6.

[52] Falk GW, Rice TW, Goldblum JR, Richter JE. Jumbo biopsy forceps protocol still misses unsuspected cancer in Barrett's esophagus with high-grade dysplasia. Gastrointest Endosc 1999;49(2):170—6.

[53] Gonzalez S, Yu WM, Smith MS, Slack KN, Rotterdam H, Abrams JA, et al. Randomized comparison of 3 different-sized biopsy forceps for quality of sampling in Barrett's esophagus. Gastrointest Endosc 2010;72 (5):935—40.

[54] Levine DS, Haggitt RC, Blount PL, Rabinovitch PS, Rusch VW, Reid BJ. An endoscopic biopsy protocol can differentiate high-grade dysplasia from early adenocarcinoma in Barrett's esophagus. Gastroenterology 1993;105:40.

[55] Eloubeidi MA, Provenzale D. Does this patient have Barrett's esophagus? The utility of predicting Barrett's esophagus at the index endoscopy. Am J Gastroenterol 1999;94(4):937—43.

[56] Abela J-E, Going JJ, Mackenzie JF, McKernan M, O'Mahoney S, Stuart RC. Systematic four-quadrant biopsy detects Barrett's dysplasia in more patients than nonsystematic biopsy. Am J Gastroenterol 2008;103(4):850—5.

[57] Fitzgerald RC, Saeed IT, Khoo D, Farthing MJ, Burnham WR. Rigorous surveillance protocol increases detection of curable cancers associated with Barrett's esophagus. Dig Dis Sci 2001;46(9):1892—8.

[58] Sampliner RE, Practice Parameters Committee of the American College of Gastroenterology. Updated guidelines for the diagnosis, surveillance, and therapy of Barrett's esophagus. Am J Gastroenterol 2002;97 (8):1888—95.

[59] Levine DS, Blount PL, Rudolph RE, Reid BJ. Safety of a systematic endoscopic biopsy protocol in patients with Barrett's esophagus. Am J Gastroenterol 2000;95 (5):1152–7.

[60] Abrams JA, Kapel RC, Lindberg GM, Saboorian MH, Genta RM, Neugut AI, et al. Adherence to biopsy guidelines for Barrett's esophagus surveillance in the community setting in the United States. Clin Gastroenterol Hepatol 2009;7(7):736–42. quiz 10.

[61] Mayinger B, Oezturk Y, Stolte M, Faller G, Benninger J, Schwab D, et al. Evaluation of sensitivity and inter- and intra-observer variability in the detection of intestinal metaplasia and dysplasia in Barrett's esophagus with enhanced magnification endoscopy. Scand J Gastroenterol 2006;41(3):349–56.

[62] Ferguson DD, Devault KR, Krishna M, Loeb DS, Wolfsen HC, Wallace MB. Enhanced magnification-directed biopsies do not increase the detection of intestinal metaplasia in patients with GERD. Am J Gastroenterol 2006;101(7):1611–16.

[63] Kara MA, Peters FP, Rosmolen WD, Krishnadath KK, ten Kate FJ, Fockens P, et al. High-resolution endoscopy plus chromoendoscopy or narrow-band imaging in Barrett's esophagus: a prospective randomized crossover study. Endoscopy 2005;37(10):929–36.

[64] Kara MA, Smits ME, Rosmolen WD, Bultje AC, Ten Kate FJ, Fockens P, et al. A randomized crossover study comparing light-induced fluorescence endoscopy with standard videoendoscopy for the detection of early neoplasia in Barrett's esophagus. Gastrointest Endosc 2005;61(6):671–8.

[65] Sharma P, Marcon N, Wani S, Bansal A, Mathur S, Sampliner R, et al. Non-biopsy detection of intestinal metaplasia and dysplasia in Barrett's esophagus: a prospective multicenter study. Endoscopy 2006;38 (12):1206–12.

[66] Olliver JR, Wild CP, Sahay P, Dexter S, Hardie LJ. Chromoendoscopy with methylene blue and associated DNA damage in Barrett's oesophagus. Lancet 2003;362(9381):373–4.

[67] Ngamruengphong S, Sharma VK, Das A. Diagnostic yield of methylene blue chromoendoscopy for detecting specialized intestinal metaplasia and dysplasia in Barrett's esophagus: a meta-analysis. Gastrointest Endosc 2009;69(6):1021–8.

[68] Canto M, Setrakian S, Willis J, Chak A, Petras R, Sivak M. Methylene blue staining of dysplastic and nondysplastic Barrett's esophagus: an in vivo and ex vivo study. Endoscopy 2001;33(5):391–400.

[69] Guelrud M, Herrera I. Acetic acid improves identification of remnant islands of Barrett's epithelium after endoscopic therapy. Gastrointest Endosc 1998;47 (6):512–15.

[70] Lambert R, Rey JF, Sankaranarayanan R. Magnification and chromoscopy with the acetic acid test. Endoscopy 2003;35(5):437–45.

[71] Longcroft-Wheaton G, Duku M, Mead R, Poller D, Bhandari P. Acetic acid spray is an effective tool for the endoscopic detection of neoplasia in patients with Barrett's esophagus. Clin Gastroenterol Hepatol 2010;8 (10):843–7.

[72] Sano Y. New diagnostic method based on color imaging using narrow-band imaging (NBI) system for gastrointestinal tract. Gastrointest Endosc 2001;53:AB125.

[73] Gono K, Yamazaki K, Doguchi N, Nonami T, Obi T, Yamaguchi M, et al. Endoscopic observation of tissue by narrowband illumination. Optical Rev 2003;10 (4):211–15.

[74] Folkman J. Tumor angiogenesis: therapeutic implications. N Engl J Med 1971;285(21):1182–6.

[75] Alvarez Herrero L, Weusten BL, Bergman JJ. Autofluorescence and narrow band imaging in Barrett's esophagus. Gastroenterol Clin North Am 2010;39(4):747–58.

[76] Sharma P, Bansal A, Mathur S, Wani S, Cherian R, McGregor D, et al. The utility of a novel narrow band imaging endoscopy system in patients with Barrett's esophagus. Gastrointest Endosc 2006;64(2):167–75.

[77] Kara MA, Ennahachi M, Fockens P, ten Kate FJ, Bergman JJ. Detection and classification of the mucosal and vascular patterns (mucosal morphology) in Barrett's esophagus by using narrow band imaging. Gastrointest Endosc 2006;64(2):155–66.

[78] Sharma P, Hawes RH, Bansal A, Gupta N, Curvers W, Rastogi A, et al. Standard endoscopy with random biopsies versus narrow band imaging targeted biopsies in Barrett's oesophagus: a prospective, international, randomised controlled trial. Gut 2013;62(1):15–21.

[79] Mannath J, Subramanian V, Hawkey CJ, Ragunath K. Narrow band imaging for characterization of high grade dysplasia and specialized intestinal metaplasia in Barrett's esophagus: a meta-analysis. Endoscopy 2010;42(5):351–9.

[80] Qumseya BJ, Wang H, Badie N, Uzomba RN, Parasa S, White DL, et al. Advanced imaging technologies increase detection of dysplasia and neoplasia in patients with Barrett's esophagus: a meta-analysis and systematic review. Clin Gastroenterol Hepatol 2013;11 (12):1562–70 e1–2

[81] Lao-Sirieix P, Fitzgerald RC. Biomarker for Barrett's oesophagus. US Patent 20,150,004,622; 2015.

[82] Moinova H, Leidner RS, Ravi L, Lutterbaugh J, Barnholtz-Sloan JS, Chen Y, et al. Aberrant vimentin methylation is characteristic of upper gastrointestinal pathologies. Cancer Epidemiol Biomarker Prevent 2012;21(4):594–600.

[83] Bird-Lieberman EL, Neves AA, Lao-Sirieix P, O'Donovan M, Novelli M, Lovat LB, et al. Molecular imaging using fluorescent lectins permits rapid endoscopic identification of dysplasia in Barrett's esophagus. Nat Med 2012;18(2):315–21.

[84] Gora MJ, Sauk JS, Carruth RW, Gallagher KA, Suter MJ, Nishioka NS, et al. Tethered capsule endomicroscopy enables less invasive imaging of gastrointestinal tract microstructure. Nat Med 2013;19(2):238–40.

[85] Kerkhof M, Van Dekken H, Steyerberg E, Meijer G, Mulder A, De Bruïne A, et al. Grading of dysplasia in Barrett's oesophagus: substantial interobserver variation between general and gastrointestinal pathologists. Histopathology 2007;50(7):920–7.

[86] Sharma P, Falk GW, Weston AP, Reker D, Johnston M, Sampliner RE. Dysplasia and cancer in a large multicenter cohort of patients with Barrett's esophagus. Clin Gastroenterol Hepatol 2006;4(5):566–72.

[87] Montgomery E, Bronner MP, Goldblum JR, Greenson JK, Haber MM, Hart J, et al. Reproducibility of the diagnosis of dysplasia in Barrett esophagus: a reaffirmation. Hum Pathol 2001;32(4):368–78.

[88] Reid BJ, Levine DS, Longton G, Blount PL, Rabinovitch PS. Predictors of progression to cancer in Barrett's esophagus: baseline histology and flow cytometry identify low- and high-risk patient subsets. Am J Gastroenterol 2000;95(7):1669–76.

[89] Montgomery E, Goldblum JR, Greenson JK, Haber MM, Lamps LW, Lauwers GY, et al. Dysplasia as a predictive marker for invasive carcinoma in Barrett esophagus: a follow-up study based on 138 cases from a diagnostic variability study. Hum Pathol 2001;32(4):379–88.

[90] Skacel M, Petras RE, Gramlich TL, Sigel JE, Richter JE, Goldblum JR. The diagnosis of low-grade dysplasia in Barrett's esophagus and its implications for disease progression. Am J Gastroenterol 2000;95 (12):3383–7.

[91] Curvers WL, ten Kate FJ, Krishnadath KK, Visser M, Elzer B, Baak LC, et al. Low-grade dysplasia in Barrett's esophagus: overdiagnosed and underestimated. Am J Gastroenterol 2010;105(7):1523–30.

[92] Srivastava A, Hornick JL, Li X, Blount PL, Sanchez CA, Cowan DS, et al. Extent of low-grade dysplasia is a risk factor for the development of esophageal adenocarcinoma in Barrett's esophagus. Am J Gastroenterol 2007;102(3):483–93.

[93] Reid BJ, Haggitt RC, Rubin CE, Roth G, Surawicz CM, Van Belle G, et al. Observer variation in the diagnosis of dysplasia in Barrett's esophagus. Hum Pathol 1988;19(2):166–78.

[94] Younes M, Lauwers GY, Ertan A, Ergun G, Verm R, Bridges M, et al. The significance of "indefinite for dysplasia" grading in Barrett metaplasia. Arch Pathol Lab Med 2011;135(4):430–2.

[95] Lomo LC, Blount PL, Sanchez CA, Li X, Galipeau PC, Cowan DS, et al. Crypt dysplasia with surface maturation: a clinical, pathologic, and molecular study of a Barrett's esophagus cohort. Am J Surg Pathol 2006;30 (4):423–35.

[96] Odze RD, Maley CC. Neoplasia without dysplasia: lessons from Barrett esophagus and other tubal gut neoplasms. Arch Pathol Lab Med 2010;134 (6):896–906.

[97] Brown IS, Whiteman DC, Lauwers GY. Foveolar type dysplasia in Barrett esophagus. Modern Pathology 2010;23(6):834–43.

[98] Mahajan D, Bennett AE, Liu X, Bena J, Bronner MP. Grading of gastric foveolar-type dysplasia in Barrett's esophagus. Modern Pathol 2010;23(1):1–11.

[99] Patil DT, Bennett AE, Mahajan D, Bronner MP. Distinguishing Barrett gastric foveolar dysplasia from reactive cardiac mucosa in gastroesophageal reflux disease. Hum Pathol 2013;44(6):1146–53.

[100] Odze RD. Diagnosis and grading of dysplasia in Barrett's oesophagus. J Clin Pathol 2006;59 (10):1029–38.

[101] Rucker-Schmidt RL, Sanchez CA, Blount PL, Ayub K, Li X, Rabinovitch PS, et al. Non-adenomatous dysplasia in Barrett's esophagus; a clinical, pathologic and DNA content flow cytometric study. Am J Surg Pathol 2009;33(6):886.

[102] Foveolar and serrated dysplasia are rare high-risk lesions in Barrett's esophagus: a prospective outcome analysis of 214 patients. In: Srivastava A, Sanchez C, Cowan D, Odze R, editors. Laboratory Investigation. New York, NY: Nature Publishing Group; 2010.

[103] Khor TS, Alfaro EE, Ooi EM, Li Y, Srivastava A, Fujita H, et al. Divergent expression of MUC5AC, MUC6, MUC2, CD10, and CDX-2 in dysplasia and intramucosal adenocarcinomas with intestinal and foveolar morphology: is this evidence of distinct gastric and intestinal pathways to carcinogenesis in Barrett esophagus?. Am J Surg Pathol 2012;36(3):331–42.

[104] Demicco EG, Farris III AB, Baba Y, Agbor-Etang B, Bergethon K, Mandal R, et al. The dichotomy in carcinogenesis of the distal esophagus and esophagogastric junction: intestinal-type vs cardiac-type mucosa-associated adenocarcinoma. Modern Pathol 2011;24 (9):1177–90.

[105] Rabinovitch PS, Longton G, Blount PL, Levine DS, Reid BJ. Predictors of progression in Barrett's esophagus III: baseline flow cytometric variables. Am J Gastroenterol 2001;96(11):3071–83.

[106] Reid BJ, Prevo LJ, Galipeau PC, Sanchez CA, Longton G, Levine DS, et al. Predictors of progression in

Barrett's esophagus II: baseline 17p (p53) loss of heterozygosity identifies a patient subset at increased risk for neoplastic progression. Am J Gastroenterol 2001;96(10):2839—48.

[107] Rygiel AM, Milano F, Ten Kate FJ, de Groot JG, Peppelenbosch MP, Bergman JJ, et al. Assessment of chromosomal gains as compared to DNA content changes is more useful to detect dysplasia in Barrett's esophagus brush cytology specimens. Genes Chromosomes Cancer 2008;47(5):396—404.

[108] Schulmann K, Sterian A, Berki A, Yin J, Sato F, Xu Y, et al. Inactivation of p16, RUNX3, and HPP1 occurs early in Barrett's-associated neoplastic progression and predicts progression risk. Oncogene 2005;24(25):4138—48.

[109] Cestari R, Villanacci V, Rossi E, Della Casa D, Missale G, Conio M, et al. Fluorescence in situ hybridization to evaluate dysplasia in Barrett's esophagus: a pilot study. Cancer Lett 2007;251(2):278—87.

Screening and Surveillance of Barrett's Esophagus

Nikhiel B. Rau[1] and Tyler M. Berzin[2]

[1]Division of Gastroenterology, Beth Israel Deaconess Medical Center, Boston, MA, United States
[2]Division of Gastroenterology, Center for Advanced Endoscopy, Department of Medicine,
Beth Israel Deaconess Medical Center, Harvard Medical School, Boston, MA, United States

6.1 OVERVIEW

Five decades after the British surgeon Norman Barrett first described the condition that bears his name, the true incidence of Barrett's esophagus (BE) in the general population and the risk of progression to cancer continue to be areas of uncertainty, debate, and controversy [1]. While the pathologic definition of BE has evolved over time, our current screening and surveillance recommendations are based in large on data derived from recent treatment trials, where clear endoscopic definitions of landmarks are recorded and expert pathologists performed biopsy analysis [1].

In this chapter, we will examine the rational behind screening and surveillance of BE. This will include the epidemiology, risk factors for developing Barrett's, risk stratification for screening, and how the presence of dysplasia affects surveillance strategies, cost-effectiveness of screening and surveillance as well as potential technological advances which may affect screening/surveillance approaches in the future.

While the defining features of BE have been covered in detail in earlier chapters, its definition is key to understanding the approach to screening and surveillance in BE. The American Gastroenterological Association defines BE as "the condition in which any extent of metaplastic columnar epithelium that predisposes to cancer development replaces the stratified squamous epithelium that normally lines the distal esophagus." Presently in the United States, intestinal metaplasia (the presence of columnar epithelium with goblet cells) is required for the diagnosis of BE because intestinal metaplasia is the predominant, while not the only, type of esophageal columnar epithelium that has malignant potential [1]. An international agreement was achieved for the first time on a definition of Barrett's in 2015 by the BOB CAT (Benign Barrett's and CAncer Taskforce) consensus group which broadens the description as "Barrett's is defined by the presence of columnar mucosa in the esophagus and it should be stated whether intestinal metaplasia is present above the gastroesophageal junction" [2].

D. Pleskow & T. Erim (Eds): Barrett's Esophagus.
DOI: http://dx.doi.org/10.1016/B978-0-12-802511-6.00006-5

It has been estimated that 5.6% of adults in the United States have BE [3]. The metaplastic columnar mucosa of BE can silently evolve into adenocarcinoma [3]. The prognosis of adenocarcinoma remains extremely poor, with a 5-year survival rate of 16% in the United States, highlighting the importance of screening and surveillance in BE [4].

6.2 PATHOGENESIS OF BARRETT'S ESOPHAGUS

One of the many challenges BE presents is that it is not entirely clear which precursor cells are responsible for metaplastic change. Metaplasia, the process wherein one mature cell type replaces another, can be the result of chronic tissue injury [3]. In patients with chronic esophageal injury from gastroesophageal reflux disease (GERD), Barrett's metaplasia develops when mucus-secreting columnar cells replace reflux-damaged esophageal squamous cells. While the mechanism of this process is not fully understood, it has been proposed that GERD might induce alterations in the expression of key developmental transcription factors, causing mature esophageal squamous cells to change into columnar cells, a process known as transdifferentiation or causing immature esophageal progenitor cells to undergo columnar rather than squamous differentiation a process known as transcommitment [3,5,6]. Interestingly, when rat models of reflux esophagitis have been studied, it has been observed that the metaplasia develops from bone marrow stem cells that enter the circulation and settle in the reflux-damaged esophagus [7]. Studies in mouse models have suggested that metaplasia might result from upward migration of stem cells from the gastric cardia [8] or from the proximal expansion of embryonic-type cells at the gastroesophageal junction [9]. It is not clear which of these processes contribute to the pathogenesis of BE in humans.

6.3 RISK FACTORS FOR BARRETT'S ESOPHAGUS

There are multiple risk factors that can predispose an individual to BE. One of the most common conditions that can propagate BE is gastroesophageal reflux disease or GERD. In individual patients, the extent of Barrett's metaplasia may be related to the severity of underlying GERD [10]. Untreated patients with long-segment BE typically have severe GERD with erosive esophagitis, whereas short-segment BE may not be associated with GERD symptoms or endoscopic signs of reflux esophagitis [11,12]. Presumably, short-segment BE develops as a consequence of protracted acid reflux involving only the most distal portion of the esophagus, a phenomenon that can be documented in apparently healthy persons [13]. Short-segment BE was not widely recognized until 1994 [14], and earlier studies generally involved patients with long-segment disease exclusively. More recent studies have involved varying proportions of patients with long-segment and short-segment BE, and the proportion can profoundly influence the frequency of associated GERD symptoms and complications.

Proposed risk factors for BE are listed in Table 6.1. BE is believed to be more common in Caucasian patients 50 years of age or older. BE is two to three times as common in men as in women, it is uncommon in blacks and Asians and is rare in children [15,16]. Other important risk factors include obesity (with a predominantly intraabdominal fat distribution) and cigarette smoking, and there is a familial form of BE, which accounts for 7–11% of all cases [17]. Most conditions associated with Barrett's metaplasia are also risk factors for esophageal adenocarcinoma (EAC) [18]. Conversely, factors that might provide protection against BE include the use of nonsteroidal anti-inflammatory drugs, *Helicobacter pylori*, and consumption of a diet high in fruits and vegetables.

TABLE 6.1 Proposed Risk Factors and Protective Factors for Barrett's Esophagus and Esophageal Adenocarcinoma[a]

Factors	Risk Factor for Barrett's Esophagus	Risk Factor for Esophageal Adenocarcinoma
Older age	Yes	Yes
White race	Yes	Yes
Male sex	Yes	Yes
Chronic heartburn	Yes	Yes
Age < 30 years at onset of GERD symptoms	Yes	—
Hiatal hernia	Yes	Yes
Erosive esophagitis	Yes	Yes
Obesity with intraabdominal fat distribution	Yes	Yes
Metabolic syndrome	Yes	Yes
Tobacco use	Yes	Yes
Family history of GERD, Barrett's esophagus, or esophageal adenocarcinoma	Yes	Yes
Obstructive sleep apnea	Yes	—
Low birth weight for gestational age	Yes	No
Preterm birth	No	Yes
Consumption of red meat and processed meat	Yes	Yes
Human papillomavirus infection	No	Yes
	Protective Factor for Barrett's Esophagus	Protective Factor for Esophageal Adenocarcinoma
Use of nonsteroidal anti-inflammatory drugs	Yes	Yes
Use of statins	Yes	Yes
Helicobacter pylori infection	Yes	Yes
Diet high in fruits and vegetables	Yes	Yes
Exposure to ambient ultraviolet radiation	—	Yes
Breast feeding for parous women	—	Yes
Tall height	Yes	Yes

[a] A dash indicates that studies have not addressed the question of whether the specified factor is associated with an increased risk or has a protective effect. Citations for the information in this table are provided in the Supplementary Appendix, available at NEJM.org. GERD denotes gastroesophageal reflux disease.

Used with permission from NEJM and American Gastroenterological Association, Spechler SJ, Sharma P, Souza RF, Inadomi JM, Shaheen NJ. American Gastroenterological Association Medical Position Statement on the management of Barrett's esophagus. Gastroenterology 2011;140(3):1084–91.

No single risk factor yet identified can account for the profound increase in the incidence of EAC in Western countries during the past 40 years, a period when GERD and BE appear to have increased only modestly in frequency [19,20]. There has been a steep rise in the frequency of central obesity, which might contribute to Barrett's carcinogenesis by promoting GERD and by increasing the production of hormones that promote cell proliferation such as leptin and insulin-like growth factors [21,22]. *H. pylori* infection, which may protect the esophagus from GERD by causing a gastritis that reduced gastric acid production, has declined in frequency during the same period when EAC has risen in developed countries [23]. Another hypothesis links the rising incidence of EAC with increased dietary intake of nitrates, which has resulted from the widespread use of nitrate-based fertilizers [23].

Estimates of the annual incidence of EAC among patients with nondysplastic Barrett's esophagus (NDBE) have ranged from 0.1% to 2.9%, with the highest estimates in studies with evidence of publication bias [24]. Recent better-quality studies suggest that the risk of EAC in the general populations of patients with NDBE is only 0.1−0.3% per year [25−28]. However, a number of factors influence the risk of cancer for individual patients. For example, cancer risk among men with BE is approximately twice that among women [27], the risk is greater with longer segment of Barrett's metaplasia [29], and the risk is especially high among persons with certain familial forms of BE [30]. In addition, the risk appears to decrease with follow-up endoscopies showing no progression to dysplasia [31]. While the incidence and epidemiology are still being elucidated for Barrett's, it is a key component to establishing screening and surveillance guidelines. Refer to Chapter 2: "Fluctuating Risk Factors and Epidemiology" for additional environmental and genetic risk factors in BE.

6.4 SCREENING FOR BARRETT'S ESOPHAGUS

For the past few decades, the primary strategy for preventing mortality from EAC has been to screen patients with GERD symptoms for BE with the use of upper endoscopy and, for patients with established Barrett's, to perform regular endoscopic surveillance to detect curable dysplasia and neoplasia [3,32]. With an annual cancer incidence of only 0.1−0.3%, there have been no randomized trials to prove that screening and surveillance is an effective strategy. There is data, however, from observational studies suggesting that for patients with BE, cancers diagnosed by means of surveillance endoscopy are more likely to be earlier stage tumors with higher survival rates than tumors discovered secondary to symptoms such as weight loss and dysphagia [3,33,34].

Despite the lack of high-quality evidence to support the practice, medical societies currently recommend endoscopic screening for BE in patients with chronic GERD symptoms who have at least one additional risk factor for EAC (Table 6.1). These risk factors include age of 50 years or older, male sex, white race, hiatal hernia, elevated body mass index, intraabdominal body fat distribution, or tobacco use [32,35−37]. If the screening examination does not reveal BE, no further endoscopic screening is recommended [36,37].

Nevertheless, there are a number of reasons to question this approach to screening for BE. The screening prerequisite of GERD symptoms limits the usefulness of the practice, because patients with short-segment BE often have no GERD symptoms, and approximately 40% of patients with EAC report no history of GERD [38]. Studies have shown that less than 10% of patients with EAC have a prior diagnosis of BE, suggesting that current screening practices may be ineffective [39,40].

6.5 SURVEILLANCE OF BARRETT'S ESOPHAGUS

6.5.1 Surveillance of Nondysplastic Barrett's Esophagus

Endoscopic surveillance of BE is used to identify patients at earlier curable stages of esophageal cancer and is endorsed by major gastroenterology societies including those in the United States and in the United Kingdom [41,42] despite fairly limited evidence that it confers a survival advantage [43,44]. Because the risk of EAC varies based on the grade of dysplasia, surveillance guidelines also vary depending on histology.

The rate of progression from NDBE to EAC is estimated to be as high as 0.6% per year [45] or as low as 0.21% per year [28]. A recent multicenter study showed a rate of progression to EAC of 0.27% per year and a rate of progression to HGD (high-grade dysplasia) of 0.48% per year [26]. In this study, 97.1% of patients with NDBE were cancer free at 10 years. A recent population-based study showed the incidence of HGD and EAC in patients with BE to be 0.38% per year [27]. Sharma et al. [46] found that half of patients who developed HGD or EAC demonstrated only NDBE on a previous biopsy, suggesting that not all cases of EAC develop in a stepwise fashion from NDBE to LGD (low-grade dysplasia) to HGD and then finally to EAC. Nevertheless, studies of patients whose EAC was detected through surveillance have consistently demonstrated improved survival over patients whose EAC was not detected through surveillance, although this observation likely represents lead time bias [34,47,48].

Systematic biopsy protocols for NDBE surveillance have been evaluated in patients with NDBE and four-quadrant biopsies obtained every 2 cm appear to be more sensitive for detection of dysplasia than random biopsies [47]. A recent study demonstrated similar rates of adequate specimens with large capacity forceps (2.8 mm) compared with jumbo forceps (3.2 mm) [49]. The standard approach to NDBE has been outlined in Fig. 6.1.

Recently, endoscopic ablation has been proposed as an alternative to surveillance for NDBE. A multicenter study of radiofrequency ablation (RFA) of NDBE achieved complete eradication of BE in 98.4% of patients at 2.5 years and 92% at 5 years, with no patients progressing past NDBE during follow-up [50]. Endoscopic ablation therapy as an alternative to surveillance of NDBE has been suggested to be cost effective in a cost utility model [51] and may be a preferred management option in select patients with NDBE, such as those with a family history of EAC [52]. Additional research evaluating this management strategy is awaited. The development of biomarkers to identify patients at highest risk of developing dysplasia or cancer may also change the need strategy for surveillance versus ablation in NDBE [53].

6.5.2 Cost-Effectiveness of NDBE Surveillance

By using current estimates of the malignant potential of BE in the wider population versus those reported in surveillance program audits, surveillance of all patients with NDBE may not be cost effective. Further work to identify high-risk individuals, perhaps in the future by using a biomarker-based strategy, might enable endoscopy surveillance to be tailored to high-risk individuals and thereby improve the economic acceptability of endoscopy-based surveillance for BE [54]. Some computer modeling studies have concluded that screening and surveillance can be cost effective under certain circumstances, but such studies are not definitive [55,56].

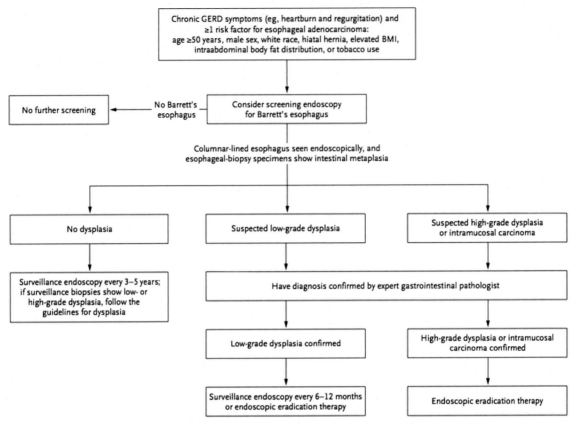

FIGURE 6.1   Algorithm for screening, surveillance, and management of Barrett's esophagus. Endoscopy for patients with dysplasia or intramucosal carcinoma should include four-quadrant biopsy sample at 1-cm intervals and endoscopic resection of mucosal irregularities. If dysplasia or intramucosal carcinoma is discovered and these procedures have not been performed, then repeat endoscopy is recommended before endoscopic eradication therapy is initiated. BMI denotes body mass index and GERD denotes gastroesophageal reflux disease. *Used with permission from NEJM and American Gastroenterological Association, Spechler SJ, Sharma P, Souza RF, Inadomi JM, Shaheen NJ. American Gastroenterological Association Medical Position Statement on the management of Barrett's esophagus. Gastroenterology 2011;140(3):1084–91.*

6.5.3 Surveillance of Barrett's Esophagus with Low-Grade Dysplasia

The natural history of BE with LGD is not well understood, but available data indicate that LGD carries a higher risk of progression to EAC rather than NDBE. The diagnosis of LGD should be confirmed by an expert GI pathologist as the recognized rate of progression may be higher in cases for which two expert GI pathologists have confirmed the presence of LGD. A large Dutch cohort study demonstrated a rate of progression from LGD to EAC of 0.77% [57]. A recent meta-analysis found similar rates of progression in studies of patients in surveillance programs: 0.7% per year in the United Kingdom, 0.7% per year in the United States, and 0.8% per year in Europe [58]. However, in studies where LGD was confirmed by two expert pathologists, the risk of progression is much higher: 9.1% per year in confirmed patients vs 0.6% in those downgraded to NDBE. Multifocal and persistent LGD confer an even higher risk of progression to neoplasia [59–62].

The American Gastroenterological Association and American College of Gastroenterology still advocate biannual to annual surveillance for patients with LGD [1,41]. Published biopsy protocols involving LGD typically follow the Seattle protocol with targeted plus four-quadrant biopsies every 1–2 cm along the length of the BE [63,64]. An approach to LGD surveillance is summarized in Fig. 6.1.

Endoscopic ablation is becoming accepted in the setting of LGD. A recent multicenter, sham-controlled trial of RFA achieved complete eradication of dysplasia in 90.5% of patients and complete eradication of BE in 81% of patients with LGD with 2-year follow-up data demonstrating complete eradication of dysplasia and BE in 98% of patients [65]. The annual rate of neoplastic progression in this study was 1 per 73 patient-years; however, no subjects (sham or ablation) progressed from LGD to cancer [66]. It should be noted that the length of follow-up was short and the development of cancer would not have been expected in this cohort. Comprehensive large studies in this population will be challenging because of the requisite long-term follow-up. There are scant published clinical data available to direct surveillance protocols after successful ablation of LGD; therefore, surveillance strategies after endoscopic ablation for LGD should be individualized [67].

6.5.4 Surveillance of Barrett's Esophagus with High-Grade Dysplasia

The gold-standard management option for BE with HGD is endoscopic ablative treatments, with or without endomucosal resection. The risk of cancer progression from HGD is unacceptably high to recommend. However, for medical or personal reasons, some patients may not be able to undergo definitive therapy for HGD, in which case surveillance should be offered.

As with LGD, we recommend that any biopsy demonstrating initial evidence of HGD requires review and confirmation by a second expert GI pathologist. Once again it is imperative to use the Seattle protocol which involves targeted biopsies of mucosal abnormalities, such as nodules and ulcers, plus four-quadrant biopsies [68]. In HGD, the endoscopist should obtain samples every 1 cm by using large capacity forceps for the length of the BE segment [68]. A less intensive protocol that uses four-quadrant biopsies every 1–2 cm with regular or large capacity forceps found a similar rate of missed cancers compared with the Seattle protocol in patients with HGD undergoing esophagectomy [69]. An algorithm for endoscopic surveillance and eradication therapy in patients with BE is outlined in Fig. 6.1.

6.6 BARRETT'S ESOPHAGUS AND ITS IMPACT ON QUALITY OF LIFE

BE patients have repeatedly been shown to have substantially lower scores when generic and organ-specific quality-of-life measures are compared to the normal population. Subjects with BE consistently report their utility of living with the disease lower than without it, and the decrease in utility correlates with the degree of dysplasia in Barrett's epithelium. A diagnosis of BE appears to cause psychological stress and may be associated with substantial but incompletely understood additional costs such as increased life and health insurance premiums [1]. Taking these psychosocial considerations into account highlights the importance of screening and surveillance in BE as progression of this largely silent disease process has profound effects on quality of life.

6.7 THE FUTURE OF SCREENING AND SURVEILLANCE OF BARRETT'S ESOPHAGUS

Due to the dramatic rise in EAC incidence, the importance of preventative approaches, and the limitations of current screening and surveillance strategies, ongoing efforts have focused on technologies which may be less invasive and/or more accurate in risk-stratifying patients. In current clinical practice, risk stratification is based on the patient's clinical history, endoscopic findings, and histopathological grade. We believe that enhanced imaging techniques, such as endomicroscopy (see Chapter 9: Enhanced Imaging of the Esophagus: Confocal Laser Endomicroscopy) and molecular diagnostic approaches will be of importance in the future to guide risk stratification.

Advanced endoscopic imaging techniques, some of which are covered more thoroughly in other chapters, have been studied for this purpose, including dye-based chromoendoscopy, optical and digital chromoendoscopy, autofluorescence endoscopy, and confocal laser endomicroscopy [70]. With regard to molecular-based approaches to risk stratification in biopsy specimens from patients with Barrett's metaplasia, abnormalities in p53 expression and in cellular DNA content on flow cytometry have been associated with neoplastic progression [53,71]. Cytogenetic abnormalities detected by means of fluorescence in situ hybridization (FISH) and biomarker panels that identify multiple abnormalities in DNA content, gene expression, and DNA methylation have shown promise as predictors of cancer risk, as have some risk stratification models that incorporate a variety of clinical, histologic, and molecular features [53,72–75]. Further, nonendoscopic, less invasive methods such as Cytosponge, combining molecular biomarkers with histopathology, may play an important role in screening and surveillance programs in the future [76]. While biomarkers have the potential to predict risk of progression to cancer and may allow us to identify patients most likely to respond to endoscopic treatment, clinical applications have yet to enter the mainstream. Both molecular analysis and novel imaging techniques have the potential in the future to radically improve our management of higher risk patients, while minimizing the need for surveillance in low-risk patients.

References

[1] American Gastroenterological Association, Spechler SJ, Sharma P, Souza RF, Inadomi JM, Shaheen NJ. American Gastroenterological Association Medical Position Statement on the management of Barrett's esophagus. Gastroenterology 2011;140(3):1084–91.

[2] Bob Cat Review AJG 110. May 2015. p. 662–682.

[3] Spechler SJ, Souza RF. Barrett's esophagus. New Engl J Med. 2014;371(9):836–45.

[4] Jemal A, Siegel R, Xu J, Ward E. Cancer statistics, 2010. CA Cancer J Clin 2010;60(5):277–300.

[5] Burke ZD, Tosh D. Barrett's metaplasia as a paradigm for understanding the development of cancer. Curr Opin Genet Develop. 2012;22(5):494–9.

[6] Wang DH, Clemons NJ, Miyashita T, et al. Aberrant epithelial–mesenchymal Hedgehog signaling characterizes Barrett's metaplasia. Gastroenterology 2010;138 (5):1810–22.

[7] Sarosi G, Brown G, Jaiswal K, et al. Bone marrow progenitor cells contribute to esophageal regeneration and metaplasia in a rat model of Barrett's esophagus. Dis Esophagus 2008;21(1):43–50.

[8] Quante M, Bhagat G, Abrams JA, et al. Bile acid and inflammation activate gastric cardia stem cells in a mouse model of Barrett-like metaplasia. Cancer Cell 2012;21(1):36–51.

[9] Wang X, Ouyang H, Yamamoto Y, et al. Residual embryonic cells as precursors of a Barrett's-like metaplasia. Cell. 2011;145(7):1023–35.

[10] Fass R, Hell RW, Garewal HS, et al. Correlation of oesophageal acid exposure with Barrett's oesophagus length. Gut 2001;48(3):310–13.

[11] Hayeck TJ, Kong CY, Spechler SJ, Gazelle GS, Hur C. The prevalence of Barrett's esophagus in the US: estimates from a simulation model confirmed by SEER data. Dis Esophagus 2010;23(6):451–7.

[12] Taylor JB, Rubenstein JH. Meta-analyses of the effect of symptoms of gastroesophageal reflux on the risk of Barrett's esophagus. Am J Gastroenterol. 2010;105 (8):1729, 1730–7.

[13] Fletcher J, Wirz A, Henry E, McColl KE. Studies of acid exposure immediately above the gastro-oesophageal squamocolumnar junction: evidence of short segment reflux. Gut 2004;53(2):168–73.

[14] Spechler SJ, Zeroogian JM, Antonioli DA, Wang HH, Goyal RK. Prevalence of metaplasia at the gastro-oesophageal junction. Lancet 1994;344(8936):1533–6.

[15] Wang A, Mattek NC, Holub JL, Lieberman DA, Eisen GM. Prevalence of complicated gastroesophageal reflux disease and Barrett's esophagus among racial groups in a multi-center consortium. Digest Dis Sci 2009;54(5):964—71.

[16] El-Serag HB, Gilger MA, Shub MD, Richardson P, Bancroft J. The prevalence of suspected Barrett's esophagus in children and adolescents: a multicenter endoscopic study. Gastrointest Endosc. 2006;64 (5):671—5.

[17] Orloff M, Peterson C, He X, et al. Germline mutations in MSR1, ASCC1, and CTHRC1 in patients with Barrett esophagus and esophageal adenocarcinoma. JAMA 2011;306(4):410—19.

[18] Lagergren J, Lagergren P. Recent developments in esophageal adenocarcinoma. CA Cancer J Clin 2013;63 (4):232—48.

[19] El-Serag HB. Time trends of gastroesophageal reflux disease: a systematic review. Clin Gastroenterol Hepatol 2007;5(1):17—26.

[20] Coleman HG, Bhat S, Murray LJ, McManus D, Gavin AT, Johnston BT. Increasing incidence of Barrett's oesophagus: a population-based study. Eur J Epidemiol. 2011;26(9):739—45.

[21] El-Serag H. The association between obesity and GERD: a review of the epidemiological evidence. Digest Dis Sci. 2008;53(9):2307—12.

[22] Greer KB, Thompson CL, Brenner L, et al. Association of insulin and insulin-like growth factors with Barrett's oesophagus. Gut 2012;61(5):665—72.

[23] Iijima K, Henry E, Moriya A, Wirz A, Kelman AW, McColl KE. Dietary nitrate generates potentially mutagenic concentrations of nitric oxide at the gastroesophageal junction. Gastroenterology 2002;122(5):1248—57.

[24] Shaheen NJ, Crosby MA, Bozymski EM, Sandler RS. Is there publication bias in the reporting of cancer risk in Barrett's esophagus? Gastroenterology 2000;119(2):333—8.

[25] Desai TK, Krishnan K, Samala N, et al. The incidence of oesophageal adenocarcinoma in non-dysplastic Barrett's oesophagus: a meta-analysis. Gut 2012;61 (7):970—6.

[26] Wani S, Falk G, Hall M, et al. Patients with nondysplastic Barrett's esophagus have low risks for developing dysplasia or esophageal adenocarcinoma. Clin Gastroenterol Hepatol 2011;9(3):220—7 quiz e226.

[27] Bhat S, Coleman HG, Yousef F, et al. Risk of malignant progression in Barrett's esophagus patients: results from a large population-based study. J Natl Cancer Inst. 2011;103(13):1049—57.

[28] Hvid-Jensen F, Pedersen L, Drewes AM, Sorensen HT, Funch-Jensen P. Incidence of adenocarcinoma among patients with Barrett's esophagus. N Engl J Med. 2011;365(15):1375—83.

[29] Anaparthy R, Gaddam S, Kanakadandi V, et al. Association between length of Barrett's esophagus and risk of high-grade dysplasia or adenocarcinoma in patients without dysplasia. Clin Gastroenterol Hepatol 2013;11(11):1430—6.

[30] Munitiz V, Parrilla P, Ortiz A, Martinez-de-Haro LF, Yelamos J, Molina J. High risk of malignancy in familial Barrett's esophagus: presentation of one family. J Clin Gastroenterol. 2008;42(7):806—9.

[31] Gaddam S, Singh M, Balasubramanian G, et al. Persistence of nondysplastic Barrett's esophagus identifies patients at lower risk for esophageal adenocarcinoma: results from a large multicenter cohort. Gastroenterology 2013;145(3):548—53.

[32] Spechler SJ. Barrett esophagus and risk of esophageal cancer: a clinical review. JAMA 2013;310(6):627—36.

[33] Fountoulakis A, Zafirellis KD, Dolan K, Dexter SP, Martin IG, Sue-Ling HM. Effect of surveillance of Barrett's oesophagus on the clinical outcome of oesophageal cancer. Br J Surg. 2004;91(8):997—1003.

[34] Corley DA, Levin TR, Habel LA, Weiss NS, Buffler PA. Surveillance and survival in Barrett's adenocarcinomas: a population-based study. Gastroenterology 2002;122(3):633—40.

[35] Spechler SJ, Sharma P, Souza RF, Inadomi JM, Shaheen NJ, American Gastroenterological Association. American Gastroenterological Association technical review on the management of Barrett's esophagus. Gastroenterology 2011;140(3):e18—52 quiz e13.

[36] Committee ASoP, Evans JA, Early DS, et al. The role of endoscopy in Barrett's esophagus and other premalignant conditions of the esophagus. Gastrointest Endosc. 2012;76(6):1087—94.

[37] Shaheen NJ, Weinberg DS, Denberg TD, et al. Upper endoscopy for gastroesophageal reflux disease: best practice advice from the clinical guidelines committee of the American College of Physicians. Ann Int Med. 2012;157(11):808—16.

[38] Chak A, Faulx A, Eng C, et al. Gastroesophageal reflux symptoms in patients with adenocarcinoma of the esophagus or cardia. Cancer. 2006;107 (9):2160—6.

[39] Dulai GS, Guha S, Kahn KL, Gornbein J, Weinstein WM. Preoperative prevalence of Barrett's esophagus in esophageal adenocarcinoma: a systematic review. Gastroenterology 2002;122(1):26—33.

[40] Bhat SK, McManus DT, Coleman HG, et al. Oesophageal adenocarcinoma and prior diagnosis of Barrett's oesophagus: a population-based study. Gut 2015;64(1):20—5.

[41] Wang KK, Sampliner RE, Practice Parameters Committee of the American College of Gastroenterology. Updated guidelines 2008 for the

diagnosis, surveillance and therapy of Barrett's esophagus. Am J Gastroenterol. 2008;103(3):788–97.

[42] Shaheen NJ, Richter JE. Barrett's oesophagus. Lancet 2009;373(9666):850–61.

[43] Cooper GS, Kou TD, Chak A. Receipt of previous diagnoses and endoscopy and outcome from esophageal adenocarcinoma: a population-based study with temporal trends. Am J Gastroenterol. 2009;104(6):1356–62.

[44] Quera R, O'Sullivan K, Quigley EM. Surveillance in Barrett's oesophagus: will a strategy focused on a high-risk group reduce mortality from oesophageal adenocarcinoma? Endoscopy 2006;38(2):162–9.

[45] Yousef F, Cardwell C, Cantwell MM, Galway K, Johnston BT, Murray L. The incidence of esophageal cancer and high-grade dysplasia in Barrett's esophagus: a systematic review and meta-analysis. Am J Epidemiol. 2008;168(3):237–49.

[46] Sharma P, Falk GW, Weston AP, Reker D, Johnston M, Sampliner RE. Dysplasia and cancer in a large multicenter cohort of patients with Barrett's esophagus. Clin Gastroenterol Hepatol 2006;4(5):566–72.

[47] Abela JE, Going JJ, Mackenzie JF, McKernan M, O'Mahoney S, Stuart RC. Systematic four-quadrant biopsy detects Barrett's dysplasia in more patients than nonsystematic biopsy. Am J Gastroenterol. 2008;103(4):850–5.

[48] Wong T, Tian J, Nagar AB. Barrett's surveillance identifies patients with early esophageal adenocarcinoma. Am J Med. 2010;123(5):462–7.

[49] Gonzalez S, Yu WM, Smith MS, et al. Randomized comparison of 3 different-sized biopsy forceps for quality of sampling in Barrett's esophagus. Gastrointest Endosc. 2010;72(5):935–40.

[50] Fleischer DE, Overholt BF, Sharma VK, et al. Endoscopic radiofrequency ablation for Barrett's esophagus: 5-year outcomes from a prospective multicenter trial. Endoscopy 2010;42(10):781–9.

[51] Inadomi JM, Somsouk M, Madanick RD, Thomas JP, Shaheen NJ. A cost–utility analysis of ablative therapy for Barrett's esophagus. Gastroenterology 2009;136 (7):2101–14, e2101–2106.

[52] Sampliner RE. Management of nondysplastic Barrett esophagus with ablation therapy. Gastroenterol Hepatol 2011;7(7):461–4.

[53] Fritcher EG, Brankley SM, Kipp BR, et al. A comparison of conventional cytology, DNA ploidy analysis, and fluorescence in situ hybridization for the detection of dysplasia and adenocarcinoma in patients with Barrett's esophagus. Human Pathol. 2008;39(8):1128–35.

[54] Gordon LG, Mayne GC, Hirst NG, et al. Cost-effectiveness of endoscopic surveillance of non-dysplastic Barrett's esophagus. Gastrointest Endosc. 2014;79(2):242–56.

[55] Sonnenberg A, Soni A, Sampliner RE. Medical decision analysis of endoscopic surveillance of Barrett's oesophagus to prevent oesophageal adenocarcinoma. Aliment Pharmacol Therapeut 2002;16(1):41–50.

[56] Inadomi JM, Sampliner R, Lagergren J, Lieberman D, Fendrick AM, Vakil N. Screening and surveillance for Barrett esophagus in high-risk groups: a cost-utility analysis. Ann Int Med. 2003;138(3):176–86.

[57] de Jonge PJ, van Blankenstein M, Looman CW, Casparie MK, Meijer GA, Kuipers EJ. Risk of malignant progression in patients with Barrett's oesophagus: a Dutch Nationwide Cohort Study. Gut 2010;59(8):1030–6.

[58] Thomas T, Abrams KR, De Caestecker JS, Robinson RJ. Meta analysis: cancer risk in Barrett's oesophagus. Aliment Pharmacol Therap 2007;26(11–12):1465–77.

[59] Phoa KN, van Vilsteren FG, Weusten BL, et al. Radiofrequency ablation vs endoscopic surveillance for patients with Barrett esophagus and low-grade dysplasia: a randomized clinical trial. JAMA 2014;311:1209–17.

[60] Shaheen NJ, Sharma P, Overholt BF, et al. Radiofrequency ablation in Barrett's esophagus with dysplasia. N Engl J Med 2009;360:2277–88.

[61] Abdalla M, Dhanekula R, Greenspan M, et al. Dysplasia detection rate of confirmatory EGD in non-dysplastic Barrett's esophagus. Dis Esophagus 2014;27:505–10.

[62] Thota PN, Lee HJ, Goldblum JR, et al. Risk stratification of patients with Barrett's esophagus and low-grade dysplasia or indefinite for dysplasia. Clin Gastroenterol Hepatol 2015;13:459–65.

[63] Wani S, Falk GW, Post J, et al. Risk factors for progression of low-grade dysplasia in patients with Barrett's esophagus. Gastroenterology 2011;141(4):1179–86 1186–1171.

[64] Srivastava A, Hornick JL, Li X, et al. Extent of low-grade dysplasia is a risk factor for the development of esophageal adenocarcinoma in Barrett's esophagus. Am J Gastroenterol. 2007;102(3):483–93.

[65] Shaheen NJ, Sharma P, Overholt BF, et al. Radiofrequency ablation in Barrett's esophagus with dysplasia. N Engl J Med. 2009;360(22):2277–88.

[66] Shaheen NJ, Overholt BF, Sampliner RE, et al. Durability of radiofrequency ablation in Barrett's esophagus with dysplasia. Gastroenterology 2011;141 (2):460–8.

[67] Hur C, Choi SE, Rubenstein JH, et al. The cost effectiveness of radiofrequency ablation for Barrett's esophagus. Gastroenterology 2012;143(3):567–75.

[68] Levine DS, Haggitt RC, Blount PL, Rabinovitch PS, Rusch VW, Reid BJ. An endoscopic biopsy protocol can differentiate high-grade dysplasia from early adenocarcinoma in Barrett's esophagus. Gastroenterology 1993;105(1):40–50.

[69] Kariv R, Plesec TP, Goldblum JR, et al. The Seattle protocol does not more reliably predict the detection of cancer at the time of esophagectomy than a less intensive surveillance protocol. Clin Gastroenterol Hepatol 2009;7(6):653−8.

[70] Boerwinkel DF, Swager A, Curvers WL, Bergman JJ. The clinical consequences of advanced imaging techniques in Barrett's esophagus. Gastroenterology 2014;146 (3):622−9.

[71] Kastelein F, Biermann K, Steyerberg EW, et al. Aberrant p53 protein expression is associated with an increased risk of neoplastic progression in patients with Barrett's oesophagus. Gut 2013;62(12):1676−83.

[72] Rubenstein JH, Morgenstern H, Appelman H, et al. Prediction of Barrett's esophagus among men. Am J Gastroenterol 2013;108(3):353−62.

[73] Alvi MA, Liu X, O'Donovan M, et al. DNA methylation as an adjunct to histopathology to detect prevalent, inconspicuous dysplasia and early-stage neoplasia in Barrett's esophagus. Clin Cancer Res 2013;19 (4):878−88.

[74] Jin Z, Cheng Y, Gu W, et al. A multicenter, double-blinded validation study of methylation biomarkers for progression prediction in Barrett's esophagus. Cancer Res 2009;69(10):4112−15.

[75] Bird-Lieberman EL, Dunn JM, Coleman HG, et al. Population-based study reveals new risk-stratification biomarker panel for Barrett's esophagus. Gastroenterology 2012;143(4):927−35.

[76] Zeki S, Fitzgerald RC. Targeting care in Barrett's oesophagus. Clin Med 2014;14(Suppl. 6):s78−83.

In Vivo Optical Detection of Dysplasia in Barrett's Esophagus with Endoscopic Light Scattering Spectroscopy

Le Qiu[1], Douglas K. Pleskow[2], Ram Chuttani[2], Lei Zhang[1], Vladimir Turzhitsky[1], Eric U. Yee[3], Mandeep Sawhney[2], Tyler M. Berzin[2], Fen Wang[2], Umar Khan[1], Edward Vitkin[1], Jeffrey D. Goldsmith[3], Irving Itzkan[1] and Lev T. Perelman[1,2]

[1]Center for Advanced Biomedical Imaging and Photonics, Department of ObGyn and Reproductive Biology, Beth Israel Deaconess Medical Center, Harvard Medical School, Boston, MA, United States
[2]Center for Advanced Endoscopy, Department of Medicine, Beth Israel Deaconess Medical Center, Harvard Medical School, Boston, MA, United States [3]Department of Pathology, Beth Israel Deaconess Medical Center, Harvard Medical School, Boston, MA, United States

7.1 INTRODUCTION

The incidence of adenocarcinoma of the esophagus is increasing more rapidly than any other type of carcinoma in the United States [1]. Almost 100% of cases occur in patients with Barrett's esophagus (BE) [2], a benign condition in which metaplastic columnar epithelium replaces the normal squamous epithelium of the esophagus. Although the prognosis of patients diagnosed with adenocarcinoma is poor, the chances of successful treatment increase significantly if the disease is detected at the dysplastic stage. The surveillance for dysplasia of patients with BE is challenging in three respects. First, dysplasia in nonnodular BE is not visible during routine endoscopy [3]. Thus, numerous systematically located biopsy specimens are taken in a prescribed pattern. Although systematic, this procedure is referred to in the literature as "random," and we will also use this term. Second, the histopathologic diagnosis of dysplasia is problematic because there is poor interobserver agreement on the classification of a particular specimen, even among expert gastrointestinal pathologists [4,5]. Third, reliance on histology imposes a time delay between endoscopy and diagnosis, requiring patient recall if disease is discovered as opposed to prompt treatment during surveillance.

Once BE has been identified, most gastroenterologists will enroll the patient in an endoscopy/biopsy surveillance program, presuming that the patient is a candidate for ablative therapy [6—8] should low-grade or high-grade dysplasia be detected. Although the cost effectiveness of this type of surveillance program has not been validated in prospective studies, the lack of such studies does not preclude its potential usefulness. Patients with BE who have esophageal carcinoma detected as part of such a surveillance program are more likely to have curable disease and have an improved 5-year survival as compared with those whose cancer was detected outside of a surveillance program.

Dysplasia in the gastrointestinal tract is defined as neoplastic epithelium confined within an intact basement membrane. Dysplasia in BE can be classified as low or high grade, based on the criteria originally defined for dysplasia in inflammatory bowel disease [9]. Low-grade dysplasia (LGD) is defined primarily by cytological abnormalities, including nuclear enlargement, crowding, stratification, hyperchromasia, mucin depletion, and mitoses in the upper portions of the crypts. These abnormalities extend to the mucosal surface. High-grade dysplasia (HGD) is characterized by even more pronounced cytological abnormalities, as well as glandular architectural abnormalities including villiform configuration of the surface, branching and lateral budding of the crypts, and formation of the so-called back-to-back glands. When there is any doubt as to the significance of histological abnormalities because of inflammation, ulceration, or histological processing artifacts, the findings may be classified as "indefinite for dysplasia" (IND) in order to prevent unnecessary clinical consequences.

Not all patients with BE progress to adenocarcinoma. The majority live their entire lives without undergoing malignant or neoplastic transformation. Others demonstrate a rapid progression to carcinoma and will die of esophageal cancer if it is not diagnosed and treated in a timely manner. Several recent attempts at identifying molecular markers that can predict which patients with BE will progress to esophageal cancer have not been proven effective in clinical trials. For example, anti-p53 antibodies have been shown to develop in patients with BE and adenocarcinoma, and may predate the clinical diagnosis of malignancy [10].

At the present time, the standard of care for surveillance of patients with BE remains debated. Although periodic endoscopic surveillance of patients with BE has been shown to detect carcinoma in its earlier stages, surveillance has significant limitations. Dysplastic and early carcinomatous lesions arising in nonnodular BE are not visible macroscopically; therefore, surveillance requires extensive random biopsies of the esophagus and histologic examination of the excised tissue for dysplasia. Random biopsy is prone to sampling error (missed dysplastic lesions) and significantly increases the cost and risk of surveillance. There is also significant interobserver disagreement between pathologists in diagnosing dysplasia. A large 10-year observational study in 409 patients with BE published in the British Medical Journal [11] concluded that the current random biopsy endoscopic surveillance strategy has very limited value.

Optical imaging offers great promise for the detection and characterization of precancers in the esophagus. Endoscopic laser-induced fluorescence (LIF) spectroscopy based imaging is believed to measure the abnormal concentrations of certain endogenous fluorophores such as porphyrins in malignant tissue. Promising results for esophagus have been obtained by Vo-Dinh et al. [12], von Holstein et al. [13], Stepp et al. [14], Messmann et al. [15], Braichotte et al. [16], and Georgakoudi et al. [17]. Another native contrast optical imaging approach, optical coherence tomography (OCT), is a method that provides two-dimensional cross-sectional images of the gastrointestinal tract. Like endoscopic ultrasound, OCT provides true anatomic images corresponding to the layers of the gastrointestinal tract [18–23], however, by using light the resolution of OCT is nearly 10-fold finer than that of high-frequency endoscopic ultrasound. Recently, Tearney et al. used high-speed OCT technology, termed optical frequency domain imaging (OFDI), to demonstrate the feasibility of a large area imaging of the entire distal esophagus in patients [24]. Fujimoto et al. [25] demonstrated the feasibility of using endoscopic ultra-high resolution optical coherence tomography (UHR OCT) to locating esophageal adenocarcinoma. Wax et al. developed an interferometric light scattering technique called angle-resolved low-coherence interferometry (a/LCI) capable of measuring nuclear morphology as a function of depth in epithelial tissue [26–28] and demonstrated its ability to detect nuclear atypia in esophageal dysplastic epithelial tissues. Bigio et al. developed a minimally invasive diagnostic light scattering technique called elastic scattering spectroscopy (ESS) and applied it to detect early cancer in colon [29,30], bladder [31], and esophagus [32]. The study in esophagus demonstrated that ESS has the potential to target conventional biopsies in Barrett's surveillance saving significant endoscopist and pathologist time.

Over a decade ago, we pioneered a new diagnostic optical technique, biomedical light scattering spectroscopy (LSS), and demonstrated that LSS can measure subcellular nuclear morphology in esophageal tissue in vivo and be directly correlated with histopathology [33]. Clinical feasibility of LSS for detection of epithelial precancers (dysplasia) in several organs was later demonstrated [34] and then extended to large field-of-view imaging [35]. Recently, we demonstrated that LSS can be employed to scan an entire esophagus during routine endoscopy, successfully guiding biopsy and detecting and mapping sites of invisible dysplasia missed by the current standard of care [36].

This article reviews the principles of LSS and describes application of endoscopic LSS for detecting invisible dysplasia in BE during routine clinical procedures.

7.2 LIGHT SCATTERING FROM CELLS

An electromagnetic wave, interacting with a dielectric or conducting particle (by a particle we will mean any bounded region with a complex refractive index different from the refractive index of the surrounding medium), induces oscillations of bound and free charges in that particle, which in turn generates electromagnetic waves inside and outside of the particle. Thus, the problem of light scattering by a single particle can be formulated in the following way: given a particle of known structure, illuminated by a plane wave of particular polarization, find the electromagnetic field inside and outside the particle. Usually, however, the simplified formulation of the problem suffices: find the approximate values of electromagnetic field at large distances from the particle.

For example, a well-known van de Hulst approximation [37] enables one to obtain the wavelength and size-dependent scattering cross section for a large particle of an arbitrary shape

$$\sigma_s(\lambda, a) \approx 2\pi a^2 \left\{ 1 - \frac{\sin 2x}{x} + \left(\frac{\sin x}{x} \right)^2 \right\} \quad (7.1)$$

where λ is a wavelength, a is a size of the particle, m is the relative refractive index averaged over the volume of the particle, and $x = 2\pi(m - 1)a/\lambda$ is the size parameter. The above expression demonstrates that scattering spectra of large particles are quite different from the monotonic $1/\lambda^4$ dependent Rayleigh scattering spectra of particles with sizes much smaller than a wavelength. Here the spectra are nonmonotonic functions exhibiting oscillations with the wavelength, with the frequency of these oscillations increasing with the particle size and its refractive index. This fact is quite important for understanding light scattering spectra of individual cells and cellular layers.

Though there are hundreds of cell types, the subcellular compartments in different cells are rather similar and are limited in number [38]. Any cell is bounded by a membrane, a phospholipid bilayer approximately 10 nm in thickness. Two major cell compartments are the nucleus which has a size of $7-10\,\mu m$ and the surrounding cytoplasm. The cytoplasm contains various other organelles and inclusions. One of the most common organelle (and the largest after the nucleus) is a mitochondrion which has the shape of a prolate spheroid. The large dimension of a mitochondrion may range from 1 to $5\,\mu m$ and the diameter typically varies between 0.2 and $0.8\,\mu m$. Other smaller organelles include lysosomes which are $250-800$ nm in size and of various shapes and peroxisomes which are 200 nm to $1.0\,\mu m$ spheroidal bodies of lower densities than the lysosomes. Peroxisomes are more abundant in metabolically active cells such as hepatocytes where they are counted in the hundreds.

Since the size of the majority of cellular organelles is comparable to the wavelength of light and their refractive index ranges from 1.38 to 1.42 [39−41] with the refractive index of the cytoplasm being within 1.34−1.36 range, the relative refractive index of the cellular organelles, m, is close to unity. At the same time, the phase shift across the particle $4\pi a(m-1)/\lambda$ is small. The above two conditions mean that for the majority of small organelles the Rayleigh−Gans approximation is satisfied [42], which means that total intensity of light scattered by these small organelles increases with the increase of its refractive index as $(m-1)^2$ and with its size as a^6.

Unfortunately, the Rayleigh−Gans approximation cannot be applied to the cell nucleus, whose size is significantly larger than that of optical wavelengths. Here we can either use the intuitively transparent but approximate van de Hulst solution from Eq. (7.1) or cumbersome but exact Mie solution.

Sizes and refractive indices of major cellular and subcellular structures are presented in

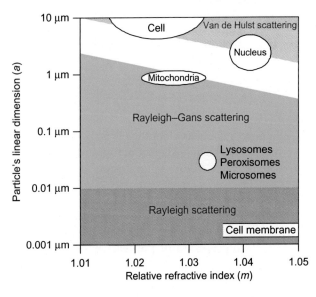

FIGURE 7.1 Optical properties of cellular and subcellular structures and relevant approximations that can be used to describe light scattering from those objects. The uncolored region represents a parameter space which requires one of the exact solutions for accurate results.

Fig. 7.1. On this figure, we also provide the information on the relevant approximations that can be used to describe light scattering from these objects when interacting with light in the visible range.

7.3 LIGHT TRANSPORT IN SUPERFICIAL TISSUES

Light transport in biological tissue is dominated by elastic scattering. The primary scattering centers are thought to be the collagen fiber network of the extracellular matrix, the mitochondria, and other intracellular substructures, all with dimensions smaller than the optical wavelength. However, larger structures, such as cell nuclei, typically 5–15 μm in diameter, also scatter light. Because of the ubiquity of light scattering, its effects are enmeshed with those of absorption, making interpretation of tissue spectra difficult. On the other hand, this feature can actually enrich the information provided by spectroscopic techniques. For example, the strong dependence of the scattering cross section (7.1) on size and refractive index of the scatterer, such as the cell nucleus, as well as on the wavelength suggests that it should be possible to design a spectroscopic technique which can differentiate cellular tissues by the sizes of the nuclei. Indeed, the hollow organs of the body are lined with a thin, highly cellular surface layer of epithelial tissue, which is supported by underlying, relatively acellular connective tissue. There are four main types of epithelial tissue: squamous, cuboidal, columnar, and transitional, which can be found in different organs of the human body. Depending on type, the epithelium consists of either a single layer of cells or several cellular layers. Here, to make the treatment of the problem more apparent, we consider epithelial layers consisting of a single well-organized layer of cells, such as simple columnar epithelium or simple squamous epithelium. For example, in healthy columnar epithelial tissues,

the epithelial cells often have an en-face diameter of 10–20 μm and height of 25 μm. In dysplastic epithelium, the cells proliferate and their nuclei enlarge and appear darker (hyperchromatic) when stained [43]. LSS can be used to measure these changes. The details of the method have been published by Perelman et al. [33] and will only be briefly summarized here.

Consider a beam of light incident on an epithelial layer of tissue. A portion of this light is backscattered from the epithelial nuclei, while the remainder is transmitted to deeper tissue layers, where it undergoes multiple scattering and becomes randomized before returning to the surface. Epithelial nuclei can be treated as spheroidal Mie scatters with a refractive index which is higher than that of the surrounding cytoplasm [38,40,41]. Normal nuclei have a characteristic size of 4–7 μm. In contrast, the size of dysplastic nuclei varies widely and can be as large as 20 μm, occupying almost the entire cell volume. In the visible range, where the wavelength is much smaller than the size of the nuclei, the van de Hulst approximation (7.1) can be used to describe the elastic scattering cross section of the nuclei. Eq. (7.1) reveals a component of the scattering cross section which varies periodically with inverse wavelength. This, in turn, gives rise to a periodic component in tissue reflectance. Since the frequency of this variation (in the inverse wavelength space) is proportional to the particle size, the nuclear size distribution can be obtained from that periodic component.

However, single scattering events cannot be measured directly in biological tissue. Because of multiple scattering, information about tissue scatterers is randomized as light propagates into the tissue, typically over one effective scattering length (0.5–1 mm, depending on the wavelength). Nevertheless, the light in the thin layer at the tissue surface is not completely randomized. In this thin region, the details of the elastic scattering process are preserved. The total signal reflected from tissue can be divided into two parts: single backscattering from the uppermost tissue structures such as cell nuclei and a background of diffusely scattered light. To analyze the single scattering component of the reflected light, the diffusive background must be removed. This can be achieved either by modeling using diffuse reflectance spectroscopy [17,44] or by other techniques such as polarization background subtraction [45,46] and coherence gating method [47].

There are several techniques that can be employed to obtain the nuclear size distribution from the remaining single scattering component of the backreflected light, which can be called the LSS spectrum. A good approximation for the nuclear size distribution can be obtained from the Fourier transform of the periodic component as described in Ref. [33]. A more advanced technique introduced by Fang et al. [48] and described here is based on linear least squares with a nonnegativity constraints algorithm.

The experimentally measured reflectance spectrum consists of a large diffusive background plus the component of forward scattered and backscattered light from the nuclei in the epithelial layer. For a thin slab of epithelial tissue containing nuclei with size distribution (number of nuclei per unit area (mm^2) and per unit interval of nuclear diameter (μm), the approximate solution of the transport equation for the backscattered component is a linear combination of the backscattering spectra of the nuclei of different sizes:

$$S(\lambda) = \int_0^\infty I(\lambda, \delta) N(\delta) d\delta + \varepsilon(\lambda) \qquad (7.2)$$

where $I(\lambda, \delta)$ is the LSS spectrum of a single scatterer with diameter δ and $\varepsilon(\lambda)$ is the experimental noise. It is convenient to write this in a matrix form $\hat{S} = \hat{I} \cdot \hat{N} + \hat{E}$, where $\hat{S}$ is the experimental spectrum measured at discrete wavelength points, $\hat{N}$ is a discreet nuclear size distribution, $\hat{I}$ is the LSS spectrum of a single scatterer with diameter δ and $\hat{E}$ is the experimental noise.

Since the LSS spectrum $\hat{I}$ is a highly singular matrix and a certain amount of noise is present in the experimental spectrum $\hat{S}$, it is not feasible to calculate the size distribution $\hat{N}$ by directly inverting the matrix $\hat{I}$. Instead we can multiply both sides of the equation $\hat{S} = \hat{I} \cdot \hat{N} + \hat{E}$ by the transpose matrix $\hat{I}^{\mathrm{T}}$ and introduce the matrix $\hat{C} = \hat{I}^{\mathrm{T}} \cdot \hat{I}$. We can now compute matrix C eigenvalues $\alpha_1, \alpha_2, \ldots$ and sequence them from large to small. This can be done because $\hat{C}$ is a square symmetric matrix. Then, we use the linear least squares with nonnegativity constraints algorithm [49] to solve the set of equations

$$\hat{I}^{\mathrm{T}}\hat{S} - (\hat{C} + \alpha_k \hat{H})\hat{N} \rightarrow \min$$
$$\hat{N} \geq 0 \qquad (7.3)$$

where $\alpha_k \hat{H} \hat{N}$ is the regularization term and matrix $\hat{H}$ represents the second derivative of the spectrum. The use of the nonnegativity constraint and the regularization procedure is critical to find the correct distribution $\hat{N}$. By using this algorithm, we can accurately reconstruct the nuclear size distribution.

As we already mentioned, to analyze the single scattering component of the reflected light the diffusive background must be removed. Here we discuss in greater detail one of the techniques used to remove the multiple scattered light in LSS measurements and improve the extraction of the single scattered light component. This technique is sometimes called polarized LSS.

In the polarized LSS technique, the tissue is illuminated with a polarized light. The light backscattered from the superficial epithelial layer retains its polarization, ie, it is polarized parallel to the incoming light. The light backscattered from the deeper tissues becomes depolarized and contains equal amounts of parallel and perpendicular polarizations. By subtracting the two, one can cancel out the contribution of the deeper tissues and the resulting signal is proportional only to the signal from the superficial epithelial layer, which contains information about early precancerous changes. The residual of the parallel and perpendicular components $I_{LSS}(\lambda) = I_{II}(\lambda) - I_{\perp}(\lambda)$ can be related to the properties of scatterers in the superficial epithelial layer using relation (7.1) and then processed using the algorithm described in the previous section.

To verify this, Backman et al. [45] employed an instrument that delivers collimated polarized light to the tissue and separates the two orthogonal polarizations of backscattered light. In this system, light from a broadband source is collimated and then refocused at a small solid angle onto the sample, using lenses and an aperture. Studies have shown that the unpolarized component of the reflected light can be canceled by subtracting the perpendicular spectral component from the parallel component allowing the single scattering signal to be extracted. These residual spectra were fitted to the model based on Mie theory.

Experiments were performed with monolayers of normal intestinal epithelial cells and T84 intestinal malignant cells placed above a thick layer of gel containing blood and $BaSO_4$, placed underneath to simulate underlying tissue. For normal intestinal epithelial cells, the best fit was obtained using $d = 5.0\ \mu m$, $\Delta d = 0.5\ \mu m$, and $n = 1.035$. For T84 intestinal malignant cells, the corresponding values were $d = 9.8\ \mu m$, $\Delta d = 1.5\ \mu m$, and $n = 1.04$. To check these results, the distribution of the average size of the cell nuclei was measured by morphometry on identical cell preparations that were processed in parallel for light microscopy. The nuclear sizes and their standard deviations were found to be in very good agreement with the parameters extracted from Mie theory (Fig. 7.2). In addition, a larger value of n was obtained for T84 intestinal malignant cells, which is in agreement with the hyperchromaticity of cancer cell nuclei observed in conventional histopathology of stained tissue sections.

7.4 IMAGING DYSPLASIA IN BARRETT'S ESOPHAGUS WITH ENDOSCOPIC POLARIZED LIGHT SCATTERING SPECTROSCOPY

LSS-based detection of dysplasia in BE has been demonstrated successfully using a simple proof-of-principle single point instrument [33,34]. This instrument was capable of collecting data at randomly selected sites by manually positioning the probe. The sites were then biopsied, the data were processed off-line, and a comparison with biopsy results was made when the pathology information became available. The high correlation between spectroscopic results and pathology was sufficiently promising to justify the development of the clinical device, which is reviewed herein.

The clinical endoscopic polarized LSS instrument [36] is compatible with existing endoscopes (Fig. 7.3). It scans large areas of the esophagus chosen by the physician and has the software and algorithms necessary to obtain quantitative, objective data about tissue structure and composition, which can be translated into diagnostic information in real time. This enables the physician to take confirming biopsies at suspicious sites and minimize the number of biopsies taken at nondysplastic sites.

The instrument detects polarized light coming primarily from the epithelial layer. Although principally using the polarization technique to extract diagnostic information about dysplasia, the endoscopic polarized LSS instrument can also sum the two polarizations to permit the use of diffuse reflectance spectroscopy, which also can provide information about early stages of adenocarcinoma [28].

The endoscopic polarized LSS instrument is a significant advance over the single point fiber optic instrument in that: (1) it scans the esophagus and has the software and algorithms necessary to obtain quantitative, objective data about tissue structure and composition, which can be translated into diagnostic information

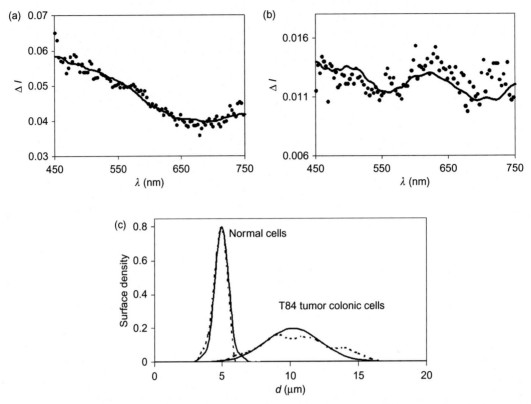

FIGURE 7.2   Spectra of polarized components of backscattered light. Spectrum of (a) normal intestinal cells, (b) T84 intestinal malignant cells, and (c) corresponding nuclear size distributions. In each case, the solid line is the distribution extracted from the data, and the dashed line is the distribution measured using light microscopy.

and guide biopsy in real time; (2) it employs collimated illumination and collection optics, which enables the instrument to generate maps of epithelial tissue not affected by the distance between the probe tip and the mucosal surface, making it dramatically less sensitive to peristaltic motion; (3) it incorporates both the polarization technique for removing the unwanted background in the LSS signal and single backscattering in the diffuse reflectance spectroscopy signal; (4) it integrates the data analysis software with the instrument in order to provide the physician with real-time diagnostic information; (5) it combines LSS information with diffuse reflectance spectroscopy information measured by the same instrument, thereby improving the diagnostic assessment capability.

The instrument makes use of commercially available gastroscopes and video processors. A standard PC is adapted to control the system. Commercially available spectrometers are also employed.

For use during endoscopy, the polarized scanning fiber optic probe is inserted into the working channel of a standard gastroendoscope (eg, Olympus GIF-H180 used in the procedures reported below) and the gastroenterologist introduces the endoscope through the mouth. Spectroscopy of the entire Barrett's segment is performed by scanning adjacent sections, 2 cm in length, with the polarized

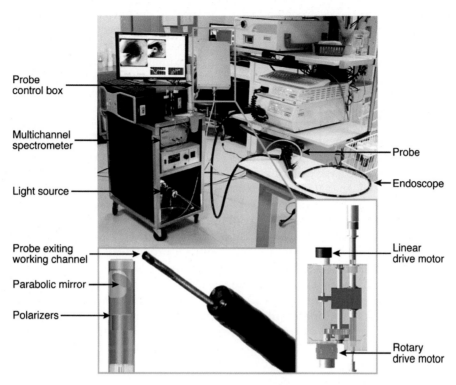

FIGURE 7.3 Clinical endoscopic scanning light scattering spectroscopy instrument. The instrument is shown in the endoscopy suite before the clinical procedure with the scanning probe inserted into the working channel of an endoscope. The insets show details of the scanning probe tip and the control box.

scanning probe as follows. The endoscope tip is positioned and the probe is extended 2 cm beyond the endoscope tip, placing it at the distal boundary of a BE region chosen for examination. One complete rotary scan of the esophageal wall is completed. The probe is withdrawn linearly 2 mm back into the endoscope tip and another rotary scan is completed. This is repeated for 10 rotary scans, so that an entire 2 cm length of BE is scanned; then, the endoscope tip is withdrawn 2 cm and the next length of BE is examined. Currently, the instrument collects 30 data points for each rotary scan and performs 10 steps during a linear scan (2 mm per step), collecting 300 data points in 2 minutes for each 2 cm segment of BE. We estimate that the scanning time can be reduced

to as little as 20 seconds by utilizing a more efficient scanning mechanism.

We performed in vivo measurements using endoscopic polarized LSS during 10 routine clinical endoscopic procedures for patients with suspected dysplasia at the Center for Advanced Endoscopy (CAE) at Beth Israel Deaconess Medical Center (BIDMC). Patients reporting to the CAE at BIDMC had undergone initial screening at other institutions and were referred with confirmed BE and suspicion of dysplasia. Our protocols were reviewed and approved by the BIDMC Institutional Review Board.

Patients reporting for routine screening of BE who had consented to participate in our study were examined. The LSS polarized fiber optic probe was inserted into the working channel of the gastroendoscope and the gastroenterologist introduced the endoscope through the mouth. The endoscopic polarized LSS instrument performed optical scanning of each complete, continuous region of the luminal esophageal wall chosen for examination by the gastroenterologist. Data from the optical scans were recorded for each linear and angular position of the probe tip as parallel and perpendicular polarization reflectance spectra, corrected for light source intensity and lineshape. The backscattering spectrum at each individual spatial location was extracted by subtracting perpendicular from parallel polarized reflectance spectra. The backscattering spectra were then normalized to remove amplitude variations due to peristalsis. The mean of the normalized spectra was calculated. The difference from the mean for each site was calculated, squared, and summed over all spectral points. A site was considered likely to be dysplastic if this parameter was greater than 10 percent of the summed mean squared. No data points are needed for calibration of this simple diagnostic rule. This analysis is straightforward and can be done in real time. By

extracting the nuclear size distributions from the backscattering spectra for each individual spatial location, we found that this simple rule is approximately equivalent to a contribution of greater than 25 percent from enlarged nuclei over 10 microns in diameter (Fig. 7.4).

Two observations support the clinical feasibility of this method. First, spectroscopic data collected during clinical procedures confirm that the polarization technique is very effective in removing unwanted background signals. Second, the issue of peristaltic motion is addressed in the endoscopic polarized LSS instrument. During a procedure, it is difficult to maintain a fixed distance between the optical probe head and the esophageal surface, due to peristaltic motion and other factors. Therefore, an important feature of the endoscopic polarized LSS instrument is its ability to collect spectra of epithelial tissue that are not affected by the orientation or distance of the distal probe tip to the mucosal surface. This is achieved with collimated illumination and collection optics. Analysis of parallel polarization spectra collected at 10 BE locations

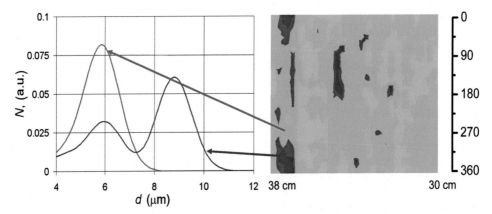

FIGURE 7.4   Nuclear size distributions for one high-grade dysplasia site and one nondysplastic site in Barrett's esophagus (BE) of one of the patients. Dark (red and pink online) regions of the map indicate areas suspicious for dysplasia based on nuclear size distributions extracted from the backscattering spectra for each individual spatial location. Nondysplastic BE sites had nuclear size distributions centered about 5–6 μm diameter while sites marked as suspicious for dysplasia have nuclear size distributions with a main peak centered from 9 to 15 μm. The arrows indicate the specific locations on the esophageal surface for which the size distributions are extracted from the polarized light scattering spectroscopy data.

during a standard clinical procedure showed that although amplitudes of the spectra differ from point to point, the spectral shape is practically unchanged and, more importantly, the oscillatory structure containing diagnostically significant information is intact.

During the initial stage of the project, we collected a total of 22,800 LSS spectra in 10 clinical procedures, covering the entire scanned regions of the esophagus. We validated the capabilities of the clinical method by comparing LSS data with subsequent pathology at each location where biopsies were taken. For the first two patients, pathology was reported per quadrant not per biopsy. For the other patients, 95 biopsies were collected at LSS locations given by their distances from the mouthpiece of the endoscope and their angles relative to the start of the LSS scan. Pathological examination revealed a total of 13 dysplastic sites out of which 9 were HGD. The rest of the sites were diagnosed as nondysplastic BE.

The diagnostic parameters for each location were extracted from the backscattering spectra, ie, the residuals of the parallel and perpendicular spectral components collected by the endoscopic polarized LSS instrument. The results are presented in the form of pseudo-color maps. Double blind comparison of the LSS maps with the biopsy reports revealed 11 true positive (TP) sites, 3 false positive (FP) sites, 80 true negative (TN) sites, and 1 false negative (FN) site. Thus LSS measurements are characterized by sensitivity of 92% and specificity of 96%.

In several BE patients who enrolled in our study and underwent routine endoscopy and biopsy with LSS, pathology revealed no dysplasia and the patients were dismissed. However, in some of these patients the LSS scan indicated probable sites of focal dysplasia, which were located in regions where biopsies had not been taken. One of the patients was recalled and biopsies were taken at the three sites indicated by LSS in addition to the standard-of-care protocol. Pathology confirmed HGD in all three LSS directed biopsies and one more HGD at a point located between two LSS indicated sites (Fig. 7.5). The latter site, considered to be a false negative, was very close to the sites indicated by LSS and was likely due to the misalignment of

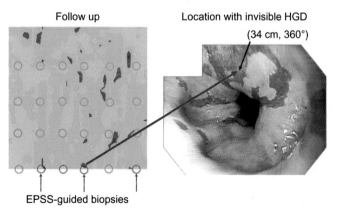

FIGURE 7.5 Biopsies taken during the initial and follow-up endoscopy procedures for patient A overlaid on the light scattering spectroscopy (LSS) map acquired during the initial procedure (left panel). Three follow-up biopsies were guided by the LSS map and pathology confirmed high-grade dysplasia (HGD) for each (indicated at 360°). High-resolution endoscopic (HRE) image of a location with invisible HGD (right panel) with narrow band imaging (NBI) enabled. Video capture was acquired in subject A at one of the locations where invisible dysplasia was missed by visual examination by HRE with NBI, but located by LSS, and later confirmed by pathology. The site is marked by an arrow. Note that the site is visually indistinguishable from the surrounding nondysplastic BE tissue.

the actual biopsy site with the LSS mapped site (a problem which will be addressed in future instrument and algorithm development). The patient was subsequently treated with radio frequency ablation (RFA).

These focal dysplasias were missed by standard-of-care procedures that blindly biopsy a tiny fraction of esophageal tissue according to a prescribed protocol, but were caught and confirmed by the capability of endoscopic polarized LSS to examine the entire esophageal epithelium millimeter by millimeter and detect dysplastic cells—thereby enabling early treatment and potentially saving future costly invasive procedures.

The frequency of dysplasia in our patient sample is consistent with that of the pre-screened patient population referred to the BIDMC CAE for confirmation and treatment but is higher than would be expected in the general BE patient population. In fact, the rarity of HGD detection in the general population of BE patients underscores the importance of having a more comprehensive and effective method for gastroesophageal cancer screening.

7.5 FUTURE DIRECTIONS

Studies described in this chapter successfully demonstrate that by analyzing spectra of light backscattered from epithelial cells one can extract information about cellular morphology in vivo and in real time—such information was not obtainable before without tissue removal. And since changes in cellular morphology are the hallmarks of precancerous and cancerous changes in most human tissues, this technique has a clear potential to provide broadly applicable means of detecting precancerous and early cancerous changes in cell-rich epithelia that line the internal surfaces throughout the body.

However, to become a useful clinical tool the clinical endoscopic polarized LSS imaging instrument described in this chapter should be capable of analyzing LSS information and translating it into histological and biochemical properties of epithelial tissue, similar to the properties observed by a pathologist, in real time during the endoscopy procedure. The instrument should also have the real-time biopsy guidance system which would enable the endoscopist to take biopsies at suspicious sites and minimize the number of biopsies taken at nondysplastic sites. Finally, the portability of the instrument should be significantly improved.

Such a clinical imaging instrument will have the potential to greatly reduce the sampling error and to eliminate sample preparation artifacts associated with endoscopy and biopsy. Because random tissue removal will not be required, it could also be used to examine extended tissue surfaces. Thus, this technique can greatly extend the utility of endoscopic techniques in detecting early cancerous changes in various organs.

Clinical endoscopic polarized LSS imaging instruments which will likely be developed and brought to practice in the next several years could open an unprecedented opportunity for screening and diagnosis of early cancer and also dramatically improve methods of monitoring cancer therapy in situ rather than in vitro.

Acknowledgments

This work was supported by US National Institutes of Health grant R01 EB003472 and US National Science Foundation grants CBET-0922876 and CBET-1402926. We would like to thank A. Turshudzhyan for helping with data collection. F. Wang is currently at the Department of Gastroenterology, Third Xiangya Hospital, Central South University, Changsha, Hunan, China.

References

[1] Blot W, Devesa SS, Kneller R, Fraumeni J. Rising incidence of adenocarcinoma of the esophagus and gastric cardia. JAMA 1991;265:1287–9.

[2] Antonioli D. The esophagus. In: Henson D, Alobores-Saavdera J, editors. The pathology of incipient neoplasia. Philadelphia: Saunders; 1993. p. 64–83.

[3] Cameron AJ. Management of Barrett's esophagus. Mayo Clin Proc 1998;73:457–61.

[4] Reid BJ, Haggitt RC, Rubin CE, Roth G, Surawicz CM, Van Belle G, et al. Observer variation in the diagnosis of dysplasia in Barrett's esophagus. Hum Pathol 1988;19:166–78.

[5] Petras RE, Sivak MV, Rice TW. Barrett's esophagus. A review of the pathologist's role in diagnosis and management. Pathol Annual 1991;26:1–32.

[6] Bergman JJGHM. Radiofrequency ablation—Great for some or justified for many? NEJM 2009;360:2353–5.

[7] Shaheen NJ, Sharma P, Overholt BF, Wolfsen HC, Sampliner RE, Wang KK, et al. Radiofrequency ablation in Barrett's esophagus with dysplasia. NEJM 2009;360:2277–88.

[8] Hur C, Wittenberg E, Nishioka NS, Gazelle GS. Quality of life in patients with various Barrett's esophagus associated health states. Health Qual Life Outcomes 2006;4:45.

[9] Riddell RH, Goldman H, Ransohoff DF, Appelman HD, Fenoglio CM, Haggitt RC, et al. Dysplasia in inflammatory bowel disease: standardized classification with provisional clinical applications. Hum Pathol 1983;14:931–86.

[10] Cawley HM, Meltzer SJ, De Benedetti VM, Hollstein MC, Muehlbauer K, Liang L, et al. Anti-p53 antibodies in patients with Barrett's esophagus or esophageal carcinoma can predate cancer diagnosis. Gastroenterology 1998;115:19–27.

[11] Macdonald CE, Wicks AC, Playford RJ. Final results from 10 year cohort of patients undergoing surveillance for Barrett's oesophagus: observational study. Br Med J 2000;321:1252–5.

[12] Vo-Dinh T, Panjehpour M, Overholt BF. Laser-induced fluorescence for esophageal cancer and dysplasia diagnosis. Ann N Y Acad Sci 1998;838:116–22.

[13] von Holstein CS, Nilsson AM, Andersson-Engels S, Willén R, Walther B, Svanberg K. Detection of adenocarcinoma in Barrett's oesophagus by means of laser induced fluorescence. Gut 1996;39:711–16.

[14] Stepp H, Sroka R, Baumgartner R. Fluorescence endoscopy of gastrointestinal diseases: basic principles, techniques, and clinical experience. Endoscopy 1998;30:379–86.

[15] Messmann H, Knüchel R, Bäumler W, Holstege A, Schölmerich J. Endoscopic fluorescence detection of dysplasia in patients with Barrett's esophagus, ulcerative colitis, or adenomatous polyps after 5 aminolevulinic acid-induced protoporphyrin IX sensitization. Gastrointest Endosc 1999;49:97–101.

[16] Braichotte DR, Wagnières GA, Bays R, Monnier P, van den Bergh HE. Clinical pharmacokinetic studies of photofrin by fluorescence spectroscopy in the oral cavity, the esophagus, and the bronchi. Cancer 1995;75:2768–78.

[17] Georgakoudi I, Jacobson BC, Van Dam J, Backman V, Wallace MB, Muller MG, et al. Fluorescence, reflectance and light scattering spectroscopies for evaluating dysplasia in patients with Barrett's esophagus. Gastroentorolgy 2001;120:1620–9.

[18] Bouma BE, Tearney GJ, Compton CC, Nishioka NS. High-resolution imaging of the human esophagus and stomach in vivo using optical coherence tomography. Gastrointest Endosc 2000;51:467–74.

[19] Sivak Jr MV, Kobayashi K, Izatt JA, Rollins AM, Ung-Runyawee R, Chak A, et al. High-resolution endoscopic imaging of the GI tract using optical coherence tomography. Gastrointest Endosc 2000;51:474–9.

[20] Izatt JA, Kulkarni MD, Wang HW, Kobayashi K, Sivak MV. Optical coherence tomography and microscopy in gastrointestinal tissues. IEEE J Select Topics Quant Electr 1996;2:1017–28.

[21] Tearney GJ, Brezinski ME, Southern JF, Bouma BE, Boppart SA, Fujimoto JG. Optical biopsy in human gastrointestinal tissue using coherence tomography. Am J Gastroenterol 1997;92:1800–4.

[22] Johanns W, Luis W, Janssen J, Kahl S, Greiner L. Argon plasma coagulation (APC) in gastroenterology: experimental and clinical experiences. Eur J Gastroenterol Hepatol 1997;9:581–7.

[23] Kobayashi K, Izatt JA, Kulkarni MD, Willis J, Sivak Jr. MV. High-resolution cross-sectional imaging of the gastrointestinal tract using optical coherence tomography: preliminary results. Gastrointest Endosc 1998;47:515–23.

[24] Suter MJ, Vakoc BJ, Yachimski PS, Shishkov M, Lauwers GY, Mino-Kenudson M, et al. Comprehensive microscopy of the esophagus in human patients with optical frequency domain imaging. Gastrointest Endosc 2008;68:745–53.

[25] Chen Y, Aguirre AD, Hsiung PL, Desai S, Herz PR, Pedrosa M, et al. Ultrahigh resolution optical coherence tomography of barrett's esophagus: preliminary descriptive clinical study correlating images with histology. Endoscopy 2007;599–605.

[26] Pyhtila JW, Chalut KJ, Boyer JD, Keener J, D'Amico T, Gottfried M, et al. In situ detection of nuclear atypia in Barrett's esophagus by using angle-resolved low-coherence interferometry. Gastrointest Endosc 2007;65:487—91.

[27] Pyhtila JW, Graf RN, Wax A. Determining nuclear morphology using an improved angle-resolved low coherence interferometry system. Opt Express 2003;11:3473—84.

[28] Amoozegar C, Giacomelli MG, Keener GD, Chalut KJ, Wax A. Appl Optics 2009;48:D20—5.

[29] Mourant JR, Bigio IJ, Boyer JD, Johnson TM, Lacey J, Bohorfoush Iii AG, et al. Elastic scattering spectroscopy as a diagnostic tool for differentiating pathologies in the gastrointestinal tract: preliminary testing. J Biomed Opt 1996;1:192—9.

[30] Dhar A, Johnson KS, Novelli MR, Bown SG, Bigio IJ, Lovat LB, et al. Elastic scattering spectroscopy for the diagnosis of colonic lesions: initial results of a novel optical biopsy technique. Gastrointest Endosc 2006;63 (2):257—61.

[31] Mourant JR, Bigio IJ, Boyer J, Conn RL, Johnson T, Shimada T. Spectroscopic diagnosis of bladder cancer with elastic light scattering. Lasers Surg Med 1995;17:350—7.

[32] Lovat LB, Johnson K, Mackenzie GD, Clark BR, Novelli MR, Davies S, et al. Elastic scattering spectroscopy accurately detects high grade dysplasia and cancer in Barrett's oesophagus. Gut 2006;55:1078—83.

[33] Perelman LT, Backman V, Wallace M, Zonios G, Manoharan R, Nusrat A, et al. Observation of periodic fine structure in reflectance from biological tissue: a new technique for measuring nuclear size distribution. Phys Rev Lett 1998;80:627—30.

[34] Backman V, Wallace MB, Perelman LT, Arendt JT, Gurjar R, Müller MG, et al. Detection of preinvasive cancer cells: early-warning changes in precancerous epithelial cells can now be spotted in situ. Nature 2000;406(6791):35—6.

[35] Gurjar RS, Backman V, Perelman LT, Georgakoudi I, Badizadegan K, Itzkan I, et al. Imaging human epithelial properties with polarized light scattering spectroscopy. Nat Med 2001;7:1245—8.

[36] Qiu L, Pleskow D, Chuttani R, Vitkin E, Leyden J, Ozden N, et al. Multispectral scanning during endoscopy guides biopsy of dysplasia in Barrett's esophagus. Nat Med 2010;16:603—66.

[37] van de Hulst HC. Light scattering by small particles. New York: Wiley; 1957.

[38] Perelman LT, Backman V. Light scattering spectroscopy of epithelial tissues: principles and applications. In: Tuchin V, editor. Handbook of Optical Biomedical Diagnostics. Bellingham: SPIE Press; 2002.

[39] Beauvoit B, Kitai T, Chance B. Contribution of the mitochondrial compartment to the optical properties of the rat-liver—a theoretical and practical approach. Biophys J 1994;67:2501—10.

[40] Beuthan J, Minet O, Helfmann J, Herrig M, Muller G. The spatial variation of the refractive index in biological cells. Phys Med Biol 1996;41:369—82.

[41] Sloot PMA, Hoekstra AG, Figdor CG. Osmotic response of lymphocytes measured by means of forward light-scattering—theoretical considerations. Cytometry 1988;9:636—41.

[42] Newton RG. Scattering theory of waves and particles. New York: McGraw Hill; 1969.

[43] Cotran RS, Robbins SL, Kumar V. Robbins pathological basis of disease. Philadelphia: W.B. Saunders Company; 1994.

[44] Zonios G, Perelman LT, Backman V, Manoharan R, Fitzmaurice M, Van Dam J, et al. Diffuse reflectance spectroscopy of human adenomatous colon polyps in vivo. Appl Opt 1999;38:6628—37.

[45] Backman V, Gurjar R, Badizadegan K, Itzkan I, Dasari RR, Perelman LT, et al. Polarized light scattering spectroscopy for quantitative measurement of epithelial cellular structures in situ. IEEE J Sel Top Quant Elect 1999;5:1019—27.

[46] Sokolov K, Drezek R, Gossage K, Richards-Kortum R. Reflectance spectroscopy with polarized light: is it sensitive to cellular and nuclear morphology. Opt Expr 1999;5:302—17.

[47] Wax A, Pyhtila JW, Graf RN, Nines R, Boone CW, Dasari RR, et al. Prospective grading of neoplastic change in rat esophagus epithelium using angle-resolved low-coherence interferometry. J Biomed Opt 2005;10:051604.

[48] Fang H, Ollero M, Vitkin E, Kimerer LM, Cipolloni PB, Zaman MM, et al. Noninvasive sizing of subcellular organelles with light scattering spectroscopy. IEEE J Sel Top Quant Elect 2003;9:267—76.

[49] Craig IJD, Brown JC. Inverse problems in astronomy: a guide to inversion strategies for remotely sensed data. USA: CRC Press; 1986.

Enhanced Imaging of the Esophagus: Optical Coherence Tomography

Michalina J. Gora[1,2] *and Guillermo J. Tearney*[1,3,4]

[1]Wellman Center for Photomedicine, Massachusetts General Hospital, Boston, MA, United States
[2]ICube Laboratory, Centre National de la Recherche Scientifique, University of Strasbourg, Strasbourg, France [3]Department of Pathology, Massachusetts General Hospital, Boston, MA, United States [4]Harvard-MIT Division of Health Sciences and Technology, Boston, MA, United States

8.1 INTRODUCTION

Optical coherence tomography (OCT) is an optical analog of ultrasound imaging that was invented in the early 1990s [1]. OCT fills a resolution gap between confocal laser endomicroscopy (CLE) and ultrasound, affording the capability of visualizing architectural microscopic morphology at a resolution of approximately $10\,\mu m$ and to a depth of several millimeters (mm). OCT is also advantageous in that it provides cross-sectional images that are similar to that of histopathology viewed under low power magnification. Similar to ultrasound, OCT measures the time delay of optical echoes backscattered from structures within tissue, capturing microstructural data as a function of depth within tissue.

8.2 PRINCIPLES OF OPTICAL COHERENCE TOMOGRAPHY

OCT uses near-infrared (NIR) light that cannot be seen by eye. The time delay of NIR light backscattered from tissue is measured using an optical method called interferometry (Fig. 8.1). After emanating from the light source, it is split so that some of it travels to the patient in the so-called sample arm through an imaging probe. The remaining light travels roughly the same path length in a reference arm where it is redirected back toward the system. Sample arm light returned from the tissue is combined with the reference arm light and detected by one or more detectors. If the distances that the light in the sample and reference arms have traveled are approximately equivalent, high and low intensities are observed by the

D. Pleskow & T. Erim (Eds): Barrett's Esophagus.
DOI: http://dx.doi.org/10.1016/B978-0-12-802511-6.00008-9

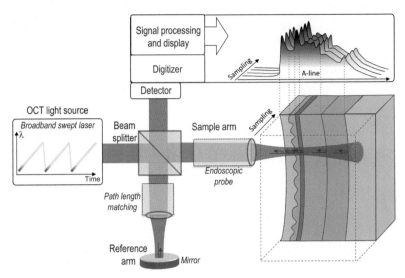

FIGURE 8.1 Schematic representation of endoscopic OCT system with a catheter placed in the sample arm. This particular OCT configuration is an SS-OCT or OFDI system, which is currently the most commonly used method for esophageal OCT imaging.

detector, a pattern known as interference. The interference pattern is analyzed by electronics inside the OCT console to measure tissue reflectance as a function of delay time or depth that the sample arm light has propagated within the tissue. This depth-dependent reflectance profile is commonly known as an A-line or A-scan. Cross-sectional images are acquired by recording A-lines as the sample arm beam is scanned across the sample by moving the optics in the probe.

There are multiple types of OCT systems that have been used for imaging the esophagus, the first-generation time-domain OCT (TD-OCT) and more recent second-generation Fourier-domain OCT (FD-OCT) systems, also known as spectral-domain OCT (SD-OCT) [2,3], or swept-source OCT (SS-OCT) [4], which has also been termed optical frequency-domain imaging (OFDI) [5]. The primary functional difference between TD-OCT and FD-OCT systems is that FD-OCT systems obtain images at a much faster rate than TD-OCT systems. This speed

advantage enables three-dimensional microscopic imaging of large regions of the esophagus, a capability that has been recently termed volumetric laser endomicroscopy (VLE) [6].

8.2.1 Optical Coherence Tomography Parameters

OCT produces cross-sectional images in depth and along a lateral dimension. As a result, there are two axes of resolution that govern the quality of the OCT image. The first, termed axial resolution, is measured along the depth axis and is proportional to the spectral bandwidth of the source. For typical OCT systems with bandwidths around 100–140 nm, the axial resolution is approximately 5–10 μm in tissue. Along the scanning trajectory that is generally perpendicular to the axis of the light beam, the lateral or transverse resolution is defined by the focusing power of the lens in the sample arm probe. Typical values for the lateral resolution are 30–50 μm.

Ranging depth is defined as the axial (depth) distance over which the image is acquired. Typical ranging depths for today's OCT systems are around 6 mm in air or 4.5 mm in tissue. Penetration depth is defined as the distance inside tissue over which an OCT signal can be obtained. This distance depends on the amount of attenuation that the light undergoes as it propagates through the esophageal wall. Generally, the penetration depth in esophageal tissues ranges from 1 to 3 mm, but the penetration depth may be smaller for example in invasive cancer, gastric cardia, or esophageal cardiac metaplasia.

Contrast in OCT comes from backscattering of different cellular and extracellular tissue constituents. The strength of the backscattering and therefore the OCT signal is based on a property of the cellular components called refractive index that governs the speed of light propagation in a medium. When refractive index differences are encountered in tissue, this interface scatters light; the greater the refractive index difference, the greater the scattering. Another factor that governs backscattering strength and thus the OCT signal is the size and concentration of scatterers. Larger scatters and higher concentrations of scatterers provide a higher OCT signal than smaller and lower concentrations of scatterers. For instance, since nuclei are large relative to other organelles and have a relatively high refractive index gradient [7], large nuclei such as those encountered in high-grade dysplasia (HGD), typically can give rise to a high OCT signal.

The speed of the OCT system is defined by the A-line rate, or the time required to acquire a single depth resolve backscattering profile. Depending on the A-line rate, the scanning mechanism, the scanning pattern, and the required sampling density, various two-dimensional frame rates and three-dimensional volume rates can be achieved. Many recent endoscopic OCT systems perform high-density radial scanning that provide circumferential cross-sectional images of the entire esophageal wall that are displayed in real time. These systems sample one or two times per lateral resolution element along the circumference, so for example for a 20 mm diameter balloon-centering catheter in esophagus, the number of A-lines required for a single cross-sectional image may be between 2000 and 4000. For VLE, a helically scanning method is used where the cross-sectional images are acquired as the rotating optics in the probe are pulled back. Typical VLE scans of the esophagus can span 6 cm, acquiring three-dimensional OCT datasets in 30–60 seconds [6,8]. More rapid scans used for positioning the probe in VLE, so called scout scans reduce this time significantly by increasing the pullback rate.

8.3 HISTORY OF OPTICAL COHERENCE TOMOGRAPHY IN THE ESOPHAGUS

8.3.1 First-Generation Esophageal Time-Domain Optical Coherence Tomography

The first OCT images of gastrointestinal (GI) tract tissues ex vivo were obtained in late 1990s, demonstrating the potential of this technology to visualize the microscopic architectural morphology of GI tract mucosa [9,10]. First internal organ imaging in vivo followed the development of a completely new concept for a fiber-based imaging probe with miniaturized focusing optics [11]. The first endoscopic OCT catheter was developed and tested in in vivo experiments in rabbits at the Massachusetts Institute of Technology (MIT) [12]. The catheter comprised a 2 mm transparent sheath capable of being inserted into a working channel of an endoscope. The sheath enclosed an optical probe that contained an optical fiber surrounded by a cable or driveshaft. The optical fiber was terminated at the distal tip by a lens and a prism for focusing and redirecting the light perpendicular to the axis of the probe [11]. Images could be obtained by rotating the driveshaft while recording sequential OCT A-lines. Images of the rabbit esophagus and trachea in this study showed remarkable detail in vivo, paving the way for future developments and human application of endoscopic OCT. The first human results with endoscopic OCT were obtained by Sergeev et al. [13] using another type of the probe that imaged in the forward direction over a field of approximately 2 mm. In this study, images were obtained in normal subjects and patients with esophageal disorders. Side-scanning probes with larger fields of view, developed by other laboratories, were subsequently demonstrated in human esophagus in vivo and reported a few years later [14,15].

Following these first in vivo human studies using endoscopic OCT, the technology entered a second phase of development focused on interpretation of OCT images obtained in human tissues both ex vivo and in vivo in comparison to histopathological diagnosis [16−20]. Performance of endoscopic OCT was also compared to a high-frequency endoscopic ultrasound (EUS) probe showing that the resolution of OCT was much higher than that of EUS [21].

In early 2000s, the first OCT image feature criteria for Barrett's esophagus (BE) and dysplasia were published [22−28] based on co-registered endoscopic OCT images and biopsy samples obtained from patients at different stages of the disease. Criteria were developed using training sets and the sensitivity and specificity of the developed criteria were tested prospectively on new images using the histopathological diagnoses as the gold standard. Using criteria (see the paragraph entitled: "Optical coherence tomography images of the esophagus" for a description of the criteria) derived from TD-OCT images, sensitivities from 81% to 97% and specificities from 57% to 92% were determined for diagnosing BE

(defined as specialized intestinal metaplasia (SIM)). Sensitivities from 54% to 83% and specificities from 72% to 75% were obtained for detection of high-grade dysplasia and intramucosal carcinoma. The large spread of results can in part be explained by the use of first-generation TD-OCT technology and research-grade prototypes, which did not provide images that have the high quality seen from second-generation systems, and by incorrect co-registration of biopsy samples caused by the need to sequentially image, remove the imaging probe, reinsert the biopsy forceps, and then biopsy the same location in the esophagus.

Along with the progress on understanding the diagnostic performance and the clinical potential of OCT for the diagnosis of BE, many research groups were focused on technical development of OCT systems to further improve quality of images. At that time, the majority of studies with TD-OCT were performed with semiconductor light sources providing axial resolution of $10-15\,\mu$m. Larger bandwidth sources based on femtosecond lasers were developed, making it possible to image with a much higher axial resolution [29,30]. In 2007, Chen et al. [31] published a study comparing standard resolution endoscopic OCT with an ultrahigh-resolution OCT system that used a femtosecond laser. While sensitivity and specificity were not reported, this group showed that the higher resolution provided better quality images and improved the visualization of finer architectural details. Because of the size and cost of these more complex lasers, most clinical endoscopic OCT systems still use semiconductor light sources.

8.3.2 Second-Generation Esophageal Fourier-Domain Optical Coherence Tomography

In comparison to the standard of care in the GI tract, OCT provides optical biopsy without need of tissue resection permitting better sampling of the organ. However, data acquisition speed of the TD-OCT systems was not sufficient for inspection of larger areas; therefore, this first-generation OCT technology was primarily limited to point sampling. Introduction of FD-OCT significantly increased OCT imaging speed and opened up new possibilities for imaging the entire distal esophagus. In order to image the entire esophageal circumference, the optical probe had to be positioned near the center of the lumen in order to keep the tissue in focus. In 2006, the OCT group at Massachusetts General Hospital [32] developed a balloon-centering probe to assure correct location of the tissue with respect to the probe [8,33] and, using a driveshaft for rotation and translation of the optics, conducted helical scanning to capture three-dimensional FD-OCT images over a 6 cm length of esophagus. New OFDI/SS-OCT technology, including a high-speed wavelength swept laser [5], was developed to enable high A-line rates (up to 54 kHz) and decrease acquisition time further [32,33]. This technique was called comprehensive volumetric microscopy [32,33], which subsequently has become synonymous with VLE [6]. Balloon-catheter OFDI was then conducted in patients undergoing endoscopy, demonstrating that this technology was capable of providing high-quality, three-dimensional OCT images of the entire distal esophagus in living patients [8]. A portion of this dataset was used by Sauk et al. [34] to test intraobserver agreement for the diagnosis of BE with VLE. Ten readers were first trained using BE diagnostic criteria [22,24] previously developed in TD-OCT studies and then asked to score another set of images. An excellent agreement for differentiation of intestinal metaplasia versus non-BE was achieved with a κ value of 0.811 (95% confidence interval (CI) 0.73−0.89; $p < 0.0001$) [34]. Subsequently, in 2011, NinePoint Medical

commercialized an OCT VLE system and a balloon-centering catheter that can be inserted into the accessory port of the endoscope. Following FDA approval in 2012, the device has been utilized in hundreds of patients to date. Most recently, a report by Wolfsen et al. [6] from multicenter experience with 100 patients has been published showing that VLE is safe and feasible for introduction into a clinical practice.

During the development of OCT VLE, a question was raised as to whether or not the pressure of the balloon on the esophageal wall influenced the appearance of tissue architecture seen by OCT. This question was investigated by Kang et al. [35] in 2010, who developed a double balloon catheter, where images were obtained either in between two balloons or through one of the balloons. In a preliminary in vivo study in swine esophagus, it was found that topographic information from mucosal surface was sometimes lost due to balloon compression. It was also noted that the presence of the balloon decreased visibility of certain internal structures like vessels but increased imaging penetration depth so that the muscularis propria could be more easily seen [35].

In parallel to the development of the balloon-centering catheters, other groups continued to develop and utilize small diameter, radially scanning probes that are inserted in the endoscope accessory port [36] and placed in close contact with the esophageal wall prior to imaging. These probes provide partial coverage of the esophageal circumference in a single longitudinal pullback; greater coverage can be achieved by repositioning the probe using the endoscope. While a smaller tissue region is imaged using these devices, the much shorter working distance of these probes facilitates image generation with higher transverse resolution providing very high-quality images of various esophageal structures and diseases [37−39].

8.4 OPTICAL COHERENCE TOMOGRAPHY IMAGES OF THE ESOPHAGUS

8.4.1 Normal Esophageal Squamous Mucosa

Five anatomic layers of the normal esophagus can be easily distinguished in OCT images (Fig. 8.2a,e): the epithelium, lamina propria, muscularis mucosa, submucosa, and muscularis propria. The intensity of the OCT signal alternates between the layers due to difference in their optical properties, which allows for clear delineation of the stratified morphology. Of note, the muscularis mucosa are distinct in images of squamous mucosa obtained with endoscopic probes used in TD-OCT (Fig. 8.2a)

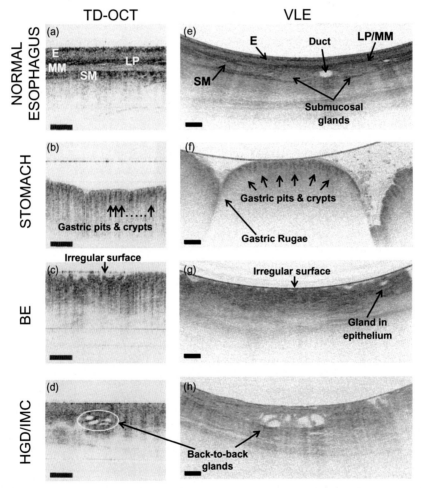

FIGURE 8.2   OCT images from a time domain OCT system [22] in the left column and from a VLE system in the right column (*courtesy of Nine Point Medical*) representing normal esophagus (a, e), normal stomach (b, f), Barrett's esophagus (c, g), and high-grade dysplasia (HGD)/intramucosal carcinoma (IMC) (d, h). *E*, epithelium; *LP*, lamina propria; *MM*, muscularis mucosa; *SM*, submucosa. Scale bars 0.5 mm.

whereas oftentimes, it is merged with the lamina propria in images acquired with balloon-centering VLE catheters. Additional structures like submucosal glands, mucin ducts, or vessels can be also found in OCT images.

8.4.2 Stomach

When VLE technology was introduced, it enabled volumetric imaging of the whole distal esophagus including the folds of the gastric cardia. It is a common approach for diagnosing BE in the volumetric OCT dataset to first localize cardia as a distal landmark and then inspect consecutive, more proximal images for presence of specialized intestinal metaplasia. Normal gastric tissue has a very typical appearance in OCT images (Fig. 8.2b,f) with the following features: (1) fine, superficial vertical structures of alternating low and high OCT signal corresponding to gastric pits, (2) much lower penetration depth, and (3) gastric rugae.

8.4.3 Barrett's Esophagus

Even though conventional OCT does not provide high enough resolution to resolve goblet cells, studies have shown that this technology is capable of differentiating SIM [22] from other upper GI tissue types. In a study by Poneros et al. [22], 166 OCT images were compared to histology of co-registered biopsy specimens. Using this data, it was determined that at least two of the following characteristic features had to be present for differentiating SIM (Fig. 8.2c,g) from normal squamous epithelium and normal gastric mucosa: (1) lack of layered morphology of normal squamous tissue and absence of "pit and crypt" structures of normal gastric mucosa, (2) heterogeneous, disorganized intensity of the OCT signal within the tissue and irregular surface topography, and (3) presence of glands below the epithelial surface. The lack of normal squamous or gastric morphology and

heterogeneous tissue backscattering were published as the most important criteria. In another study published by Sauk et al. [34], heterogeneity and irregular surface topography most frequently lead 10 readers (5 gastroenterologist, 1 pathologist, 4 OCT experts) to render a diagnosis of BE.

8.4.4 Dysplasia in Barrett's Esophagus

The most comprehensive study on identifying dysplasia in BE (Fig. 8.2d,h) has been published by Evans et al. [26,40]. A scoring system for dysplasia progression was developed and tested in 177 biopsy-correlated TD-OCT images. For each OCT image, a dysplasia index was calculated based on a sum of scores for the following two features: (1) surface maturation and (2) gland architecture. Poor surface maturation is a histological feature related to an increase in the nuclear to cytoplasmic ratio of the surface epithelium. Because refractive indices of the cytoplasm and chromatin are significantly different, an increase in nuclear size and density increases the intensity of the OCT signal at the surface of the tissue. If the OCT signal from the surface was weaker than the signal from the subsurface, it had a score 0; if equivalent, the score was 1; and if the surface OCT signal was stronger, then the score was 2. A similar scoring was defined for appearance of glands in the OCT image: a normal glandular architecture with typical linear structures with alternating intensity of OCT signal and minimal number of smooth dilated glands/ducts was given a score of 0; more irregular glandular architecture and an increased number of dilated glands/ducts had a score of 1; and high glandular irregularity with glands that were back-to-back or had highly asymmetric shapes or contained debris were given a score of 2. A summed dysplasia index that was equal or larger than 2 was found to be 83.3% sensitive (95% CI 70—93) and 75.0% specific (95% CI 68—84) for the diagnosis of high-grade dysplasia or intramucosal carcinoma.

Since FD-OCT and TD-OCT images are similar in appearance, it is probable that a form of these TD-OCT criteria will apply to VLE data but this remains to be demonstrated in the literature in vivo, as validation of VLE has only been conducted ex vivo to date. Leggett et al. [41] investigated capabilities of VLE for detection of dysplasia associated with BE. In this study, endoscopic mucosal resection (EMR) specimens from 27 patients with BE were used to determine the sensitivity and specificity of VLE compared to histology. The findings of this study showed a sensitivity of 86% (95% CI 69–96), specificity of 88% (95% CI 60–99) and an overall diagnostic accuracy of 87% (95% CI 86–88) [41]. In another study, difficulties with one-to-one correlation between VLE and histology using EMR specimens was addressed by Swager et al. [42] where various combinations of marker placement and tissue block sectioning were investigated for improved correlation.

8.4.5 Response to Esophageal Ablation

Following esophageal ablation, squamous epithelium can be found overlying Barrett's epithelium [43], which may be of concern if this residual Barrett's has neoplastic potential. OCT has considerable advantages over video endoscopy with respect to the assessment of subsquamous BE as it provides microscopic information deep into the esophageal wall (Fig. 8.3). OCT images of buried Barrett's were first reported in vivo in a case study by Adler et al. [44]. Application of OCT for detection of subsquamous BE was further investigated by Cobb et al. [45] where esophagectomy specimens were collected from 14 patients and immediately imaged with an ultrahigh-resolution benchtop OCT system before histological processing. Co-registered OCT and histology data showed subsquamous BE that

appeared by OCT as glands underneath neosquamous epithelium. In this study, it was noted that it could be difficult to differentiate Barrett's glands that appeared to have a thick double-walled appearance from vessels with thinner walls [45]. Later, Zhou et al. [37] performed a study where OCT volumetric data was obtained from 18 patients before complete eradication of intestinal metaplasia and 16 patients post-eradication. In their analysis of OCT cross sections, buried glands with irregular size and shape and sparse distribution were differentiated from tube-like vessels and from well-organized esophageal glands below the lamina propria/muscularis mucosa. A total number of 620 buried glands were found in 72% of patients (13/18) before complete eradication of intestinal metaplasia, this number decreased to 114 buried glands found in 63% of patients (10/16) after the procedure; however, no significant change in size or distribution was found. However, due to the lack of histological correlation, it was impossible to definitively confirm that the structures identified were buried BE glands. In the most recent study with a VLE system published by Swager et al. [46], many post-ablation buried Barrett's regions identified with the VLE corresponded to normal histological structures like dilated glands and blood vessels. In this study, the VLE system was first used to identify areas of buried Barrett's during EGD procedure (found in 13 out of 17 patients), and in a second step, the tissue was imaged ex vivo after the area was excised. Only one area identified by VLE as positive was also found to contain buried Barrett's on corresponding histology (Fig. 8.3). While these studies have confirmed that OCT can be used to visualize subsquamous microscopic structures, further research is necessary to develop criteria specific to subsquamous Barrett's that can be useful for surveillance of post-ablation patients.

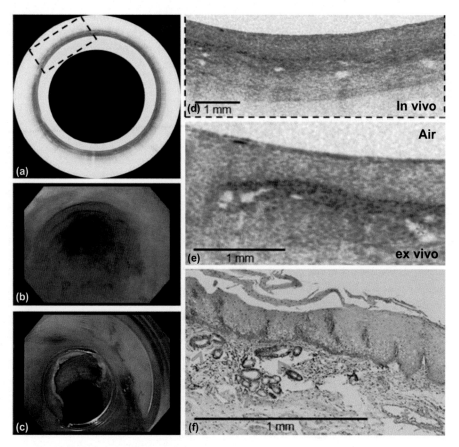

FIGURE 8.3   In vivo volumetric laser endomicroscopy (VLE) scan (a) showing the entire esophageal cross section. Narrow-band imaging showing no residual Barrett's within the neosquamous epithelium (b) and white light endoscopy of the site after endoscopic resection (c). Buried Barrett's glands seen by VLE in vivo (d), ex vivo VLE (e), and corresponding histology (f). Arrowheads indicate buried Barrett's glands; asterisk indicates a blood vessel [46].

8.5 NEW ADVANCES IN VOLUMETRIC LASER ENDOMICROSCOPY

8.5.1 Optical Coherence Tomography Image Guided Biopsy

In order to utilize OCT information available in real time to improve biopsy yield, a VLE image guided biopsy concept was introduced by Suter et al. [47]. The new platform for guided biopsy was developed with a higher power laser (output power of 400 mW and wavelength of 1450 nm) added to the OCT system for creating superficial cautery marks visible in both white light endoscopy and OCT. Each mark was created by a 2-second pulse of the marking laser light that was delivered to the tissue without any modifications to the optical fiber or focusing optics of the balloon catheter. Safety and feasibility of the new guided biopsy platform was tested in swine in vivo and a high accuracy of 97.07% (95% CI 89.8−99.7) for hitting the target location was demonstrated [47]. In first in human studies with the same system, it was shown that regions of interest seen by VLE could be accurately identified and marked with transverse error (along the circumference) of 1.2±1.3 mm and longitudinal error (along the length of the esophagus) of 0.5±0.9 mm [48]. After optimizing laser marking parameters in the first 12 subjects, the marking procedure was performed in 10 new subjects where 3 target sites were selected in real time during VLE procedure for each subject. All marking sites were clearly visible endoscopically and were successfully biopsied for histopathological diagnosis. For each site collected, OCT data and the endoscopy were reviewed by independent blinded readers and the OCT and endoscopy diagnoses were compared to histopathological analysis of the biopsy sites. Diagnostic accuracies of 67% (95% CI 47−83) for white light endoscopy and 93% (95% CI 78−99) for OCT discrimination of columnar-lined esophagus was measured, demonstrating that OCT image guided biopsy is a feasible concept. In this study, the imaging needed to be stopped and the laser activated to create each mark. Newer superficial laser cautery lasers are being developed that are capable of marking the tissue much faster so that marks can be placed on the esophagus on in real time [49].

8.5.2 En Face Visualization in Optical Coherence Tomography

Three-dimensional or volumetric OCT data collected and stored during the imaging procedure can be reprocessed offline in order to obtain the best visualization of key mucosal features. In addition to the standard cross-sectional view, longitudinal recuts are common and helpful in VLE systems. Another recut of the three-dimensional dataset, the so-called *en face* projection (Fig. 8.4c,b) [37−39,44] has been shown to provide efficient visualization of glandular and vascular patterns. An *en face* two-dimensional image is computed from the OCT volumetric

dataset, where one axis corresponds to the probe scan angle and the other axis to the longitudinal position along the esophageal lumen. Each pixel in the image is generated by summation of a portion of the A-line for every point in the cylindrical scan. Such a representation emphasizes transverse features, for example, the pit and crypt pattern in the cardia or glandular features of BE [37−39]. Additionally, summation can be performed for only selected range of depths at any depth of the tissue to differentiate epithelial structures from those in deeper layers. The quality of *en face* images strongly depends on sampling density in both circumferential and longitudinal directions and the uniformity of the rotation and pullback of the OCT imaging optics [50]. A new endoscopic OCT system developed at MIT, using an ultrahigh-speed wavelength swept vertical cavity surface emitting laser, increased the A-line acquisition rate up to 1-mHz, which significantly enhances sampling density [51]. In order to take the full advantage of the ultrafast acquisition speed, a much faster scanning mechanism than proximally effectuated rotation and pullback of the distal optics was used. Similar to previously published studies [52,53], the MIT group implemented a micromotor design where helical scanning of the optical beam is performed by a micromotor at the distal end of the probe that deflects the light to the side of the catheter. Following preliminary results obtained in rabbits in vivo with the 1 MHz system [51], results from human studies with a 600 kHz system were reported [54,55], showing the capability of this higher speed technology to provide *en face* views of BE mucosa that resemble surface patterns seen with high-magnification NBI (Fig. 8.4). High-quality *en face* endoscopic OCT images of normal and diseased esophagus have shown its potential for improving information content captured by OCT; however, a more systematic study is still needed in order to compare how much the *en face* view improves the accuracy of OCT diagnosis compared to cross-sectional data alone.

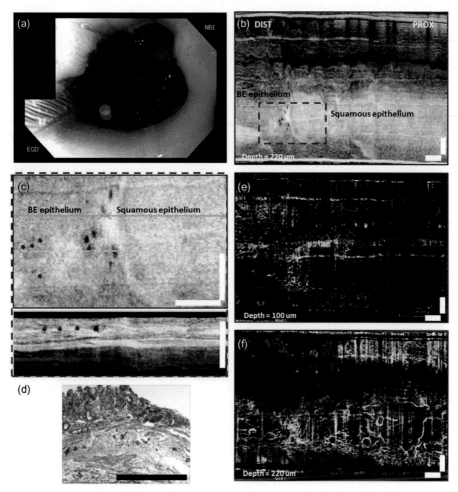

FIGURE 8.4 Endoscopic optical coherence tomography (OCT) and angiography of Barrett's esophagus (BE). (a) Endoscopic view of BE using narrow-band imaging showing dark red and irregular surface features. (b) En face OCT image at 220 mm depth (lamina propria layer). (c) Enlarged en face OCT image and the corresponding cross-sectional OCT image from the dashed red region in (b). (d) Corresponding histology from the imaged region. (e) En face OCT angiogram at 100 mm depth showing surface vasculature in the BE region. (f) En face OCT angiogram at 220 mm depth showing high density of microvasculature along the squamocolumnar junction. *Red arrows*, BE glands; *white bar*, 1 mm. *Source: Caption and figure from Tsai T-H, Ahsen OO, Lee H-C, Liang K, Figueiredo M, Tao YK, et al. Endoscopic optical coherence angiography enables 3-dimensional visualization of subsurface microvasculature. Gastroenterology 2014;147:1219–21.*

8.5.3 Tethered Capsule Endomicroscopy

A number of studies with OCT have shown the clinical potential of this technology for improving diagnostic accuracy and overcoming random sampling errors during surveillance by conducting microscopic imaging over a large field of view in a short period of time. However, all of these endoscopic OCT probe designs required the assistance of endoscopy, making them undesirable for screening for Barrett's via OCT in unsedated patients. In a study published by Gora et al. [56], a new implementation of OCT technology in a form of swallowable tethered capsule was introduced. The tethered capsule endomicroscopy (TCE) device comprised a small, rigid capsule (11 mm × 24 mm) attached to a distal end of a string-like tether. After the capsule was swallowed by an unsedated patient, OCT images were collected while the capsule traversed down the esophagus via peristalsis or up when pulled by the tether [57]. The catheter worked in a similar manner to the endoscope where operator can navigate the device inside of the lumen. During a standard imaging procedure that lasted for approximately 5 minutes, the entire esophagus was imaged by OCT several times. After the procedure, the capsule was removed and disinfected for the next use, which brought down the per-utilization cost of the device. A clinical experience with TCE from 26 subjects has recently been published, including 10 BE patients [57]. Only two healthy volunteers were unable to swallow the capsule and 5 patients experienced a weak gag reflex during the procedure. High-quality images covering more than 50% of the circumference were obtained in total of 93.7% of frames in subjects without hiatal hernia and 89% in patients with hiatal hernia. Future technical advances in TCE technology have been reported and demonstrated in swine where an integrated micromotor and linear translation mechanism have been incorporated in the capsule to enable visualization of high-quality *en face* views of the epithelium [58].

8.6 NEW OPTICAL COHERENCE TOMOGRAPHY TECHNOLOGY DEVELOPMENT

Advances in OCT technology and efforts in combining OCT with other diagnostic techniques are focused on providing additional contrast or functional information to increase the amount of clinically relevant information that can be provided by this technology. Because

they are earlier in the development cycle, many of the studies and techniques presented below are still in a preclinical validation or preliminary clinical testing stage.

8.6.1 Vasculature Imaging

Extraction of functional information about vasculature and blood flow from the OCT data was first published for application in the human eye. The first catheter-based Doppler endoscopic OCT system was developed and tested in human by Yang et al. in 2003 [59] where blood flow in very small subsurface vessels was detected. In a follow-on publication, the same endoscopic OCT system was used to investigate differences in microvasculature in normal and pathological GI tissues [60] and to monitor microvascular tissue response during photodynamic therapy in an animal model of BE [61]. In both studies, prominent changes in vasculature of pathological tissues were found, including: (1) absence of larger blood vessels situated in the submucosa and (2) diffuse patterns of small blood vessels in the epithelium. Advances in OCT esophageal vasculature imaging followed using the OFDI form of FD-OCT [2,5], published by Vakoc et al. [33]. The increased acquisition speed of the OFDI system allowed for the generation of a three-dimensional vasculature map [33] showing vessels and capillaries in a swine esophagus, rendered from VLE data that was obtained in vivo. In 2014, Tsai et al. [54] published three-dimensional vasculature maps in normal volunteers and patients with BE (Fig. 8.4e,f). Results showed that endoscopic OCT angiography provides information about subsurface vascular patterns, which may be a marker of disease severity/progression [54] (Fig. 8.5).

8.6.2 Polarization-Sensitive Optical Coherence Tomography

Polarization-sensitive OCT (PS-OCT) is an advanced form of this technology, where

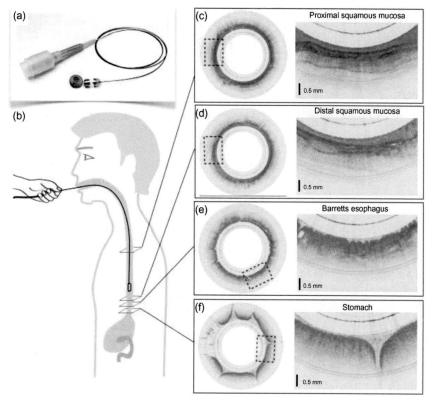

FIGURE 8.5   (a) Tethered capsule endomicroscope photograph and (b) schematic representation of the procedure with corresponding TCE data obtained in a subject with diagnosis of BE representing examples of (c) proximal squamous mucosa, (d) distal squamous mucosa, (e) Barrett's esophagus, and (f) stomach.

birefringence of the tissue, related to collagen and smooth muscle content [62–64], can be measured in addition to standard structural information. In one study published by Wang et al. [65], a simple and robust PS-OCT system was developed, allowing implementation of PS-OCT through an endoscope-compatible probe. In swine esophageal data acquired ex vivo, a significant difference was found between the birefringence of the epithelium and lamina propria/muscularis mucosa. It is possible that such additional contrast can be useful in assessing epithelial thickness, which could be a predictive parameter for determining therapy response in dysplastic BE.

8.6.3 Angle-Resolved Low-Coherence Interferometry

In angle-resolved low-coherence interferometry (A/LCI), one OCT A-line is collected many times at different angles with respect to the incoming light beam. This data is analyzed to resolve the mean size and average density of cell nuclei [66]. A/LCI technology was implemented in a fiber-based catheter for endoscopic surveillance of BE. With the A/LCI probe, nuclear morphology was measured at 172 sites in 46 patients. These sites were subsequently biopsied for histopathological diagnosis. Findings from this study showed that A/LCI measurements could differentiate dysplastic tissue from non-dysplastic tissue with an accuracy of 86% [67]. Interestingly, this study found that only the size of nuclei in the deepest layer of the epithelium provided significant diagnostic value [67].

8.6.4 Simultaneous Optical Coherence Tomography and Fluorescence Imaging

Combining morphological OCT data with chemical/molecular information about the tissue has been also investigated for further improvement of BE diagnosis. This concept was first successfully tested ex vivo where the tissue was illuminated with an excitation wavelength (eg, 400 nm) and the fluorescence carried either endogenous biochemical information (eg, reduced nicotinamide adenine dinucleotide, collagen, or fluorescent proteins) or signal from exogenous contrast. Because OCT and laser-induced fluorescence (LIF) used different light wavelengths, it was necessary to significantly modify the imaging probe to deliver and collect OCT and LIF light independently. Multifiber configurations such as this can be challenging to implement because with conventional OCT, the inner optical probe rotates in order to effectuate circumferential scanning of the tissue. This problem was overcome by Mavadia et al. [68] where OCT and LIF were integrated in an endoscopy-compatible catheter with a mini motor in the distal end. Similar to previously published approaches [69,70], this group used double-clad fiber in the optical probe. This imaging system, tested in a rabbit's esophagus in vivo, has significantly decreased complexity of the multimodality systems, facilitating clinical translation of this multimodality technology.

8.6.5 Optical Coherence Tomography—Guided Ablation

Since OCT provides detailed information about internal structures of the tissues up to 2 mm, it can be a powerful tool for real-time monitoring of treatment [71]. Moreover, if combined with high-power lasers, the same catheter can be used for monitoring and therapy at the same time. Lasers operating in tissue water absorption picks at the wavelength of 1.44 [72,73] and 1.9 μm [74] are preferred for tissue coagulation and ablation. In 2007, Vakoc et al. [72] presented a method for visualizing the laser ablation therapy process in porcine esophageal specimens ex vivo, by showing that fine changes in the OCT signal were related to temperature and ablation depth. In this study with a continuous-wave laser with an output power of 320 mW at a wavelength of 1450 nm, a thermal damage of approximately 0.7 mm in depth was achieved in 1.2 seconds. In order to increase the coagulation speed for on-the-fly ablation, Baac et al. [73] developed a high-power pulsed Raman fiber laser at a wavelength of 1436 nm where visible thermal damage marks were obtained in porcine esophagus ex vivo with pulses as short as 900 μs for pulse energies greater than approximately 7 mJ [73]. Utilization of short pulse lasers allows for tissue coagulation with much lower energy pulses, because laser energy is delivered faster than the thermal diffusion of the tissue. The pulsed Raman fiber laser was later combined with an OCT system using a double-clad coupler and fiber for simultaneous and co-localized OCT and laser tissue coagulation, reported by Beaudette et al. [49]. Results obtained in a porcine esophagus specimen ex vivo showed the potential of this technology to coagulate tissue on-the-fly with real-time OCT guidance and monitoring [49].

8.6.6 Automated Optical Coherence Tomography Image Analysis

Application of OCT for guiding biopsy or treatment would require clinicians to perform tissue diagnosis in real time during the imaging procedure. Especially for VLE where the amount of OCT image data is large, it would be desirable to develop algorithms that can support clinicians in image interpretation. The most important steps for automatic analysis of the OCT data are image segmentation, removal of artifacts, and extraction of image features related to the condition to be diagnosed. Among others [31,40,75,76], the OCT group at Case Western Reserve University focused on quantitative and automatic analysis of GI tissues for computer-aided diagnosis (CAD) of dysplasia in BE. In the study by Qi et al. [40], a total of 96 correlated OCT images with biopsy were analyzed. The developed CAD system achieved accuracy of 84% for classification of nondysplastic versus dysplastic BE tissue, showing the potential of this technology to support clinicians' diagnoses of esophageal OCT images.

8.7 FUTURE OUTLOOK

In recent years, OCT has received much attention as a new technology that can potentially improve current standard of care for screening and surveillance of BE. Commercialized VLE devices are now being used clinically during the surveillance of BE patients. With the addition of guided biopsy features, this technology can ensure the acquisition of more meaningful biopsies, which should improve the management of patients with BE. In parallel, new technological advancements are in the development pipeline that add functional measurements like blood flow or additional contrast that should improve the diagnostic capacity of OCT further. The development of new, less invasive probes, such as tethered OCT capsules, is another promising new area of OCT technology development. Such devices can be used in unsedated patients and therefore have the potential to be more cost-effective and better tolerated than endoscopy, making them more appropriate for population-based screening. The field overall is still in its early phases; much more work needs to be done validating the accuracy of OCT and VLE for BE and dysplasia diagnosis. In addition, training clinicians to interpret the OCT datasets remains a challenge. Because VLE provides a vast amount of detailed image data, approaches should be investigated for visualizing and interpreting these images in a short period of time during the procedure. Finally, as with all new technologies, demonstrating that OCT improves the outcomes of patients with esophageal diseases will ultimately be required for widespread adoption.

DISCLOSURES

Massachusetts General Hospital has a licensing arrangement with NinePoint Medical. Drs Gora and Tearney can receive royalties from this licensing arrangement.

References

[1] Huang D, Swanson EA, Lin CP, Schuman JS, Stinson WG, Chang W, et al. Optical coherence tomography. Science 1991;254:1178–81.

[2] Leitgeb R, Hitzenberger C, Fercher A. Performance of Fourier domain vs. time domain optical coherence tomography. Opt Express 2003;11:889–94.

[3] De Boer JF, Cense B, Park BH, Pierce MC, Tearney GJ, Bouma BE. Improved signal-to-noise ratio in spectral-domain compared with time-domain optical coherence tomography. Opt Lett 2003;28:2067–9.

[4] Choma M, Sarunic M, Yang C, Izatt J. Sensitivity advantage of swept source and Fourier domain optical coherence tomography. Opt Express 2003;11: 2183–9.

[5] Yun S, Tearney G, de Boer J, Iftimia N, Bouma B. High-speed optical frequency-domain imaging. Opt Express 2003;11:2953–63.

[6] Wolfsen HC, Sharma P, Wallace MB, Leggett C, Tearney G, Wang KK. Safety and feasibility of volumetric laser endomicroscopy in patients with Barrett's esophagus (with videos). Gastrointest Endosc 2015.

[7] Drezek R, Guillaud M, Collier T, Boiko I, Malpica A, Macaulay C, et al. Light scattering from cervical cells throughout neoplastic progression: influence of nuclear morphology, DNA content, and chromatin texture. J Biomed Opt 2003;8:7−16.

[8] Suter MJ, Vakoc BJ, Yachimski PS, Shishkov M, Lauwers GY, Mino-Kenudson M, et al. Comprehensive microscopy of the esophagus in human patients with optical frequency domain imaging. Gastrointest Endosc 2008;68:745−53.

[9] Izatt J, Kulkarni MD, Wang H-W, Kobayashi K, Sivak Jr MV. Optical coherence tomography and microscopy in gastrointestinal tissues. Sel Top Quantum Electron IEEE J 1996;2:1017−28.

[10] Tearney GJ, Brezinski ME, Southern JF, Bouma BE, Boppart SA, Fujimoto JG. Optical biopsy in human gastrointestinal tissue using optical coherence tomography. Am J Gastroenterol 1997;92:1800−4.

[11] Tearney G, Brezinski M, Fujimoto J, Weissman N, Boppart S, Bouma B, et al. Scanning single-mode fiber optic catheter−endoscope for optical coherence tomography. Opt Lett 1996;21:543−5.

[12] Tearney GJ, Brezinski ME, Bouma BE, Boppart SA, Pitris C, Southern JF, et al. In vivo endoscopic optical biopsy with optical coherence tomography. Science 1997;276:2037−9.

[13] Sergeev A, Gelikonov V, Gelikonov G, Feldchtein F, Kuranov R, Gladkova N, et al. In vivo endoscopic OCT imaging of precancer and cancer states of human mucosa. Opt Express 1997;1:432−40.

[14] Rollins AM, Ung-Arunyawee R, Chak A, Wong RC, Kobayashi K, Sivak MV, et al. Real-time in vivo imaging of human gastrointestinal ultrastructure by use of endoscopic optical coherence tomography with a novel efficient interferometer design. Opt Lett 1999;24:1358−60.

[15] Li XD, Boppart SA, Van Dam J, Mashimo H, Mutinga M, Drexler W, et al. Optical coherence tomography: advanced technology for the endoscopic imaging of Barrett's esophagus. Endoscopy 2000;32:921−30.

[16] Kobayashi K, Izatt JA, Kulkarni MD, Willis J, Sivak MV. High-resolution cross-sectional imaging of the gastrointestinal tract using optical coherence tomography: preliminary results. Gastrointest Endosc 1998; 47:515−23.

[17] Pitris C, Jesser C, Boppart SA, Stamper D, Brezinski ME, Fujimoto JG. Feasibility of optical coherence tomography for high-resolution imaging of human gastrointestinal tract malignancies. J Gastroenterol 2000;35:87−92.

[18] Bouma BE, Tearney GJ, Compton CC, Nishioka NS. High-resolution imaging of the human esophagus and stomach in vivo using optical coherence tomography. Gastrointest Endosc 2000;51:467−74.

[19] Jackle S, Gladkova N, Feldchtein F, Terentieva A, Brand B, Gelikonov G, et al. In vivo endoscopic optical coherence tomography of esophagitis, Barrett's esophagus, and adenocarcinoma of the esophagus. Endoscopy 2000;32:750−5.

[20] Sivak MV, Kobayashi K, Izatt JA, Rollins AM, Ung-Runyawee R, Chak A, et al. High-resolution endoscopic imaging of the GI tract using optical coherence tomography. Gastrointest Endosc 2000;51:474−9.

[21] Das A, Sivak MV, Chak A, Wong RC, Westphal V, Rollins AM, et al. High-resolution endoscopic imaging of the GI tract: a comparative study of optical coherence tomography versus high-frequency catheter probe EUS. Gastrointest Endosc 2001;54:219−24.

[22] Poneros JM, Brand S, Bouma BE, Tearney GJ, Compton CC, Nishioka NS. Diagnosis of specialized intestinal metaplasia by optical coherence tomography. Gastroenterology 2001;120:7−12.

[23] Zuccaro G, Gladkova N, Vargo J, Feldchtein F, Zagaynova E, Conwell D, et al. Optical coherence tomography of the esophagus and proximal stomach in health and disease. Am J Gastroenterol 2001;96:2633−9.

[24] Poneros JM, Nishioka NS. Diagnosis of Barrett's esophagus using optical coherence tomography. Gastrointest Endosc Clin N Am 2003;13:309−23.

[25] Isenberg G, Sivak MV, Chak A, Wong RC, Willis JE, Wolf B, et al. Accuracy of endoscopic optical coherence tomography in the detection of dysplasia in Barrett's esophagus: a prospective, double-blinded study. Gastrointest Endosc 2005;62:825−31.

[26] Evans JA, Poneros JM, Bouma BE, Bressner J, Halpern EF, Shishkov M, et al. Optical coherence tomography to identify intramucosal carcinoma and high-grade dysplasia in Barrett's esophagus. Clin Gastroenterol Hepatol 2006;4:38−43.

[27] Evans JA, Bouma BE, Bressner J, Shishkov M, Lauwers GY, Mino-Kenudson M, et al. Identifying intestinal metaplasia at the squamocolumnar junction by using optical coherence tomography. Gastrointest Endosc 2007;65:50−6.

[28] Evans JA, Nishioka NS. The use of optical coherence tomography in screening and surveillance of Barrett's esophagus. Clin Gastroenterol Hepatol 2005;3:S8−11.

[29] Bouma B, Brezinski M, Fujimoto J, Tearney G, Boppart S, Hee M. High-resolution optical coherence tomographic imaging using a mode-locked Ti: Al$_2$O$_3$ laser source. Opt Lett 1995;20:1486−8.

[30] Drexler W, Morgner U, Kärtner F, Pitris C, Boppart S, Li X, et al. In vivo ultrahigh-resolution optical coherence tomography. Opt Lett 1999;24:1221−3.

[31] Chen Y, Aguirre AD, Hsiung PL, Desai S, Herz PR, Pedrosa M, et al. Ultrahigh resolution optical coherence tomography of Barrett's esophagus: preliminary

descriptive clinical study correlating images with histology. Endoscopy 2007;39:599—605.

[32] Yun SH, Tearney GJ, Vakoc BJ, Shishkov M, Oh WY, Desjardins AE, et al. Comprehensive volumetric optical microscopy in vivo. Nat Med 2006;12:1429—33.

[33] Vakoc BJ, Shishko M, Yun SH, Oh W-Y, Suter MJ, Desjardins AE, et al. Comprehensive esophageal microscopy by using optical frequency-domain imaging (with video). Gastrointest Endosc 2007;65:898—905.

[34] Sauk J, Coron E, Kava L, Suter M, Gora M, Gallagher K, et al. Interobserver agreement for the detection of Barrett's esophagus with optical frequency domain imaging. Dig Dis Sci 2013;58:2261—5.

[35] Kang W, Wang H, Pan Y, Jenkins MW, Isenberg GA, Chak A, et al. Endoscopically guided spectral-domain OCT with double-balloon catheters. Opt Express 2010;18:17364—72.

[36] Tsai T-H, Fujimoto JG, Mashimo H. Endoscopic optical coherence tomography for clinical gastroenterology. Diagnostics 2014;4:57—93.

[37] Zhou C, Tsai TH, Lee HC, Kirtane T, Figueiredo M, Tao YKK, et al. Characterization of buried glands before and after radiofrequency ablation by using 3-dimensional optical coherence tomography (with videos). Gastrointest Endosc 2012;76:32—40.

[38] Adler D, Zhou C, Tsai T-H, Lee H-C, Becker L, Schmitt J, et al. Three-dimensional optical coherence tomography of Barrett's esophagus and buried glands beneath neo-squamous epithelium following radiofrequency ablation. Endoscopy 2009;41:773.

[39] Zhou C, Kirtane T, Tsai T-H, Lee H-C, Adler DC, Schmitt JM, et al. Cervical inlet patch-optical coherence tomography imaging and clinical significance. World J Gastroenterol 2012;18:2502.

[40] Qi X, Pan Y, Sivak MV, Willis JE, Isenberg G, Rollins AM. Image analysis for classification of dysplasia in Barrett's esophagus using endoscopic optical coherence tomography. Biomed Opt Express 2010;1:825—47.

[41] Leggett CL, Gorospe EC, Chan DK, Muppa P, Owens V, Smyrk TC, et al. Comparative diagnostic performance of volumetric laser endomicroscopy and confocal laser endomicroscopy in the detection of dysplasia associated with Barrett's esophagus. Gastrointest Endosc 2015.

[42] Swager A, Boerwinkel DF, de Bruin DM, Weusten BL, Faber DJ, Meijer SL, et al. Volumetric laser endomicroscopy in Barrett's esophagus: a feasibility study on histological correlation. Dis Esophagus 2015.

[43] Mashimo H. Subsquamous intestinal metaplasia after ablation of Barrett's esophagus: frequency and importance. Curr Opin Gastroenterol 2013;29:454—9.

[44] Adler DC, Zhou C, Tsai T-H, Schmitt J, Huang Q, Mashimo H, et al. Three-dimensional endomicroscopy of the human colon using optical coherence tomography. Opt Express 2009;17:784.

[45] Cobb MJ, Hwang JH, Upton MP, Chen Y, Oelschlager BK, Wood DE, et al. Imaging of subsquamous Barrett's epithelium with ultrahigh-resolution optical coherence tomography: a histologic correlation study. Gastrointest Endosc 2010;71:223—30.

[46] Swager AF, Boerwinkel DF, de Bruin DM, Faber DJ, van Leeuwen TG, Weusten BL, et al. Detection of buried Barrett's glands after radiofrequency ablation with volumetric laser endomicroscopy. Gastrointest Endosc 2016;83:80—8.

[47] Suter MJ, Jillella PA, Vakoc BJ, Halpern EF, Mino-Kenudson M, Lauwers GY, et al. Image-guided biopsy in the esophagus through comprehensive optical frequency domain imaging and laser marking: a study in living swine. Gastrointest Endosc 2010;71:346—53.

[48] Suter MJ, Gora MJ, Lauwers GY, Arnason T, Sauk J, Gallagher KA, et al. Esophageal-guided biopsy with volumetric laser endomicroscopy and laser cautery marking: a pilot clinical study. Gastrointest Endosc 2014;79:886—96.

[49] Beaudette K, Baac HW, Madore W-J, Villiger M, Godbout N, Bouma BE, et al. Laser tissue coagulation and concurrent optical coherence tomography through a double-clad fiber coupler. Biomed Opt Express 2015;6:1293—303.

[50] Ahsen OO, Lee H-C, Giacomelli MG, Wang Z, Liang K, Tsai T-H, et al. Correction of rotational distortion for catheter-based en face OCT and OCT angiography. Opt Lett 2014;39:5973—6.

[51] Tsai T-H, Potsaid B, Tao YK, Jayaraman V, Jiang J, Heim PJ, et al. Ultrahigh speed endoscopic optical coherence tomography using micromotor imaging catheter and VCSEL technology. Biomed Opt Express 2013;4:1119—32.

[52] Tran PH, Mukai DS, Brenner M, Chen Z. In vivo endoscopic optical coherence tomography by use of a rotational microelectromechanical system probe. Opt Lett 2004;29:1236—8.

[53] Pan Y, Xie H, Fedder GK. Endoscopic optical coherence tomography based on a microelectromechanical mirror. Opt Lett 2001;26:1966—8.

[54] Tsai T-H, Ahsen OO, Lee H-C, Liang K, Figueiredo M, Tao YK, et al. Endoscopic optical coherence angiography enables 3-dimensional visualization of subsurface microvasculature. Gastroenterology 2014;147:1219—21.

[55] Tsai T-H, Lee H-C, Ahsen OO, Liang K, Giacomelli MG, Potsaid BM, et al. Ultrahigh speed endoscopic optical coherence tomography for gastroenterology. Biomed Opt Express 2014;5:4387—404.

[56] Gora MJ, Sauk JS, Carruth RW, Gallagher KA, Suter MJ, Nishioka NS, et al. Tethered capsule endomicroscopy

enables less invasive imaging of gastrointestinal tract microstructure. Nat Med 2013;19:238—40.

[57] Gora MJ, Sauk JS, Carruth RW, Lu W, Carlton DT, Soomro A, et al. Imaging the upper gastrointestinal tract in unsedated patients using tethered capsule endomicroscopy. Gastroenterology 2013;145:723—5.

[58] Liang K, Traverso G, Lee H-C, Ahsen OO, Wang Z, Potsaid B, et al. Ultrahigh speed en face OCT capsule for endoscopic imaging. Biomed Opt Express 2015;6:1146—63.

[59] Yang VX, Gordon M, Tang S-j, Marcon N, Gardiner G, Qi B, et al. High speed, wide velocity dynamic range Doppler optical coherence tomography (Part III): in vivo endoscopic imaging of blood flow in the rat and human gastrointestinal tracts. Opt Express 2003;11:2416—24.

[60] Yang VX, Tang S-j, Gordon ML, Qi B, Gardiner G, Cirocco M, et al. Endoscopic Doppler optical coherence tomography in the human GI tract: initial experience. Gastrointest Endosc 2005;61:879—90.

[61] Standish BA, Yang VX, Munce NR, Song L-MWK, Gardiner G, Lin A, et al. Doppler optical coherence tomography monitoring of microvascular tissue response during photodynamic therapy in an animal model of Barrett's esophagus. Gastrointest Endosc 2007;66:326—33.

[62] Islam MS, Oliveira MC, Wang Y, Henry FP, Randolph MA, Park BH, et al. Extracting structural features of rat sciatic nerve using polarization-sensitive spectral domain optical coherence tomography. J Biomed Opt 2012;17:056012-1—056012-9.

[63] Matcher SJ, Winlove CP, Gangnus SV. The collagen structure of bovine intervertebral disc studied using polarization-sensitive optical coherence tomography. Phys Med Biol 2004;49:1295.

[64] Chin L, Yang X, McLaughlin RA, Noble PB, Sampson DD. En face parametric imaging of tissue birefringence using polarization-sensitive optical coherence tomography. J Biomed Opt 2013;18:066005.

[65] Wang Z, Lee HC, Ahsen OO, Lee B, Choi W, Potsaid B, et al. Depth-encoded all-fiber swept source polarization sensitive OCT. Biomed Opt Express 2014; 5:2931—49.

[66] Pyhtila JW, Chalut KJ, Boyer JD, Keener J, D'Amico T, Gottfried M, et al. In situ detection of nuclear atypia in Barrett's esophagus by using angle-resolved low-coherence interferometry. Gastrointest Endosc 2007;65:487—91.

[67] Zhu Y, Terry NG, Wax A. Angle-resolved low-coherence interferometry: an optical biopsy technique for clinical detection of dysplasia in Barrett's esophagus. Expert Rev Gasteroenterol Hepatol 2012;6:37—41.

[68] Mavadia J, Xi J, Chen Y, Li X. An all-fiber-optic endoscopy platform for simultaneous OCT and fluorescence imaging. Biomed Opt Express 2012;3:2851—9.

[69] Yoo H, Kim JW, Shishkov M, Namati E, Morse T, Shubochkin R, et al. Intra-arterial catheter for simultaneous microstructural and molecular imaging in vivo. Nat Med 2011;17:1680—4.

[70] Ryu SY, Choi HY, Na J, Choi ES, Lee BH. Combined system of optical coherence tomography and fluorescence spectroscopy based on double-cladding fiber. Opt Lett 2008;33:2347—9.

[71] Boppart SA, Herrmann J, Pitris C, Stamper DL, Brezinski ME, Fujimoto JG. High-resolution optical coherence tomography-guided laser ablation of surgical tissue. J Surg Res 1999;82:275—84.

[72] Vakoc BJ, Tearney GJ, Bouma BE. Real-time microscopic visualization of tissue response to laser thermal therapy. J Biomed Opt 2007;12:020501—020501-3.

[73] Baac HW, Uribe-Patarroyo N, Bouma BE. High-energy pulsed Raman fiber laser for biological tissue coagulation. Opt Express 2014;22:7113—23.

[74] Villiger M, Soroka A, Tearney GJ, Bouma BE, Vakoc BJ. Injury depth control from combined wavelength and power tuning in scanned beam laser thermal therapy. J Biomed Opt 2011;16:118001—1180019.

[75] Qi X, Sivak MV, Isenberg G, Willis JE, Rollins AM. Computer-aided diagnosis of dysplasia in Barrett's esophagus using endoscopic optical coherence tomography. J Biomed Opt 2006;11:044010.

[76] Garcia-Allende PB, Amygdalos I, Dhanapala H, Goldin RD, Hanna GB, Elson DS. Morphological analysis of optical coherence tomography images for automated classification of gastrointestinal tissues. Biomed Opt Express 2011;2:2821—36.

Enhanced Imaging of the Esophagus: Confocal Laser Endomicroscopy

Martin Goetz

Innere Medizin I, Universitätsklinikum Tübingen, Tübingen, Germany

9.1 INTRODUCTION

Barrett's esophagus (BE) is associated with the risk of development of esophageal adenocarcinoma. Current surveillance according to the Seattle protocol includes white light endoscopy (WLE) with the collection of random four-quadrant biopsy specimens over every 1–2 cm of the columnar-lined esophagus. The aim of such surveillance is detection of neoplasia, ideally at an (endoscopically) curable stage. This state-of-the-art approach is labor intensive and prone to sampling error since only a minor part of the mucosal surface is undergoing microscopic analysis ex vivo. Despite high definition (HD) and virtual or spraying surface contrast enhancement to augment WLE, its ability to reliably detect premalignant lesions remains suboptimal. This is partly based on the fact that high-grade intraepithelial neoplasia (IN) intramucosal cancer in BE is found in a patchy pattern side by side with lower grades of dysplasia and metaplasia.

Endomicroscopy is a technique in which the mucosa is magnified by a confocal scanner. When combined with WLE this technique can be used to highlight areas that are suspicious for dysplastic epithelium. A fluorescent agent is injected intravenously prior to image acquisition and images are viewed at real time during endoscopy. This tool is best used when examining small areas of the esophageal mucosa. Endomicroscopy is fundamentally different from light microscopy: the use of the confocal technique allows microscopy even underneath the tissue surface in intact tissue without the need to physically shine light through thin tissue sections. Such optical sectioning permits magnification to about 1000-fold and reveals subtle structural details of the mucosa on a (sub-)cellular level. Since its first description over 10 years ago, many trials have covered confocal laser endomicroscopy (CLE) in BE. This indicates a clinical need to optimize detection of Barrett's associated IN and reflects the low confidence of many gastroenterologists in untargeted quadrant biopsies to incidentally pick up preneoplastic lesions. Volumetric laser endomicroscopy (VLE) shows overlap in

D. Pleskow & T. Erim (Eds): Barrett's Esophagus.
DOI: http://dx.doi.org/10.1016/B978-0-12-802511-6.00009-0

naming and indication but relies on optical coherence tomography, a technically different cross-sectional imaging technique that is covered in Chapter 8: "Enhanced Imaging of the Esophagus: Optical Coherence Tomography."

9.2 CONFOCAL ENDOMICROSCOPY DEVICES

Two CLE systems are currently used in clinical practice, an endoscope-based system (eCLE; Pentax, Tokyo, Japan) and a probe-based system (pCLE; Mauna Kea Technologies, Paris, France) [1]. Both are point techniques that achieve very high resolution of a small mucosal area.

In eCLE, that is not marketed at this point (early 2015) but still available in many endoscopy suites, a miniaturized confocal scanning device is integrated into the distal tip of a standard resolution endoscope. The tip of the scanner protrudes slightly and is visible in the endoscopic image at the seven o'clock position so that it can be placed onto the point of interest under endoscopic guidance. The free working channel can be used for labeling, targeted biopsies, or other interventions. The scanning mechanism relies on a single optical fiber working as a pinhole, mounted into a resonant magnetic tuning fork that scans the area at high resolution at two different speeds. Lateral resolution is 1024×1024 pixels ($\sim 0.7\,\mu m$) at a frame rate of approximately $0.6\,s^{-1}$ or 512×1024 pixels at $1.2\,s^{-1}$. The depth of imaging can be actuated by the user at $7\,\mu m$ steps from surface to about $200\,\mu m$.

In probe-based CLE (pCLE), the confocal probe is fitted through the working channel of any gastrointestinal endoscope. This carries with it the advantage of using it with different types of endoscopes, including state-of-the-art HD scopes. The probes are available with different diameters that even allow use within the bile duct or through an FNA needle during endoscopic ultrasonography (which are outside the focus of this review). The probes use a fiber bundle for laser light propagation and collection of fluorescence. The imaging plane is fixed for the different probe types and can be somewhat adapted by exerting different extent of pressure with the probe. Most trials have used the Gastroflex UHD CLE probe. Some authors use a short transparent cap on the endoscope to facilitate stabilization of the probe tip on a region of interest. Resolution of pCLE is lower than with eCLE, but image acquisition faster, providing microscopic video sequences at real time.

In both systems, the excitation wavelength is 488 nm (blue light), and emitted light is captured in the green range. Imaging relies on the application of a fluorescent agent. In most trials, fluorescein is injected intravenously at $2.5-5\,mL$ of a 10% solution. Fluorescent contrast is available for tissue imaging after few seconds. Fluorescein is partly bound by plasma proteins and also extravasates into the tissue. As a result, the tissue structure becomes visible almost immediately with good resolution of the mucosal structures and the capillaries of the lamina propria. Nuclei are not discernible, but structural information of the mucosa is usually sufficient as a basis for a therapeutic decision (see later). Fluorescein is safe, and adverse events are rare [2]. While some early trials have also used topical acriflavine (which visualizes nuclei) [3,4], this is largely abandoned due to a theoretical risk of harboring mutagenic effects of the nuclear staining. Cresyl violet results in an indirect visualization of nuclei but has only been evaluated in small trials [5,6].

In CLE, the resultant image on the screen is parallel to the tissue surface, ie, perpendicular to sectioning in conventional histopathology. Interpretation of the microscopic images is usually performed online in order to make use of the advantage of having a microscopic tissue analysis available during the endoscopic session. This requires a thorough knowledge of the mucosal histopathology by the endoscopist. For starting CLE in the endoscopy unit, it might be beneficial to ask a histopathologist into the room for the first CLE sessions, however, this is not mandatory. Usually, the endoscopist bases his microscopic diagnosis on a two-step decision: the first being the differentiation of normal versus abnormal (pattern recognition) and the second being appreciation of the fine suspicious alterations in abnormal tissue (detail description). This will be explained later for Barrett's esophagus and associated neoplasia in more detail.

9.3 CONFOCAL LASER ENDOMICROSCOPY IN BARRETT'S ESOPHAGUS

9.3.1 Confocal Laser Endomicroscopy Features of Barrett's Esophagus

In BE, the squamous epithelium of the lower esophagus is replaced by metaplastic columnar-lined epithelium (Fig. 9.1). One of the hallmarks of the metaplasia is the presence of goblet cells (although this has been questioned for a smaller subset of Barrett's patients). Goblet cells are easily discernible by CLE as high columnar cells with dark mucin inclusions toward the luminal aspect. The differentiation of the glandular structure of BE against the squamous epithelium is usually very straightforward whereas delineation against the darker, cobbled-stone appearance of the cardiac epithelium may be more difficult.

In the first trial using CLE for BE [7], vessel and cellular aspects were described for the typical microscopic definition of BE: BE contains columnar-lined epithelium with goblet cells in the luminal superficial aspects of the mucosal layer. In deeper parts, villous-like glandular structures contain dark cylindrical epithelial cells. Capillaries within the Lamina propria are regular and visible in deeper optical sections. Often, a superficial double lining is visible that corresponds to the brush border

of the epithelium. In BE-associated neoplasia, fluorescein leaks from superficial and deep irregular capillaries. Nests of malignant cells are darker probably due to a lower pH—fluorescence after fluorescein is pH dependent. Neoplastic cells may pile up to multilayer structures or are no longer contained by a normal basal membrane and infiltrate from the epithelium into the lamina propria.

For pCLE, criteria have been tested and validated for detection of BE-associated neoplasia [8]: a saw-toothed epithelial surface, enlarged and pleomorphic cells, not equidistant glands unequal in size and shape, and goblet cells that are not easily identified, showed an overall accuracy in diagnosing BE-associated neoplasia of 81.5%.

9.3.2 Clinical Trials for Diagnosing Barrett's Esophagus and Barrett's Esophagus-Associated Neoplasia

Initial studies have aimed at the clinical feasibility of CLE in Barrett's esophagus and the diagnostic accuracy in comparison to histopathology. These early trials defined the criteria to diagnose BE and BE-associated neoplasia that were then evaluated and used in follow-up trials. An early study using CLE in 63 patients with BE was published in 2006 [7]. Here, based on the above-mentioned criteria eCLE was able to predict Barrett's epithelium and Barrett's associated neoplasia during endoscopy with a sensitivity of 98.1% and 92.9% and a specificity of 94.1% and 98.4%, respectively. Accuracy was 96.8% and 97.4%, respectively. Interobserver and intraobserver agreements for the prediction of histopathological diagnosis were substantial (κ, 0.843 and 0.892, respectively). In a similar setting, eCLE was performed in 50 patients referred for known BE [9]. Again, optical biopsies were performed in a circular fashion every 1−2 cm of the columnar-lined esophagus, corresponding to an "optical Seattle biopsy"

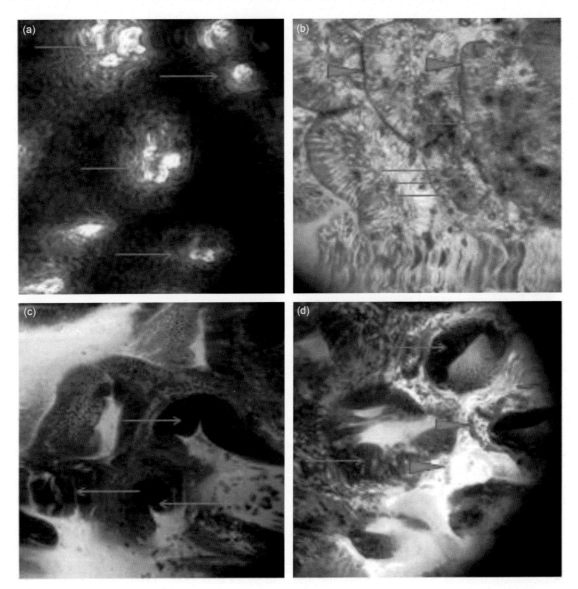

FIGURE 9.1   (a) In squamous epithelium, intrapapillary capillary loops rise vertically toward the tissue surface and are visible as cork-screw like bright fluorescein-filled vessel loops (*arrows*) within the dark squamous cells. Light fluorescein extravasation from these IPCLs highlights the intercellular spaces in-between squamous cells. (b) In Barrett's esophagus, darker mucin-filled goblet cells (*arrows*) with columnar epithelial lining are seen within the villous tissue pattern. The epithelium shows a double lining at the surface, corresponding to the brush border (*arrowheads*). (c) Darker cells pile up in slightly irregular glands (*arrows*) in a Barrett's esophagus that was diagnosed with neoplasia in random biopsies. Targeted EMR of this region confirmed low-grade intraepithelial neoplasia (IN). (d) Cellular and glandular irregularity (*arrows*) increases in this confirmed high-grade IN in Barrett's esophagus. Bright fluorescein leaks from superficial tortuous irregular vessels (*arrowheads*) as a functional sign of neoplasia-associated increased vascular permeability (a–d: eCLE, edge length 475 μm).

protocol, and on a total of three visible lesions. With targeted histology as the gold standard, BE-associated neoplasia could be predicted with an accuracy of 98.1% and a substantial agreement between endomicroscopy and histology ($\kappa = 0.76$).

Early trials using pCLE demonstrated a high negative predictive value (NPV) of 99% for BE, however, showed a PPV of only 44% [10]. Similar numbers were generated in a second trial with pCLE, where the NPV to rule out neoplasia was 95%, but PPV was only 18% [11]. There has been some speculation about the reason for the initially observed differences in the numbers of pCLE and eCLE, and one possible explanation might be higher resolution of eCLE. In fact, one small case series has compared pCLE criteria and eCLE criteria, and found an interrater agreement of 0.17 and 0.68, respectively [12]. Accordingly, the overall accuracy in detecting dysplasia was only 37% and 44.3%. However, this was an ex vivo study on only 13 specimens and thus does not provide a solid head-to-head comparison. Another useful explanation was that compromising on PPV rather than NPV makes sense in a clinical setting in order to be sure to not miss suspicious lesions. Accordingly, follow-up trials demonstrated higher sensitivity and specificity rates of 87.8% for pCLE [13]. In a post hoc analysis in a meta-analysis, the type of CLE used did not make a significant difference as to sensitivity and specificity [14].

A following set of trials compared CLE optical biopsies in a Seattle fashion (with targeted real biopsies only on suspicious areas) with random physical biopsies according to the Seattle protocol, the current standard of care. This slightly different approach makes use of the fact that a larger mucosal area is covered by in vivo endomicroscopy by moving the scanner across the columnar-lined esophagus than by taking pinpoint real biopsies. In the first of these studies [15], the yield per biopsy was almost doubled by endomicroscopic targeting. In about two-thirds of patients, the need for random biopsies would have been abolished based on normal CLE findings. This study corroborated the point that gastroenterologists are able to reliably differentiate nonneoplastic from neoplastic Barrett's epithelium and strengthened the argument that close observation with CLE would obviate the need for random biopsies in favor of CLE-targeted biopsies. In a multicenter follow-up trial, 192 BE patients were randomized to undergo screening with high-definition white light endoscopy (HD-WLE) and random biopsies versus HD-WLE followed by eCLE and targeted biopsies only [16]. Diagnosis and proposed management were documented after HD-WLE in both groups, and after eCLE in the second group. In the eCLE group, a lower number of biopsies resulted in a significantly higher yield for BE-associated neoplasia (7% in the HD-WLE group vs 34% in HD-WLE eCLE group). In a per-biopsy analysis, the use of eCLE in addition to HD-WLE resulted in a 4.8-fold decrease in the number of biopsies during endoscopy. In a per-patient analysis, eCLE had a 2.7-fold higher diagnostic yield for neoplasia. With eCLE, 34% of patients had a correct change in management. Of 26 patients without lesions, 5 had inapparent high-grade IN in flat BE detected by eCLE and 21 did not require any biopsies.

The next set of studies established the use of CLE in addition to HD-WLE. In a trial comparing HD-WLE versus HD-WLE with pCLE in a total of 100 patients [17], HD-WLE alone did not find suspicious areas, but random biopsies according to the Seattle protocol picked up incident BE-associated neoplasia in 5 of 50 patients (10%, 1 patient with high-grade neoplasia). pCLE identified areas suspicious for neoplasia in 21 of 50 patients (42%), of which 14 cases were confirmed by histopathology, 2 with high grade. This was significantly higher than in the HD-WLE only group, with numbers for sensitivity, specificity, PPV and NPV of pCLE for neoplasia of 100%, 83%, 67%, and 100%, respectively. Similar results supporting the use of pCLE in addition to HD-WLE were found in an enriched population of 101 consecutive BE patients scheduled for surveillance or treatment of BE-associated neoplasia [13]. In contrast, another study included 50 consecutive patients with known dysplastic Barrett's esophagus, and a prediction of histology was performed with HD-WLE, followed by NBI, and finally by CLE, and correlated to biopsies targeted to this spot. A total of 91 biopsy spots harbored high-grade IN or intramucosal cancer. CLE did not add to the accuracy rates of 82.8% for HD-WLE or 81.4% for NBI. In these patients with a known high pretest probability, all mucosal carcinomas could be detected by targeted biopsies guided by HD-WLE or NBI alone.

Two meta-analyses have evaluated the value of CLE in BE patients, pooling pCLE and eCLE trials. A recent meta-analysis [18] summarized results in BE patients up to 2013 in a total of 7 trials, including all prospective studies that compared the accuracy of CLE with standard four-quadrant biopsies (total of 345 patients and 3080 lesions) [7,8,10,11,13,19,20]. On a per-lesion analysis, the pooled sensitivity for the detection of high-grade IN and early carcinoma in Barrett's esophagus was 68%, the pooled specificity 88%. Respective numbers for sensitivity and specificity in a per-patient analysis were 86% and 83%. A second meta-analysis [14] with slightly different eligibility criteria on 8 trials [7,10,11,13,16,17,21,22] involving 709 patients and 4008 specimens showed comparable results in a per-lesion analysis (sensitivity 70%, specificity 91%) and per-patient analysis (sensitivity 89%, specificity 75%) for the detection of neoplasia. The authors of these meta-analyses conclude that CLE with targeted biopsies has good diagnostic accuracy for detecting high-grade IN and early carcinoma in BE. However, the data had been gathered from patients with a high overall prevalence of high-grade IN and early

carcinoma, and may therefore not be generalizable to all screening patients. Thus abandoning the Seattle protocol at this stage cannot be recommended for the broad clinical practice, despite good results in expert centers.

A further set of trials used CLE for guidance of therapy or surveillance after therapy. It has been hypothesized that CLE could be an appropriate tool to confirm completeness of ablation and to rule out residual buried Barrett's glands after radiofrequency ablation of BE. This is difficult to achieve with HD-WLE alone. A recent study assessed the usefulness of pCLE in addition to HD-WLE in a prospective, multicenter, randomized fashion [21]. After endoscopic ablation, patients were followed up with HD-WLE or HD-WLE aided by pCLE, with treatment of residual Barrett's mucosa guided by endoscopic or/and endomicroscopic findings. A total of 119 patients were included before termination of the trial at interim analysis. The addition of pCLE did not result in a higher proportion of patients with complete absence of BE at follow-up. However, only a low proportion of patients achieved complete eradication in both groups (15 of 57 patients (26%) for HD-WLE; 17/62, 27% with HD-WLE + pCLE) which might have influenced results. Another reason for this unanticipated result could be the limited depth of imaging especially in squamous epithelium, making transmission of blue laser light for CLE difficult. However, some neoplastic glands at least partially undermining squamous epithelium were diagnosed in two trials using eCLE [7,16], however, without prior ablation treatment, thus not representing true buried glands.

How long does it take for an experienced gastroenterologist and endoscopist to be confident in establishing a microscopic diagnosis based on CLE findings (rather than taking a biopsy and asking the histopathologist for a diagnosis)? A short learning curve after a structured teaching session and a set of surveilled teaching cases has been suggested by an abstract [23]. In a prospective, double-blind review of pCLE images of 40 sites of BE and matched tissue specimens as a gold standard, the sensitivity for the diagnosis of neoplasia for 11 endoscopists (CLE specialists and nonspecialists) was 88%, with a specificity of 96% and a substantial diagnostic agreement (κ, 0.72) that was even better when only pCLE specialists were evaluated (κ, 0.83) [19]. Recently, a thorough evaluation of the pCLE criteria was published [8]. After development of the above-mentioned criteria, a validation set was used for the prediction of neoplasia in BE. Overall accuracy in predicting high-grade IN and cancer was high and interobserver agreement substantial. Accuracy and agreement between experienced and nonexperienced assessors were not different after initial structured teaching, suggesting a short learning curve. These pCLE criteria also served as a basis to establish the accuracy and interobserver agreement between endoscopists and histopathologists and among histopathologists [24]. Among histopathologists, the accuracy for the diagnosis of all grades of neoplasia was 77.8% in 90 videos on BE and rose to 93.8% when pathologists had "high confidence" in their assessment of the videos, with substantial interobserver agreement. Comparable results were achieved by endoscopists in the same set of videos, suggesting that endoscopists can be trained to achieve a sound diagnosis in BE based on predefined criteria. This trial also underlines the need for high quality microscopic images to allow for a confident (and more accurate) diagnosis.

9.4 CONFOCAL LASER ENDOMICROSCOPY IN BARRETT'S PATIENTS

When using CLE in clinical practice, the following points may help to optimize results. First, CLE at the GI junction is more prone to movement artifacts than CLE in most other regions of the gastrointestinal tract. Spasmolytics and deep sedation usually support good imaging, but of course do not influence moving artifacts by heartbeat or respiration. Even in expert hands and under study conditions, good quality images were only reported in 38% of the obtained images from the distal esophagus [7]. With pCLE, some authors prefer using a cap in order to facilitate stable positioning of the probe. Proton pump inhibitors should be administered before CLE to treat inflammation and to minimize inflammatory changes mimicking or hiding neoplasia. If eCLE is used, prior examination with an HD endoscope should be considered, as well as use of pCLE in conjunction with HD-WLE (and potentially virtual chromoendoscopy) to optimize macroscopic evaluation of the distal esophagus before using CLE as a point technique. Contrast enhancement with acetic acid results in blurred CLE images and should not be used prior to endomicroscopy. If biopsies are to be taken or suspicious areas to be resected, imaging should be completed before the intervention since bleeding (with spillout of fluorescein) interferes with good imaging. Suspicious areas can be labeled with an APC or suction mark for later identification if they cannot be reidentified by macroscopic landmarks. This is important since with eCLE, the endoscope may have to be changed for mucosal resection (not for taking biopsies—the working channel is free), and with pCLE, the probe has to be taken out of the working channel to allow passage of endoscopic instrumentarium. We often use CLE to reidentify endoscopically inapparent neoplasia detected by random biopsies. In such a setting, relocalization is followed by immediate endoscopic resection, and the rest of the columnar-lined mucosa is microscopically screened in the same session. A similar protocol was followed in previous trials [15,16] and resulted in a change of the management plan in 34% of patients, including the decision to initiate endoscopic treatment at the time of endoscopy in patients with indeterminate lesions and flat neoplasia [16].

The clinical role of CLE in Barrett's esophagus has been subject of critical discussion on the background of a slightly shifted clinical appreciation of BE that has been found to result in a lower incidence of Barrett-associated carcinoma in recent population-based studies and also with the advent of a change in therapy of invisible neoplasia by circumferential ablation. One of the fundamental points is that CLE does not permit deep tissue imaging with the use of blue laser excitation so that submucosal invasion cannot be confirmed or excluded. However, this may be true for many of the pinch biopsies as well. Other risk factors (microvessel or lymphatic vessel invasion, tumor grading) of cancerous lesions cannot be determined with CLE at present, so certain risk factors to promote an à priori surgical approach cannot be appreciated. However, if in doubt CLE at least allows for a targeted diagnostic EMR of such areas. Another point raised is that lesions that mandate EMR are mostly elevated and endoscopically visible in most patients, whereas all other lesions can be effectively addressed by radiofrequency ablation [25]. However, this also requires prior diagnosis for indication, and Barrett's epithelium was not completely eradicated in many of the ablation trials even under the highly controlled setting of clinical trials. Whether this represents true buried glands or a false positive diagnosis of buried glands is matter of ongoing discussions [26].

In most studies, CLE is compared to either random biopsies or targeted biopsies from the spot imaged before. However, random biopsies according to the Seattle protocol are associated with sampling error [27], and histopathologic interpretation (especially of low-grade dysplasia) is inconsistent with a high rate of interobserver variability even among expert histopathologists [28]. Even though the stringent definition of histopathology as a predefined gold standard is absolutely necessary for valid clinical trials, in clinical routine CLE is able to decrease the sampling error by being able to optically sample a larger area of the mucosa by moving the scanner across the tissue surface and may thus increase the diagnostic yield for high-grade IN and intramucosal cancer. As a consequence, patients can be treated immediately by EMR after in vivo microscopic diagnosis.

9.5 PERSPECTIVE

In perspective, any advanced imaging technique for the surveillance of Barrett's esophagus needs to address enhanced detection, optimized characterization and prediction of treatment options. For facilitating detection of suspicious lesions, a red flag technique would be helpful. Many trials have evaluated virtual chromoendoscopy techniques such as NBI or contrast enhancement (eg, with acetic acid) and also autofluorescence imaging, in this setting. Molecular imaging uses fluorescently labeled probes to specifically highlight neoplastic lesions based on their molecular fingerprint upon interaction with the molecular target [29]. Detection of lesion-specific fluorescence can be obtained by wide field or point endoscopy techniques such as CLE.

Within the esophagus, a cathepsin-sensing probe was evaluated in an orthotopic xenograft mouse model of esophageal adenocarcinoma. Molecular wide field imaging was performed with a dual channel endoscope combining WLE and near infrared light [30]. Periostin is overexpressed in patients with invasive esophageal squamous cell carcinoma. In a murine model, labeled antibodies to periostin detected neoplasia with a fluorescence endoscope [31]. In humans, the lack of expression of cell surface glycans in Barrett's associated neoplasia was used in a proof-of-principle trial on four esophagectomy specimens after topical application of fluorescently labeled lectins [32]. These trials, although using fluorescence imaging, have not been linked to CLE so far. For CLE and molecular imaging, a recent trial has demonstrated a good interobserver agreement and accuracy in EMR specimens of Barrett's associated IN. Ex vivo topical contrast with a fluorescent glucose analog with preferential uptake in neoplastic mucosa doubled the accuracy rates of pCLE and eCLE, however, against a background of low accuracy rates of 37% (pCLE) and 44% (eCLE) [12]. First in vivo data on molecular imaging of dysplasia in Barrett's esophagus was recently reported after construction of fluorescently labeled peptides with specificity for esophageal neoplasia that were selected using phage display technology [33,34]. A preliminary technical report documented selective binding of the probe to Barrett's associated neoplasia after topical application of 5 mL of a specific probe during upper endoscopy and in vivo incubation time of 5 min. Imaging was performed with pCLE [35], and further trials are awaited. While none of these approaches are currently ready for clinical practice, they highlight the search for optimal markers and imaging modalities including CLE for a molecular detection of neoplasia in the esophagus. Binding analysis by enhanced molecular imaging by CLE could even be used to predict response to targeted chemotherapy in malignant disease [36]. Analysis with CLE provides a point analysis at present. This could be combined with wide field fluorescent endoscopes to rapidly provide a suspicious area that can then be targeted by a point technique such as CLE to detect and characterize suspicious lesions on a molecular basis.

References

[1] Goetz M, Malek NP, Kiesslich R. Microscopic imaging in endoscopy: endomicroscopy and endocytoscopy. Nat Rev Gastroenterol Hepatol 2014;11:11–18.

[2] Wallace MB, Meining A, Canto MI, Fockens P, Miehlke S, Roesch T, et al. The safety of intravenous fluorescein for confocal laser endomicroscopy in the gastrointestinal tract. Aliment Pharmacol Ther 2010;31:548–52.

[3] Kiesslich R, Burg J, Vieth M, Gnaendiger J, Enders M, Delaney P, et al. Confocal laser endoscopy for diagnosing intraepithelial neoplasias and colorectal cancer in vivo. Gastroenterology 2004;127:706–13.

[4] Polglase AL, McLaren WJ, Skinner SA, Kiesslich R, Neurath MF, Delaney PM. A fluorescence confocal endomicroscope for in vivo microscopy of the upper- and the lower-GI tract. Gastrointest Endosc 2005;62:686–95.

[5] George M, Meining A. Cresyl violet as a fluorophore in confocal laser scanning microscopy for future in-vivo histopathology. Endoscopy 2003;35:585–9.

[6] Goetz M, Toermer T, Vieth M, Dunbar K, Hoffman A, Galle PR, et al. Simultaneous confocal laser endomicroscopy and chromoendoscopy with topical cresyl violet. Gastrointest Endosc 2009;70:959–68.

[7] Kiesslich R, Gossner L, Goetz M, Dahlmann A, Vieth M, Stolte M, et al. In vivo histology of Barrett's esophagus and associated neoplasia by confocal laser endomicroscopy. Clin Gastroenterol Hepatol 2006;4:979–87.

[8] Gaddam S, Mathur SC, Singh M, Arora J, Wani SB, Gupta N, et al. Novel probe-based confocal laser endomicroscopy criteria and interobserver agreement for the detection of dysplasia in Barrett's esophagus. Am J Gastroenterol 2011;106:1961–9.

[9] Trovato C, Sonzogni A, Ravizza D, Fiori G, Tamayo D, De Roberto G, et al. Confocal laser endomicroscopy for in vivo diagnosis of Barrett's oesophagus and associated neoplasia: a pilot study conducted in a single Italian centre. Dig Liver Dis 2013;45:396–402.

[10] Pohl H, Rosch T, Vieth M, Koch M, Becker V, Anders M, et al. Miniprobe confocal laser microscopy for the detection of invisible neoplasia in patients with Barrett's oesophagus. Gut 2008;57:1648–53.

[11] Bajbouj M, Vieth M, Rosch T, Miehlke S, Becker V, Anders M, et al. Probe-based confocal laser endomicroscopy compared with standard four-quadrant biopsy for evaluation of neoplasia in Barrett's esophagus. Endoscopy 2010;42:435–40.

[12] Gorospe EC, Leggett CL, Sun G, Anderson MA, Gupta M, Penfield JD, et al. Diagnostic performance of two confocal endomicroscopy systems in detecting Barrett's dysplasia: a pilot study using a novel bioprobe in ex vivo tissue. Gastrointest Endosc 2012;76:933–8.

[13] Sharma P, Meining AR, Coron E, Lightdale CJ, Wolfsen HC, Bansal A, et al. Real-time increased detection of neoplastic tissue in Barrett's esophagus with probe-based confocal laser endomicroscopy: final results of an international multicenter, prospective, randomized, controlled trial. Gastrointest Endosc 2011;74:465–72.

[14] Wu J, Pan YM, Wang TT, Hu B. Confocal laser endomicroscopy for detection of neoplasia in Barrett's esophagus: a meta-analysis. Dis Esophagus 2014;27:248−54.

[15] Dunbar KB, Okolo 3rd P, Montgomery E, Canto MI. Confocal laser endomicroscopy in Barrett's esophagus and endoscopically inapparent Barrett's neoplasia: a prospective, randomized, double-blind, controlled, crossover trial. Gastrointest Endosc 2009;70:645−54.

[16] Canto MI, Anandasabapathy S, Brugge W, Falk GW, Dunbar KB, Zhang Z, et al. In vivo endomicroscopy improves detection of Barrett's esophagus-related neoplasia: a multicenter international randomized controlled trial (with video). Gastrointest Endosc 2014;79:211−21.

[17] Bertani H, Frazzoni M, Dabizzi E, Pigo F, Losi L, Manno M, et al. Improved detection of incident dysplasia by probe-based confocal laser endomicroscopy in a Barrett's esophagus surveillance program. Dig Dis Sci 2013;58:188−93.

[18] Gupta A, Attar BM, Koduru P, Murali AR, Go BT, Agarwal R. Utility of confocal laser endomicroscopy in identifying high-grade dysplasia and adenocarcinoma in Barrett's esophagus: a systematic review and meta-analysis. Eur J Gastroenterol Hepatol 2014;26:369−77.

[19] Wallace MB, Sharma P, Lightdale C, Wolfsen H, Coron E, Buchner A, et al. Preliminary accuracy and interobserver agreement for the detection of intraepithelial neoplasia in Barrett's esophagus with probe-based confocal laser endomicroscopy. Gastrointest Endosc 2010;72:19−24.

[20] Jayasekera C, Taylor AC, Desmond PV, Macrae F, Williams R. Added value of narrow band imaging and confocal laser endomicroscopy in detecting Barrett's esophagus neoplasia. Endoscopy 2012;44:1089−95.

[21] Wallace MB, Crook JE, Saunders M, Lovat L, Coron E, Waxman I, et al. Multicenter, randomized, controlled trial of confocal laser endomicroscopy assessment of residual metaplasia after mucosal ablation or resection of GI neoplasia in Barrett's esophagus. Gastrointest Endosc 2012;76 539−547 e1.

[22] Becker V, Vieth M, Bajbouj M, Schmid RM, Meining A. Confocal laser scanning fluorescence microscopy for in vivo determination of microvessel density in Barrett's esophagus. Endoscopy 2008;40:888−91.

[23] Dunbar K, Canto M. Confocal endomicroscopy. Curr Opin Gastroenterol 2008;24:631−7.

[24] Tofteland N, Singh M, Gaddam S, Wani SB, Gupta N, Rastogi A, et al. Evaluation of the updated confocal laser endomicroscopy criteria for Barrett's esophagus among gastrointestinal pathologists. Dis Esophagus 2014;27:623−9.

[25] Boerwinkel DF, Swager A, Curvers WL, Bergman JJ. The clinical consequences of advanced imaging techniques in Barrett's esophagus. Gastroenterology 2014;146 622−629.

[26] Pouw RE, Visser M, Odze RD, Sondermeijer CM, ten Kate FJ, Weusten BL, et al. Pseudo-buried Barrett's post radiofrequency ablation for Barrett's esophagus, with or without prior endoscopic resection. Endoscopy 2014;46:105−9.

[27] Wani S, Mathur SC, Curvers WL, Singh V, Alvarez Herrero L, Hall SB, et al. Greater interobserver agreement by endoscopic mucosal resection than biopsy samples in Barrett's dysplasia. Clin Gastroenterol Hepatol 2010;8:783−8.

[28] Curvers WL, ten Kate FJ, Krishnadath KK, Visser M, Elzer B, Baak LC, et al. Low-grade dysplasia in Barrett's esophagus: overdiagnosed and underestimated. Am J Gastroenterol 2010;105:1523−30.

[29] Atreya R, Goetz M. Molecular imaging in gastroenterology. Nat Rev Gastroenterol Hepatol 2013;10:704−12.

[30] Habibollahi P, Figueiredo JL, Heidari P, Dulak AM, Imamura Y, Bass AJ, et al. Optical imaging with a cathepsin B activated probe for the enhanced detection of esophageal adenocarcinoma by dual channel fluorescent upper GI endoscopy. Theranostics 2012;2:227−34.

[31] Wong GS, Habibollahi P, Heidari P, Lee JS, Klein-Szanto AJ, Waldron TJ, et al. Optical imaging of Periostin enables early endoscopic detection and characterization of esophageal cancer in mice. Gastroenterology 2012;144(2):294−7.

[32] Bird-Lieberman EL, Neves AA, Lao-Sirieix P, O'Donovan M, Novelli M, Lovat LB, et al. Molecular imaging using fluorescent lectins permits rapid endoscopic identification of dysplasia in Barrett's esophagus. Nat Med 2012;18:315−21.

[33] Sturm MB, Joshi BP, Lu S, Piraka C, Khondee S, Elmunzer BJ, et al. Targeted imaging of esophageal neoplasia with a fluorescently labeled peptide: first-in-human results. Sci Transl Med 2013;5 184ra61.

[34] Li M, Anastassiades CP, Joshi B, Komarck CM, Piraka C, Elmunzer BJ, et al. Affinity peptide for targeted detection of dysplasia in Barrett's esophagus. Gastroenterology 2010;139:1472−80.

[35] Sturm MB, Piraka C, Elmunzer BJ, Kwon RS, Joshi BP, Appelman HD, et al. In vivo molecular imaging of Barrett's esophagus with confocal laser endomicroscopy. Gastroenterology 2013;145:56−8.

[36] Goetz M, Hoetker MS, Diken M, Galle PR, Kiesslich R. In vivo molecular imaging with cetuximab, an anti-EGFR antibody, for prediction of response in xenograft models of human colorectal cancer. Endoscopy 2013;45:469−77.

History of Ablative Therapies for Barrett's and Superficial Adenocarcinoma

Jennifer T. Higa[1] and Joo Ha Hwang[2]

[1]Department of Medicine, University of Washington School of Medicine, Seattle, WA, United States
[2]Gastroenterology Section, Harborview Medical Center, University of Washington School of Medicine, Seattle, WA, United States

10.1 INTRODUCTION

A general outline of this chapter includes a brief background followed by historical trends of endoscopic ablation (with the exception of radiofrequency ablation and cryospray ablation, which are covered in chapters 11 and 12: Radiofrequency Ablation and Cryospray Ablation) for nondysplastic Barrett's esophagus (BE), BE with low-grade dysplasia (LGD), and BE with high-grade dysplasia (HGD)/intramucosal esophageal adenocarcinoma (EA). Finally, therapeutic options for endoscopic intervention are addressed detailing technical application, eradication and recurrence rates, complications and limitations of treatment, and intermodality comparisons.

10.1.1 Background

For decades, esophagectomy was the only reported treatment for dysplastic BE or EA. This changed in the early 1990s, with the publication of data showing successful ablation of nondysplastic BE using laser therapy [1,2]. Since then, multiple modalities have evolved intended for the ablation of BE, including multipolar electrocautery (MPEC), argon plasma coagulation (APC), photodynamic therapy (PDT), and more recently cryospray and radiofrequency ablation (RFA) therapies. Unfortunately, for most of these modalities, and especially the older technologies, the current body of literature lacks rigor and primarily consists of case series with "marked heterogeneity around study endpoints, duration of follow-up, postablation surveillance protocols" [3]. Newer interventions, including radiofrequency ablation and cryotherapy, are backed by more rigorous data and are covered in separate chapters (see chapters 11 and 12: Radiofrequency Ablation and Cryospray Ablation). Furthermore, multimodal endoscopic therapy, or endoscopic resection of visibly abnormal tissue followed by endoscopic ablation of remaining BE tissue, is currently well regarded as a comprehensive and efficacious

D. Pleskow & T. Erim (Eds): Barrett's Esophagus.
DOI: http://dx.doi.org/10.1016/B978-0-12-802511-6.00010-7

treatment for BE with HGD or superficial adeno-carcinoma [4]. Historically, esophagectomy has been employed as the recommended treatment for EA with drawbacks of high short-term (post-procedural) morbidity, estimated at 6−37% [5,6]. With advancements in endoscopic therapy for mucosal disease (ie, high-grade dysplastic disease or adenocarcinoma confined to the mucosal layer), new data shows that endo-scopic ablation has comparable survival out-comes with less morbidity and is more cost effective than surgery [7−9]. The current body of research provides critical data and insight into historical practices and guides current clinical decision making and management of BE using ablative techniques, thus we will compare and examine these techniques in this chapter (Table 10.1).

10.2 ENDOSCOPIC ABLATION: HISTORY AND OVERVIEW

Endoscopic eradication of high-grade dysplastic BE or superficial EA utilizes ablative therapy to destroy dysplastic cells. The ablated neoplastic mucosa is then replaced with "neos-quamous" epithelium, which theoretically decreases the risk of malignancy. This re-epithelialization was first demonstrated in the 1990s in cases of nondysplastic BE using lasers to achieve deep tissue injury. This was particu-larly efficacious if concomitant high doses of an oral proton pump inhibitor (PPI) were administered [1,10]. Subsequent trials using PDT, APC, MPEC, RFA, and cryospray therapy followed in the footsteps of the initial laser therapy case series. Laser therapy for ablation of BE was eventually rendered obsolete by the increasing evidence for use of these other ther-apies. Expansion of endoscopic options for treatment has been met with enthusiasm as historically BE with HGD or superficial

adenocarcinoma may have been subjected to morbid surgery [11,12].

10.3 ENDOSCOPIC ABLATION AS A TREATMENT PARADIGM

The advantage of these mucosal ablation techniques over surgical treatment with eso-phagectomy is complete eradication of dys-plastic or carcinomatous tissue without the operative risk and long recovery periods asso-ciated with esophagectomy. Furthermore, it is a potentially curative option for patients who are otherwise considered nonsurgical candi-dates. There is little long-term data comparing ablative therapy with surgery; however, in two larger retrospective studies, long-term out-comes were comparable between patients receiving endoscopic therapy for mucosal EA and esophagectomy [13,14]. These authors also note an increasing trend in utilization of endo-scopic ablation over time, likely a result of improvement in endoscopic technologies and preferential avoidance of the aforementioned morbidity of esophageal resection [14].

However, with regard to the merits of endo-scopic ablation, it is important to consider the appropriate timing of this intervention within the treatment framework for BE. In contrast to mechanical resection with techniques like endo-scopic mucosal resection (EMR), endoscopic submucosal dissection (ESD), or esophagect-omy, endoscopic ablation does not produce a specimen posttreatment for pathological evalu-ation. Because of this, patients with suspected superficial adenocarcinoma should be T-staged with either endoscopic ultrasound (EUS) and/or EMR prior to undergoing ablative therapy. Meta-analysis estimates risk of progression from HGD to EA at 6% per year [15]. As such, society guidelines recommend EUS with EMR of visibly dysplastic or malignant lesions prior to ablation to assess likelihood of disease progression [16]. This recommendation is

TABLE 10.1 Summary Table of Ablative Therapies

Category of Therapy	Ablative Therapy	Eradication Rates	Advantages	Disadvantages
Photochemical	PDT	40–77%	Available RCT data	Buried glands
				Challenging to administer
			Long-term data	High cost
				Risk of photosensitivity reactions
Photothermal	Nd:Yag, KTP:Yag laser	28–100%	Wide availability	Buried glands
				Challenging to administer
				Limited long-term data
				No RCT data
Thermal	RFA	82–98%	Available RCT data	High cost
			Less buried glands	Limited availability
			Low complication rate	Limited long-term data
Thermal	APC	67–86%	Available RCT data	Buried glands
			Low cost	High regression rates
			Wide availability	Practical for very short segments of BE only
Thermal	Cryotherapy	68–88%	Low complication rate	Limited quality of evidence for use
			Low cost	Limited long-term data
Thermal	MPEC	65–77%	Low cost	Limited long-term data
			Wide availability	Limited quality of evidence for use
				Postprocedure dysphagia/odynophagia
Thermal	Heater Probe	100%[a]	Low cost	Buried glands
				Limited long-term data
				Limited quality of evidence for use
				No RCT data

[a] *Findings based on one study available in the literature.*
Abbreviations: PDT, photodynamic therapy; RCT, randomized control trials; RFA, radiofrequency ablation; APC, argon plasma coagulation; Nd, neodymium; Yag, yttrium aluminum garnet; KTP, potassium titanyl phosphate; MPEC, multipolar electrocautery.

particularly salient in patients with nodular findings who are at an increased risk of malignancy and lymph node metastases, given the higher probability of submucosal invasion, and therefore not amenable to ablative intervention which has a role only in cases where neoplasia is limited to the mucosa [17]. After EUS is obtained, patients more suitable for endoscopic ablation, in addition to meeting criteria for disease limited to the mucosal layer, include cases of BE with HGD but without esophageal cancer (EC), shallow EC smaller than 2 cm in diameter, favorable histology (moderate or well differentiated), and patient agreeability to endoscopic surveillance program.

An additional drawback for endoscopic ablation includes the posttreatment risk of developing "buried glands" defined as metaplastic glands within the lamina propria with or without dysplasia hidden beneath the neosquamous epithelium. The deep location of the buried metaplastic glands eludes macroscopic detection by conventional endoscopy and therefore presents a risk for progression to neoplastic disease. The true neoplastic potential of this phenomenon is unknown; however, this may explain the metachronous neoplasia noted in long-term follow-up studies of endoscopic ablation.

In the current state of the art, there is data that suggests improved outcomes can be achieved with complete obliteration of all metaplastic tissue even in nondysplastic areas of BE, which prevents the development of neoplastic disease [18]. Additionally, PPI therapy is likely to decrease development of dysplastic changes in BE [19]. As such, mucosal ablation in conjunction with acid suppression therapy using PPI would likely optimize the tissue environment for generation of neosquamous epithelium while simultaneously reducing the risk of recurrent dysplasia. Presently, there exist no established dosing guidelines for administration of PPIs, but a recent position statement of the American Gastroenterological Association recommends standard doses of acid reduction agents to effectively treat GERD and heal esophagitis [20]. Further, empiric PPI use is recommended for neoplastic prophylaxis in documented cases of BE, though the evidence for this indication has not been proven in long-term controlled trials. As a result, it is acknowledged that this treatment model merits discussion of risks and benefits with the patient such that therapy may proceed in an informed manner [3].

10.3.1 History of Endoscopic Therapy for Patients Without Dysplasia

The bulk of the available research regarding endoscopic therapy for nondysplastic BE examines the application of thermal ablation techniques (ie, laser, MPEC, and APC). Close to complete eradication of nondysplastic BE is likely to have occurred in the majority of these cases; however, the clinical significance of this achievement is unclear. Some studies show that BE without dysplasia still carries a risk of progression to LGD, HGD, and even EA; however, these data also indicate that prophylactic effects of complete ablation of nondysplastic tissues against EC, if any, is unknown [21,22]. Additionally, the durability of the result to prevent recurrent metaplasia is also in question. Presently, many GI societies (including the ASGE and AGA) do not recommend ablation for nondysplastic BE and instead recommend endoscopic surveillance with biopsies at a 3- to 5-year interval [3].

10.3.2 History of Endoscopic Therapy for Patients with Low-Grade Dysplasia

There exists very little data on endoscopic therapy outcomes for LGD due to lack of consensus among pathologists and gastroenterologists regarding the natural history of the disease and difficulty making a histopathologic diagnosis. In most of the available literature, patients with LGD are included as a subgroup population for larger studies that focus on non-dysplastic BE or HGD. In trials designed to compare the efficacy of PDT and APC, eradication rates were comparable for cases of LGD [23]. More recently, one study demonstrated that use of RFA significantly reduces the risk of progression from LGD to over a 3-year follow-up period compared to endoscopic surveillance alone [24]. Current society guidelines for management of BE with LGD involve either endoscopic surveillance every 6−12 months or endoscopic ablation with RFA [20,25].

Despite this, consensus regarding treatment of LGD remains difficult to achieve due to challenges inherent to the diagnosis of LGD. This is likely attributable to poor inter- and intraobserver correlations in the diagnosis of LGD, which even at the histopathologic level is difficult to distinguish from inflammatory changes secondary to chronic reflux disease. Additionally, there exists wide heterogeneity around the known natural history of LGD, which presents difficulties in determining how aggressively surveillance and treatment should be conducted [3]. A meta-analysis conducted by Almond et al. examined LGD diagnoses across 37 studies (6 controlled clinical trials, all other observational) involving 521 patients [26]. Treatments included 9 RFA, 5 PDT, 9 APC, 2 MPEC, 3 laser, and 9 combination (surgical plus ablation) therapies. The authors concluded that LGD is likely over diagnosed and endoscopic ablation is unlikely to reduce risk of neoplastic progression; however, they acknowledged heterogeneity among studies with short-term follow-up which somewhat confounds this data. Due to these challenges, the existing literature concerning the efficacy of endoscopic therapies for the treatment of LGD is limited at best. Barring new information clearly defining the likelihood that BE with LGD leads to malignancy absent medical intervention, the long-term effects of endoscopic ablation remain epistemologically indeterminate.

10.3.3 History of Endoscopic Therapy for Patients with High-Grade Dysplasia or Intramucosal Adenocarcinoma

Initial evidence to support endoscopic treatment of HGD in BE is primarily derived from a meta-analysis of 4 studies and 236 patients, which reported a weighted incidence of EA at 6.58 per 100 patient-years (95% CI 4.99−8.46) during the initial 1.5−7 years after HGD diagnosis [15]. This high rate of progression to malignant disease has since prompted more aggressive treatment of cases of BE with HGD [15].

Endoscopic therapy is curative for cases where disease is limited to the mucosa [27], and EMR has been proven to be a durable and effective treatment for remission of neoplastic changes in BE [28]. The combination of EMR followed by endoscopic ablation is particularly effective. In one large, long-term study of 1000 patients, treatment with EMR demonstrated early eradication rates of 96.3% and subsequent development of metachronous neoplasia at a rate of 14.5% was successfully treated with endoscopic retreatment in 115 of 140 patients [29]. Risk factors for cancer recurrence include patients who did not receive post-EMR ablation of remnant areas of BE, patients with poorly differentiated EA, and patients with long segment BE. ESD appears to be a feasible treatment option for patients at risk for incomplete EMR or with poor pathologic procurement with EMR. However, this alternative is limited by the number of available practitioners who perform this therapy and complications like esophageal perforation [30]. Because of this, EMR should be performed on HGD/mucosal EA cases to clarify extent of disease and provide pathologic T staging. High-resolution endoscopic exam should be performed in any biopsy confirmed cases of BE with HGD to detect visible abnormalities amenable for EMR [9]. As mentioned previously, patients with submucosal EA should undergo esophagectomy [20].

10.4 ENDOSCOPIC THERAPIES

10.4.1 Photodynamic Therapy

PDT capitalizes on the interaction between a class of photosensitive drugs and a specific wavelength of laser light that elicits a photoexcitatory reaction. The microvasculature of the targeted tissue is damaged by the photochemical reaction leading to eventual cellular necrosis (Fig. 10.1). The degree of damage incurred is related to the type and concentration of the photosensitizing drug and the wavelength and energy of light delivered endoscopically [31]. The intrinsic macromolecular structure of these drugs preferentially concentrates the drug within skin, reticuloendothelial tissues, and neoplastic growths, making them an effective agent for targeted ablation of tissues. Different formulations of chlorine, chlorophyll, and porfimer contribute to the number of therapeutic options for PDT; however, in the United States, porfimer sodium is the only approved option for systemic use. Globally, 5-aminolevulinic acid (ALA) is also employed, however, it is not approved for systemic use in the United States.

10.4.1.1 Procedure

Porfimer sodium is the only photosensitive drug currently used for PDT in the United States and activates at wavelengths 515 and 630 nm. The general therapeutic application of PDT in the United States involves reconstitution of porfimer sodium in normal saline or 5% dextrose that is typically dosed at 2 mg/kg administered intravenously over 3–5 minutes. Due to the photosensitive nature of the compound, the slurry must be protected from any light exposure once reconstituted. Light treatment is performed at 40–50 hours after intravenous infusion to allow the majority of drug to be cleared from normal tissues. Typical metabolism of the chemical occurs within 40–72 hours. Porfimer sodium is contraindicated in patients taking other photosensitizers including fluoroquinolones, griseofulvin, phenothiazines, sulfonamides, sulfonylurea, tetracyclines, and thiazide diuretics.

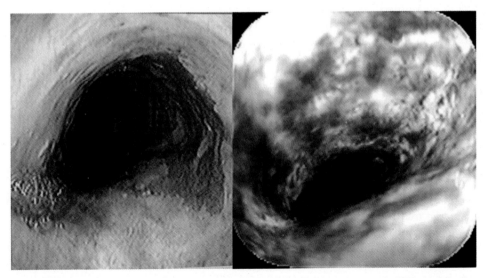

FIGURE 10.1   Barrett's esophagus pretreatment (left). Three days posttreatment with PDT (right). *Source: Images courtesy of Michael Kimmey, M.D.*

The application of PDT involves deploying a cylindrical centering balloon that stabilizes a diffusion catheter which serves as the light delivery device. The balloon is inflated within the esophageal lumen to stabilize the optical fiber to equilibrate the circumferential distribution and intensity of light. Light dosimetry (joules per centimeter) for BE with HGD is 150−200 and 300 J/cm for EA; this quantity is calculated by multiplying power output of the diffuser (watts) by treatment time (seconds) divided by diffuser length (cm). Currently, the Diomed 630 PDT Laser Model 2TUSA (Diomed Inc., Andover, MA) is the only FDA-approved light source for administering PDT for use with porfimer sodium, which simplifies the calibration process [31]. The endoscopist inputs information regarding the target organ, pathology, and fiber length from which the device automatically calibrates dosimetry (light power and duration). Treatment plans can be altered from standard guidelines to include repeat treatments for missed or "skipped" lesions, or following debridement of adenocarcinoma.

Limitations of treatment using PDT with porfimer sodium include acute toxicities of photosensitizing agents, systemic phototoxicity, and local effects from therapy. The most common postprocedural complaints include odynophagia and chest pain, abdominal pain, nausea and vomiting, fever, and pleural effusion. Rare adverse effects include anemia related to bleeding from mucosal ulceration, esophageal perforation, atrial fibrillation, and respiratory compromise. Local complications of scarring including esophageal stricture formation reported at upwards of 58% of cases typically occurring about 1−2 months after treatment (Fig. 10.2). Frequency of side effects is correlated with higher light dosage and amount of therapeutic exposure (ie, type of pretreatment, time between treatment sessions, and number of treatment sessions) [32].

PDT is contraindicated in patients with esophageal or gastric varices, esophageal ulcers >1 cm, esophageal fistulae, porphyria or sensitivity to porphyrins, or esophageal tumors with vascular invasion. Patients must be compliant with photosensitivity precautions including full skin

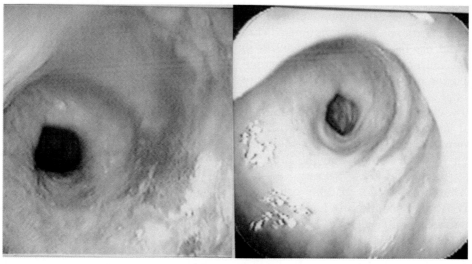

FIGURE 10.2 Healed esophagus 3 months post-PDT (left). Esophageal stricture post-PDT (right). *Source: Images courtesy of Michael Kimmey, M.D.*

coverage with hat, eyewear, complete clothing of limbs, etc. for at least 30 days postdrug administration to minimize risk of cutaneous phototoxicity.

10.4.1.2 PDT Compounds

Outside of the United States, PDT photosensitization using ALA is common. ALA is a metabolic precursor of the compound protoporphyrin IX and can be administered orally. 5-ALA PDT has demonstrated greater efficacy over porfimer sodium PDT for BE segments <6 cm [33]. These findings have been attributed to a number of factors including a higher mucosal (vs submucosal) concentration of the drug. An additional benefit of ALA over porfimer sodium is a shorter half-life of the 5-ALA compound which results in an increased rate of tissue clearance thereby decreasing the probability of adverse photosensitivity reactions and less esophageal stricturing [34,35]. Transaminase abnormality is associated in 50% of cases using ALA but is typically transient, lasting 3–4 days.

10.4.1.3 Efficacy

PDT using porfimer sodium was the first endoscopic ablation technique proven to reduce malignancy risk using randomized clinical data and has been used since the 1970s to treat other cancers [36]. However, contemporary utilization of PDT has diminished in favor of RFA after a clinical trial comparing PDT to RFA showed superior therapeutic efficacy, cost, and reduced complication rates in the RFA group [37]. PDT eradication rates are documented in the literature in the range of 77–100% based on observational studies [38,39]. One randomized placebo controlled trial demonstrated significantly higher rates of eradication using PDT with PPI (77%) versus PPI alone (39%) ($p < 0.0001$) [36]. Five-year follow-up data from the same group demonstrated significant difference in durability as evidenced by regression rates in the PDT-treated groups (15%) versus PPI treated alone (29%) [40].

Other studies demonstrate durable response to PDT but note a greater risk of buried metaplasia with PDT when compared with RFA and documented occurrences of subsquamous adenocarcinoma [41,42].

Nonrandomized control trials (RCT) of PDT using 5-ALA demonstrated efficacious treatment of HGD and early EA [43,44]. The first randomized trial of PDT using 5-ALA demonstrated significant macroscopic regression over placebo photosensitizer in treatment of BE with LGD [45].

10.5 THERMAL ABLATION

10.5.1 Argon Plasma Coagulation

APC is a noncontact, direct endoscopic application of electrocoagulation using ionized argon gas plasma. Implementation requires passage of a tungsten-tipped catheter through the working channel of the endoscope. A stream of argon gas becomes ionized at the distal catheter tip allowing conduction of electrical current once placed in close enough proximity to tissue in a grounded patient (eg, patient on whom a grounding pad has been placed). The flow of argon transfers the electrical current into the tissue causing thermal ablation (Fig. 10.3). Parameters of tissue coagulation include the following: flow rate of argon gas, electrical power of the cautery tip (generator setting), duration of application, and proximity of therapeutic application to target tissue (proxy for degree of resistance). These factors all affect the depth of tissue injury. There are multiple types of APC systems available as well as a variety of probes to customize forward, side, or circumferential application of electrocoagulation [46].

10.5.1.1 Procedure

Typical settings for APC used in neoplastic ablation are at a higher power output and

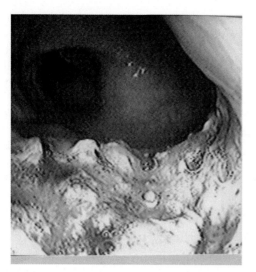

FIGURE 10.3 Immediate posttreatment with APC.
Source: Images courtesy of Michael Kimmey, M.D.

higher flow of argon gas (eg, 70–90 watts, 1–2 L/min) compared to typical settings used for control of bleeding lesions [3,47] with anticipated depth of injury of 3–4 mm. Ex vivo studies demonstrate that for esophageal ablation of BE a longer duration of treatment (3 seconds vs 1 second) had greater impact on depth of thermal injury than power settings (range 40–99 watts) [48]. However in vivo data suggests that higher power ablation is more efficacious at obliterating HGD and intramucosal EA [49]. Available literature reports vary regarding optimal operating distance of the catheter tip ranging from 2 to 8 mm [50] or a distance of 1 mm with the probe tip maintained at a 90° angle from tissue target [48]. Limitations to therapy include operator malfunction by embedding the probe within tissue which can result in deep tissue injury or submucosal emphysema with argon gas.

10.5.1.2 Efficacy

The bulk of data examining the efficacy of APC exists in the literature derived from the treatment of nondysplastic BE or BE with LGD.

There are a wide range of complete eradication rates reported by investigators and the procedure is generally regarded as safe [51] and more efficacious at destroying metaplastic tissue compared to PPI therapy alone [52]. A study in patients with nondysplastic BE with complete macroscopic ablation at time of treatment was followed for median duration of 51 months. Recurrence rates were documented at 19.7% for macroscopic evidence of BE and 12.1% with histologic evidence (3% annual relapse rate) with none of the patients developing neoplasia or EC [53]. Conflicting with this report is another long-term follow-up study, which demonstrated no protective effect against developing adenocarcinoma which developed in 9% of patients treated with APC (two cases from buried metaplasia, one from remnant BE) [54].

Recurrence rates for treatment of BE with LGD following ablation with APC is likewise less than reassuring; as one study showed recurrent metaplasia in 66% of patients during the course of a 51-month follow-up [55]. The risk of developing posttreatment neoplasia may be related to formation of buried glands postablation due to the shallow depth of tissue injury intrinsic to the modality [56]. With respect to treatment of BE with HGD, one study looking at outcomes over 7 years demonstrated reasonable eradication rates of 86% (25 patients) [57], however, the existing data regarding recurrence rates is limited and results appear mixed. Basu et al. demonstrated incomplete eradication of metaplastic tissue estimated at rates up to 68% using APC [58]. Some data suggests eradication rates can be improved using higher power settings; one such study reported eradication rates as high as 98.6% achieved using a power output of 90 watts (W) [59]. However, a recent multicenter trial of APC ablation in nondysplastic BE demonstrated complete eradication in only 77% of cases using a power output of 90 W [60]. Moreover, 17.6% of patients suffered adverse effects of treatment including retrosternal pain, odynophagia, fever, and bleeding requiring endoscopic intervention (repeat APC treatment and one requiring clipping). More serious complications of APC included formation of esophageal strictures/stenosis and esophageal perforation. For these reasons, the aforementioned disputed durability suggests the need for further study of APC before a clear clinical role can be established.

10.5.2 Multipolar Electrocautery

MPEC probes are another commonly utilized thermal ablation technique. Unlike APC which uses superheated plasma for thermal ablation and does not make physical contact to injury targets, MPEC requires direct contact with tissue to apply thermal injury (Fig. 10.4). Functionally, these devices contain a power source with two distal electrical probes arranged in an open circuit. Once the probes come in direct contact with target tissue, the circuit is completed with the tissue acting as conductor between the two probes. As the tissue presents the highest resistance in this circuit, current traveling with this tissue results in the generation of heat, causing thermal injury. This injury is localized to the path of current flow, ie, the tissue area between the distal probes. MPEC is used commonly for hemostasis in cases of GI bleeding and probes are available in the United States in bipolar or multipolar form with hydroflush capabilities. Some include retractable hollow bore needles useful for injecting saline, surgical tattoo markers, or epinephrine for the treatment of bleeding lesions.

10.5.2.1 Procedure

Parameters of tissue coagulation include the following: duration of application, power setting of the electrosurgical generator, and amount of pressure applied to the tissue. A power setting of 20 W is typical for ablative therapy. One ex vivo study associated 1 mm depth of tissue injury with 2-second application of light pressure at 20 W of power [61]. Additionally, the authors found that amount of pressure and duration of application affected the degree of thermal injury to a greater degree than electrical current settings.

Limitations to therapy with MPEC include application that is time-consuming/challenging, and lack of standardization for parameters for application (power settings, pressure, duration of application).

10.5.2.2 Efficacy

There is a great deal of variability in the literature for eradication rates of BE using MPEC for ablation, with reports indicating complete eradication in the range of 65−100% of cases [3,62]. Montes et al. examined the effectiveness of MPEC plus antireflux surgery in BE without dysplasia and observed that MPEC appeared to be effective with few complications related to the ablative treatment (ie, 10 days of self-limited dysphagia/odynophagia) [63]. In contrast, one meta-analysis comparing ablative techniques to esophagectomy found rates of patient reported odynophagia and dysphagia were highest with MPEC [64].

However, there are multicenter trials data looking at MPEC plus high dose PPI that demonstrated 78% macroscopic and histologic eradication rates leading the authors to recommend this treatment modality for short segments of nondysplastic BE [62]. While these studies suggest that MPEC is a low-cost, efficacious treatment paradigm for nondysplastic BE, long-term reports to confirm durability of results, ascertain complication rate of buried metaplasia, or prevention of EA do not exist. As such current society guidelines do not recommend use of MPEC for ablation of nondysplastic or LGD BE, given incomplete understanding of recurrence rates and optimal treatment approach. Currently, there are no

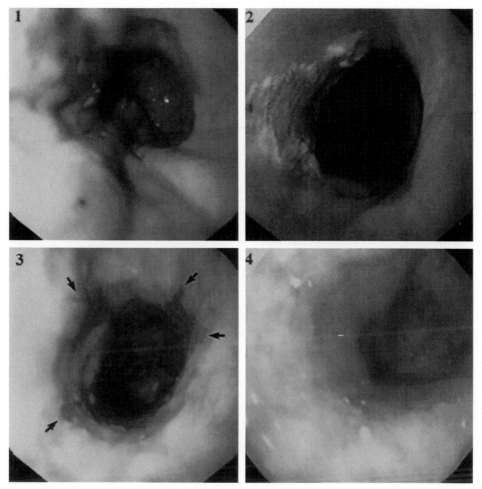

FIGURE 10.4 (Labeled 1–4). Tattooed margin of Barrett's esophagus (1), post-MPEC therapy (2), pretreatment image (3), posttreatment with seven MPEC sessions (4). *Source: Used with permission from Sampliner RE, Fennerty B, Garewal HS. Reversal of Barrett's esophagus with acid suppression and multipolar electrocoagulation: preliminary results. Gastrointest Endosc 1996;44(5):532–5.*

trials of MPEC ablation for treatment of HGD or submucosal EA.

10.5.3 KTP Laser and Nd:Yag Laser

Both KTP:Yag and Nd:Yag lasers are solid-state lasers that emit high-energy light which is then absorbed by tissue. This results in a transfer of photonic energy that is absorbed by tissue resulting in heat which causes localized

thermal injury (Figs. 10.5–10.8). Neodymium (Nd) serves as the dopant in the Nd:Yag laser and comprises a small portion of the yttrium aluminum garnet (Yag) crystal which emits collimated infrared light. Similarly, potassium titanyl phosphate (KTP) is the dopant in KTP: Yag lasers. There are a number of laser types available for endoscopic ablation used by other specialties (eg, urology, otolaryngology, and pulmonary medicine) including Nd:Yag, KTP:

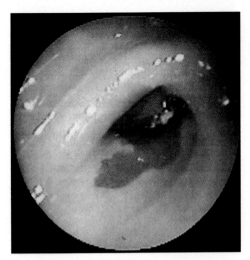

FIGURE 10.5   Barrett's esophagus pretreatment. *Source: Used with permission from Brandt LJ, Kauvar DR. Laser-induced transient regression of Barrett's epithelium. Gastrointest Endosc 1992;38(5):619−22.*

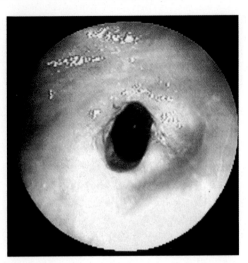

FIGURE 10.7   Six weeks posttreatment with laser therapy. *Source: Used with permission from Brandt LJ, Kauvar DR. Laser-induced transient regression of Barrett's epithelium. Gastrointest Endosc 1992;38(5):619−22.*

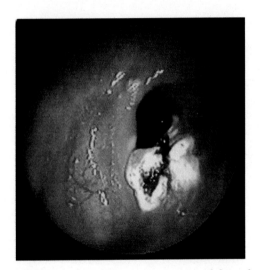

FIGURE 10.6   Immediate posttreatment with laser therapy. *Source: Used with permission from Brandt LJ, Kauvar DR. Laser-induced transient regression of Barrett's epithelium. Gastrointest Endosc 1992;38(5):619−22.*

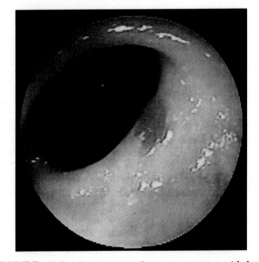

FIGURE 10.8   Fourteen weeks posttreatment with laser therapy. *Source: Used with permission from Brandt LJ, Kauvar DR. Laser-induced transient regression of Barrett's epithelium. Gastrointest Endosc 1992;38(5):619−22.*

Yag, CO_2, neodymium−holmium and diode lasers. However, only the Yag variants have been investigated within the GI literature for esophageal ablation [2]. Additionally, devices are designed as "contact" or "noncontact." Contact lasers make direct physical contact to target tissues using a sapphire interface and can be used with more tangentially located areas that can present a challenge for *en face* endoscopic approach. For noncontact devices, the laser fiber is maintained within the working channel of the endoscope with either tangential or straight-on beams for application. There is evidence to suggest that variable power contact devices using sapphire tips offer greater control with less unintentional thermal injury such as adjacent edema and charring [65].

10.5.3.1 Procedure

Parameters for tissue coagulation (and depth of laser light penetration) include type of laser (which determines the wavelength of light emitted), tissue type of target region, power settings, and duration of application. Nd:Yag lasers penetrate tissue more deeply (up to 4–6 mm) when compared with KTP:Yag laser [66].

Much of the early, noncomparative data investigating the use of endoscopic ablation of nondysplastic BE focused on laser therapy and the ability to generate neosquamous epithelium [67,68]. Laser therapy for ablation in BE has largely fallen out of favor due to challenges of application including requirement of special safety training, protective eyewear, and limited portability [51]. This has also resulted in a dearth of data regarding the efficacy of laser treatment and there are no recent studies examining the eradication rates, recurrence rates, or complications of this modality.

10.5.3.2 Heat Probe

Endoscopic ablation of BE involves application of an electric heat probe to inflict direct thermal injury. Unlike MPEC, which employs electrical current passing through the target tissue, endoscopic heat probe causes direct thermal injury similar to an iron. The degree of ablation is determined purely by the duration of application and amount of coaptive pressure applied. These variables are reliant on the skill of the operator and consistency of ablation is challenging. The probe itself consists of a Teflon-coated aluminum cylinder containing a metal coil filament that is mounted to the probe tip. The amount of heat is controlled through preset current profiles, and the device is activated by foot pedal depression.

10.5.3.3 Efficacy

One small case series in 13 patients with non-dysplastic BE examined feasibility of eradication of abnormal mucosa using a heat probe with concomitant PPI use [69]. The authors reported complete macroscopic eradication after multiple treatment sessions but 23% of patients were found to have buried metaplasia in follow-up.

10.6 INTERMODALITY COMPARISON

10.6.1 PDT Versus APC

Data comparing PDT to APC includes one randomized study which demonstrated higher eradication rates using PDT with porfimer sodium over APC in treatment of dysplastic BE [70]. The data is heterogeneous particularly with regard to treatment endpoints in cases of nondysplastic BE. One randomized trial demonstrated similar rates of buried metaplasia between the APC- and PDT-treated groups; however, APC-treated BE had 97% eradication rates significantly exceeding the 50% eradication rates demonstrated in the PDT groups ($p < 0.0001$) [71]. One long-term study published comparable efficacies between both therapies in prevention of metachronous neoplasia following EMR [72]. Although most studies of PDT are with porfimer sodium, some data exists using PDT with 5-ALA. PDT with 5-ALA seems to be more effective than APC at inducing disease regression but is commonly associated with complications of nausea/vomiting, pain during treatment and transient abnormal liver function tests. One patient in this study died 3 days postprocedure from reasons that remained unclear even after autopsy [34].

10.6.2 MPEC Versus APC

There is very little data on MPEC outcomes and only two studies comparing MPEC to other modalities. In two prospective trials between APC versus MPEC, there were no differences in eradication rates for treatment of nondysplastic BE. One randomized controlled trial required multiple treatment sessions for both APC and MPEC without a significant difference in the number of sessions required to demonstrate macroscopic eradication (75% with MPEC, 63% with APC) in patients with nondysplastic BE [73]. After 2 years, 70% of patients maintained regressed disease. Another randomized prospective trial of MPEC versus APC on PPI documented similar findings [11].

10.6.3 Others

There are no comparative trials of laser therapy to other ablative devices. As such, utilization of laser therapy for BE with HGD or submucosal EA is not endorsed as first line therapy.

No comparative data exists for heat probe with other modalities.

10.7 CONCLUSION

Historically, there have been multiple endoscopic ablative techniques developed for treatment of BE with or without neoplasia. Many of these modalities have fallen out of common use due to various factors. Of the techniques discussed in this chapter, PDT and APC have the most supportive data and historical use, however, unfavorable adverse effect profiles and greater inconvenience with administration limit their usefulness. Other methods, such as MPEC, are similarly limited by difficulty with administration and meager evidence-based efficacy, while high-energy lasers were obviated by exorbitant cost and exotic training requirements.

Finally, while selection of ablative therapy must be individualized and will be dependent on patient case, endoscopist's skillset, and clinical resource availability, efficacy remains the most important aspect of BE therapy. Discussion of these ablative therapies is primarily to provide historical perspective on endoscopic ablative therapies for BE as RFA and cryotherapy are now more commonly employed. Further detail regarding these ablative techniques is presented in the subsequent chapters.

References

[1] Berenson MM, Johnson TD, Markowitz NR, Buchi KN, Samowitz WS. Restoration of squamous mucosa after ablation of Barrett's esophageal epithelium. Gastroenterology 1993;104(6):1686–91.

[2] Brandt LJ, Kauvar DR. Laser-induced transient regression of Barrett's epithelium. Gastrointest Endosc 1992;38(5):619–22.

[3] Spechler SJ, Sharma P, Souza RF, Inadomi JM, Shaheen NJ, American Gastroenterological Association. American Gastroenterological Association technical review on the management of Barrett's esophagus. Gastroenterology 2011;140(3):e18–52 quiz e13

[4] Spechler SJ, Fitzgerald RC, Prasad GA, Wang KK. History, molecular mechanisms, and endoscopic treatment of Barrett's esophagus. Gastroenterology 2010;138 (3):854–69.

[5] Williams VA, Watson TJ, Herbella FA, Gellersen O, Raymond D, Jones C, et al. Esophagectomy for high grade dysplasia is safe, curative, and results in good alimentary outcome. J Gastrointest Surg 2007;11 (12):1589–97.

[6] Sujendran V, Sica G, Warren B, Maynard N. Oesophagectomy remains the gold standard for treatment of high-grade dysplasia in Barrett's oesophagus. Eur J Cardiothorac Surg 2005;28(5):763–6.

[7] Inadomi JM, Somsouk M, Madanick RD, Thomas JP, Shaheen NJ. A cost–utility analysis of ablative therapy for Barrett's esophagus. Gastroenterology 2009;136 (7):2101–14.

[8] Zehetner J, DeMeester SR, Hagen JA, Ayazi S, Augustin F, Lipham JC, et al. Endoscopic resection and ablation versus esophagectomy for high-grade dysplasia and intramucosal adenocarcinoma. J Thorac Cardiovasc Surg 2011;141(1):39–47.

[9] Fitzgerald RC, di Pietro M, Ragunath K, Ang Y, Kang JY, Watson P, et al. British Society of Gastroenterology guidelines on the diagnosis and management of Barrett's oesophagus. Gut 2014;63(1):7–42.

[10] Sampliner RE, Hixson LJ, Fennerty MB, Garewal HS. Regression of Barrett's esophagus by laser ablation in an anacid environment. Dig Dis Sci 1993;38(2):365—8.

[11] Dulai GS, Jensen DM, Cortina G, Fontana L, Ippoliti A. Randomized trial of argon plasma coagulation vs. multipolar electrocoagulation for ablation of Barrett's esophagus. Gastrointest Endosc 2005;61(2):232—40.

[12] Sharma VK, Wang KK, Overholt BF, Lightdale CJ, Fennerty MB, Dean PJ, et al. Balloon-based, circumferential, endoscopic radiofrequency ablation of Barrett's esophagus: 1-year follow-up of 100 patients. Gastrointest Endosc 2007;65(2):185—95.

[13] Prasad GA, Wang KK, Buttar NS, Lightdale CJ, Fennerty MB, Dean PJ, et al. Long-term survival following endoscopic and surgical treatment of high-grade dysplasia in Barrett's esophagus. Gastroenterology 2007;132(4):1226—33.

[14] Ngamruengphong S, Wolfsen HC, Wallace MB. Survival of patients with superficial esophageal adenocarcinoma after endoscopic treatment vs surgery. Clin Gastroenterol Hepatol 2013;11(11):1424—9 e1422; quiz e1481

[15] Rastogi A, Puli S, El-Serag HB, Bansal A, Wani S, Sharma P. Incidence of esophageal adenocarcinoma in patients with Barrett's esophagus and high-grade dysplasia: a meta-analysis. Gastrointest Endosc 2008;67(3):394—8.

[16] Committee ASoP, Evans JA, Early DS, Fukami N, Ben-Menachem T, Chandrasekhara V, et al. The role of endoscopy in Barrett's esophagus and other premalignant conditions of the esophagus. Gastrointest Endosc 2012;76(6):1087—94.

[17] Buttar NS, Wang KK, Sebo TJ, Riehle DM, Krishnadath KK, Lutzke LS, et al. Extent of high-grade dysplasia in Barrett's esophagus correlates with risk of adenocarcinoma. Gastroenterology 2001;120(7):1630—9.

[18] Dunbar KB. Endoscopic eradication therapy for mucosal neoplasia in Barrett's esophagus. Curr Opin Gastroenterol 2013;29(4):446—53.

[19] El-Serag HB, Aguirre TV, Davis S, Kuebeler M, Bhattacharyya A, Sampliner RE. Proton pump inhibitors are associated with reduced incidence of dysplasia in Barrett's esophagus. Am J Gastroenterol 2004;99(10):1877—83.

[20] American Gastroenterological Association, Spechler SJ, Sharma P, Souza RF, Inadomi JM, Shaheen NJ. American Gastroenterological Association Medical Position Statement on the management of Barrett's esophagus. Gastroenterology 2011;140(3):1084—91.

[21] Shaheen NJ, Crosby MA, Bozymski EM, Sandler RS. Is there publication bias in the reporting of cancer risk in Barrett's esophagus? Gastroenterology 2000;119(2):333—8.

[22] Sharma P, Falk GW, Weston AP, Reker D, Johnston M, Sampliner RE. Dysplasia and cancer in a large multicenter cohort of patients with Barrett's esophagus. Clin Gastroenterol Hepatol 2006;4(5):566—72.

[23] Ackroyd R, Kelty CJ, Brown NJ, Stephenson TJ, Stoddard CJ, Reed MW. Eradication of dysplastic Barrett's oesophagus using photodynamic therapy: long-term follow-up. Endoscopy 2003;35(6):496—501.

[24] Phoa KN, van Vilsteren FG, Weusten BL, Bisschops R, Schoon EJ, Ragunath K, et al. Radiofrequency ablation vs endoscopic surveillance for patients with Barrett esophagus and low-grade dysplasia: a randomized clinical trial. JAMA 2014;311(12):1209—17.

[25] Spechler SJ, Souza RF. Barrett's esophagus. N Engl J Med 2014;371(9):836—45.

[26] Almond LM, Hodson J, Barr H. Meta-analysis of endoscopic therapy for low-grade dysplasia in Barrett's oesophagus. Br J Surg 2014;101(10):1187—95.

[27] Buskens CJ, Westerterp M, Lagarde SM, Bergman JJ, ten Kate FJ, van Lanschot JJ. Prediction of appropriateness of local endoscopic treatment for high-grade dysplasia and early adenocarcinoma by EUS and histopathologic features. Gastrointest Endosc 2004;60(5):703—10.

[28] Konda VJ, Gonzalez Haba Ruiz M, Koons A, Hart J, Xiao SY, Siddiqui UD, et al. Complete endoscopic mucosal resection is effective and durable treatment for Barrett's-associated neoplasia. Clin Gastroenterol Hepatol 2014;12(12):2002—10.

[29] Pech O, May A, Manner H, Behrens A, Pohl J, Weferling M, et al. Long-term efficacy and safety of endoscopic resection for patients with mucosal adenocarcinoma of the esophagus. Gastroenterology 2014;146(3):652—60 e651

[30] Chevaux JB, Piessevaux H, Jouret-Mourin A, Yeung R, Danse E, Deprez PH. Clinical outcome in patients treated with endoscopic submucosal dissection for superficial Barrett's neoplasia. Endoscopy 2015;47(2):103—12.

[31] Petersen BT, Chuttani R, Croffie J, DiSario J, Liu J, Mishki D, et al. Photodynamic therapy for gastrointestinal disease. Gastrointest Endosc 2006;63(7):927—32.

[32] Wang KK, Nijhawan PK. Complications of photodynamic therapy in gastrointestinal disease. Gastrointest Endosc Clin N Am 2000;10(3):487—95.

[33] Dunn JM, Mackenzie GD, Banks MR, Mosse CA, Haidry R, Green S, et al. A randomised controlled trial of ALA vs. photofrin photodynamic therapy for high-grade dysplasia arising in Barrett's oesophagus. Lasers Medical Sci 2013;28(3):707—15.

[34] Hage M, Siersema PD, van Dekken H, Steyerberg EW, Haringsma J, van de Vrie W, et al. 5-aminolevulinic

acid photodynamic therapy versus argon plasma coagulation for ablation of Barrett's oesophagus: a randomised trial. Gut 2004;53(6):785–90.

[35] Peters F, Kara M, Rosmolen W, Aalders M, Ten Kate F, Krishnadath K, et al. Poor results of 5-aminolevulinic acid-photodynamic therapy for residual high-grade dysplasia and early cancer in barrett esophagus after endoscopic resection. Endoscopy 2005;37(5):418–24.

[36] Overholt BF, Lightdale CJ, Wang KK, Canto MI, Burdick S, Haggitt RC, et al. Photodynamic therapy with porfimer sodium for ablation of high-grade dysplasia in Barrett's esophagus: international, partially blinded, randomized phase III trial. Gastrointest Endosc 2005;62(4):488–98.

[37] Ertan A, Zaheer I, Correa AM, Thosani N, Blackmon SH. Photodynamic therapy vs radiofrequency ablation for Barrett's dysplasia: efficacy, safety and cost-comparison. World J Gastroenterol 2013;19(41):7106–13.

[38] Wolfsen HC. Photodynamic therapy for mucosal esophageal adenocarcinoma and dysplastic Barrett's esophagus. Dig Dis 2002;20(1):5–17.

[39] Overholt BF, Panjehpour M, Haydek JM. Photodynamic therapy for Barrett's esophagus: follow-up in 100 patients. Gastrointest Endosc 1999;49 (1):1–7.

[40] Overholt BF, Wang KK, Burdick JS, Lightdale CJ, Kimmey M, Nava HR, et al. Five-year efficacy and safety of photodynamic therapy with photofrin in Barrett's high-grade dysplasia. Gastrointest Endosc 2007;66(3):460–8.

[41] Gray NA, Odze RD, Spechler SJ. Buried metaplasia after endoscopic ablation of Barrett's esophagus: a systematic review. Am J Gastroenterol 2011;106 (11):1899–908.

[42] Ban S, Mino M, Nishioka NS, Puricelli W, Zukerberg LR, Shimizu M, et al. Histopathologic aspects of photodynamic therapy for dysplasia and early adenocarcinoma arising in Barrett's esophagus. Am J Surg Pathol 2004;28(11):1466–73.

[43] Gossner L, Stolte M, Sroka R, Rick K, May A, Hahn EG, et al. Photodynamic ablation of high-grade dysplasia and early cancer in Barrett's esophagus by means of 5-aminolevulinic acid. Gastroenterology 1998;114(3):448–55.

[44] Pech O, Gossner L, May A, Rabenstein T, Vieth M, Stolte M, et al. Long-term results of photodynamic therapy with 5-aminolevulinic acid for superficial Barrett's cancer and high-grade intraepithelial neoplasia. Gastrointest Endosc 2005;62(1):24–30.

[45] Ackroyd R, Brown NJ, Davis MF, Stephenson TJ, Marcus SL, Stoddard CJ, et al. Photodynamic therapy for dysplastic Barrett's oesophagus: a prospective,

double blind, randomised, placebo controlled trial. Gut 2000;47(5):612–17.

[46] Committee AT, Tokar JL, Barth BA, Banerjee S, Chauhan SS, Gottlieb KT, et al. Electrosurgical generators. Gastrointest Endosc 2013;78(2):197–208.

[47] Farin G, Grund KE. Technology of argon plasma coagulation with particular regard to endoscopic applications. Endosc Surg Allied Technol 1994;2(1):71–7.

[48] Watson JP, Bennett MK, Griffin SM, Matthewson K. The tissue effect of argon plasma coagulation on esophageal and gastric mucosa. Gastrointest Endosc 2000;52(3):342–5.

[49] Byrne JP, Armstrong GR, Attwood SE. Restoration of the normal squamous lining in Barrett's esophagus by argon beam plasma coagulation. Am J Gastroenterol 1998;93(10):1810–15.

[50] Ginsberg GG, Barkun AN, Bosco JJ, Burdick JS, Isenberg GA, Nakao NL, et al. The argon plasma coagulator: February 2002. Gastrointest Endosc 2002;55 (7):807–10.

[51] American Society for Gastrointestinal Endoscopy Technology Committee. Mucosal ablation devices. Gastrointest Endosc 2008;68(6):1031–42.

[52] Bright T, Watson DI, Tam W, Game PA, Ackroyd R, Devitt PG, et al. Prospective randomized trial of argon plasma coagulation ablation versus endoscopic surveillance of Barrett's esophagus in patients treated with antisecretory medication. Dig Dis Sci 2009;54 (12):2606–11.

[53] Madisch A, Miehlke S, Bayerdorffer E, Wiedemann B, Antos D, Sievert A, et al. Long-term follow-up after complete ablation of Barrett's esophagus with argon plasma coagulation. World J Gastroenterol 2005;11 (8):1182–6.

[54] Milashka M, Calomme A, Van Laethem JL, Wiedemann B, Antos D, Sievert A, et al. Sixteen-year follow-up of Barrett's esophagus, endoscopically treated with argon plasma coagulation. United Eur Gastroenterol J 2014;2(5):367–73.

[55] Mork H, Al-Taie O, Berlin F, Kraus MR, Scheurlen M. High recurrence rate of Barrett's epithelium during long-term follow-up after argon plasma coagulation. Scand J Gastroenterol 2007;42(1):23–7.

[56] Van Laethem JL, Jagodzinski R, Peny MO, Cremer M, Deviere J. Argon plasma coagulation in the treatment of Barrett's high-grade dysplasia and in situ adenocarcinoma. Endoscopy 2001;33(3):257–61.

[57] Attwood SE, Lewis CJ, Caplin S, Hemming K, Armstrong G. Argon beam plasma coagulation as therapy for high-grade dysplasia in Barrett's esophagus. Clin Gastroenterol Hepatol 2003;1(4):258–63.

[58] Basu KK, Pick B, Bale R, West KP, de Caestecker JS. Efficacy and one year follow up of argon plasma

coagulation therapy for ablation of Barrett's oesophagus: factors determining persistence and recurrence of Barrett's epithelium. Gut 2002;51(6):776–80.

[59] Schulz H, Miehlke S, Antos D, Schentke KU, Vieth M, Stolte M, et al. Ablation of Barrett's epithelium by endoscopic argon plasma coagulation in combination with high-dose omeprazole. Gastrointest Endosc 2000;51(6):659–63.

[60] Manner H, May A, Miehlke S, Dertinger S, Wigginghaus B, Schimming W, et al. Ablation of non-neoplastic Barrett's mucosa using argon plasma coagulation with concomitant esomeprazole therapy (APBANEX): a prospective multicenter evaluation. Am J Gastroenterol 2006;101(8):1762–9.

[61] Laine L, Long GL, Bakos GJ, Vakharia OJ, Cunningham C. Optimizing bipolar electrocoagulation for endoscopic hemostasis: assessment of factors influencing energy delivery and coagulation. Gastrointest Endosc 2008;67(3):502–8.

[62] Sampliner RE, Faigel D, Fennerty MB, Lieberman D, Ippoliti A, Lewin K, et al. Effective and safe endoscopic reversal of nondysplastic Barrett's esophagus with thermal electrocoagulation combined with high-dose acid inhibition: a multicenter study. Gastrointest Endosc 2001;53(6):554–8.

[63] Montes CG, Brandalise NA, Deliza R, Novais de Magalhaes AF, Ferraz JG. Antireflux surgery followed by bipolar electrocoagulation in the treatment of Barrett's esophagus. Gastrointest Endosc 1999;50(2):173–7.

[64] Menon D, Stafinski T, Wu H, Lau D, Wong C. Endoscopic treatments for Barrett's esophagus: a systematic review of safety and effectiveness compared to esophagectomy. BMC Gastroenterol 2010;10:111.

[65] Weston AP. Use of lasers in Barrett's esophagus. Gastrointest Endosc Clin N Am 2003;13(3):467–81.

[66] Willems PW, Vandertop WP, Verdaasdonk RM, van Swol CF, Jansen GH. Contact laser-assisted neuroendoscopy can be performed safely by using pretreated "black" fibre tips: experimental data. Lasers Surg Med 2001;28(4):324–9.

[67] Gossner L, May A, Stolte M, Seitz G, Hahn EG, Ell C. KTP laser destruction of dysplasia and early cancer in columnar-lined Barrett's esophagus. Gastrointest Endosc 1999;49(1):8–12.

[68] Barham CP, Jones RL, Biddlestone LR, Hardwick RH, Shepherd NA, Barr H. Photothermal laser ablation of Barrett's oesophagus: endoscopic and histological evidence of squamous re-epithelialisation. Gut 1997;41 (3):281–4.

[69] Michopoulos S, Tsibouris P, Bouzakis H, Sotiropoulou M, Kralios N. Complete regression of Barrett's esophagus with heat probe thermocoagulation: mid-term results. Gastrointest Endosc 1999;50(2): 165–72.

[70] Ragunath K, Krasner N, Raman VS, Haqqani MT, Phillips CJ, Cheung I. Endoscopic ablation of dysplastic Barrett's oesophagus comparing argon plasma coagulation and photodynamic therapy: a randomized prospective trial assessing efficacy and cost-effectiveness. Scand J Gastroenterol 2005;40(7): 750–8.

[71] Kelty CJ, Ackroyd R, Brown NJ, Stephenson TJ, Stoddard CJ, Reed MW. Endoscopic ablation of Barrett's oesophagus: a randomized-controlled trial of photodynamic therapy vs. argon plasma coagulation. Aliment Pharmacol Ther 2004;20(11–12): 1289–96.

[72] Pech O, Behrens A, May A, Nachbar L, Gossner L, Rabenstein T, et al. Long-term results and risk factor analysis for recurrence after curative endoscopic therapy in 349 patients with high-grade intraepithelial neoplasia and mucosal adenocarcinoma in Barrett's oesophagus. Gut 2008;57(9):1200–6.

[73] Sharma P, Wani S, Weston AP, Bansal A, Hall M, Mathur S, et al. A randomised controlled trial of ablation of Barrett's oesophagus with multipolar electrocoagulation versus argon plasma coagulation in combination with acid suppression: long term results. Gut 2006;55(9):1233–9.

Radiofrequency Ablation

Kamar Belghazi, Jacques J. Bergman and Roos E. Pouw

Department of Gastroenterology and Hepatology, Academic Medical Center, Amsterdam,
The Netherlands

11.1 INTRODUCTION

Barrett's esophagus (BE) occurs when an abnormal, intestinal-type epithelium called "specialized intestinal metaplasia" replaces the stratified squamous epithelium that normally lines the distal esophagus. The condition develops as a consequence of chronic gastroesophageal reflux disease and predisposes to the development of adenocarcinoma of the esophagus [1].

Traditionally, high-grade dysplasia (HGD) and intramucosal carcinoma (IMC) arising from BE were treated with esophagectomy, while nondysplastic BE and BE with low-grade dysplasia (LGD) were managed with endoscopic surveillance. Problems associated with these approaches included significant morbidity and mortality from esophagectomy [2–4], and the risk of missed or interval development of cancer in patients undergoing surveillance [5]. Over the past decades, much research has focused on endoscopic imaging and treatment techniques to improve endoscopic detection and treatment of early esophageal neoplasia.

Radiofrequency ablation (RFA) is an endoscopic treatment modality for eradication of BE.

Primary circumferential ablation is performed using a balloon-based bipolar electrode, while secondary treatment of residual Barrett's mucosa is performed using an endoscope-mounted bipolar electrode on an articulated platform [6]. Studies suggest that this ablation technique is highly effective in removing Barrett's mucosa and associated dysplasia and in preventing progression of disease, while minimizing the known drawbacks of other ablation techniques (eg, photodynamic therapy (PDT), argon plasma coagulation (APC)) such as esophageal stenosis and subsquamous foci of intestinal metaplasia (IM, "buried Barrett's") [7].

11.2 INDICATIONS FOR RADIOFREQUENCY ABLATION

11.2.1 Barrett's Esophagus with Visible Lesions Containing High-Grade Dysplasia or Intramucosal Carcinoma

Patients with BE and visible abnormalities containing HGD or IMC may be treated with

D. Pleskow & T. Erim (Eds): Barrett's Esophagus.
DOI: http://dx.doi.org/10.1016/B978-0-12-802511-6.00011-9

RFA, but only after endoscopic resection (ER) of all macroscopic lesions.

ER provides a relatively large tissue specimen that allows for histopathological staging of a lesion, enabling optimal selection of patients who are eligible for further endoscopic management [8–10]. Patients found to have submucosal invading lesions on histology (>T1sm1) have a 15–30% risk of positive local lymph nodes and should be referred for surgery. However, the risk of lymph node involvement is minimal in patients with HGD and IMC limited to the mucosa, with good to moderate differentiation, no signs of lymphatic/vascular invasion, which is radically resected, making these patients candidates for endoscopic management [11,12]. In addition to staging a lesion, ER also renders the mucosa flat prior to RFA, which helps to ensure that

the ablation reaches the muscularis mucosae (Fig. 11.1).

Additional RFA of all remaining Barrett's mucosa after focal ER of visible lesions is advised to prevent metachronous lesions arising from the residual Barrett's mucosa after focal ER [13].

11.2.2 Barrett's Esophagus with Flat High-Grade Dysplasia

Patients with BE and flat HGD seem to be ideal candidates for RFA, since successful eradication of their dysplastic BE prevents the development of cancer [14]. However, proper patient selection is critical. To ensure that only patients with flat HGD are being treated with RFA monotherapy, several studies have

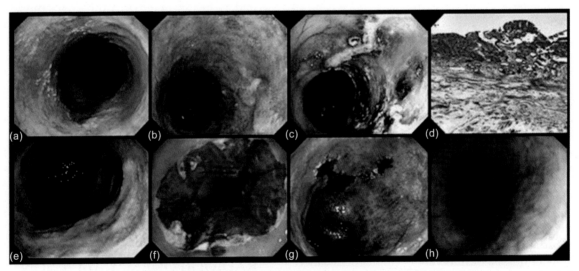

FIGURE 11.1   Endoscopic and histological images of long-segment Barrett's esophagus (BE) with early cancer treated with a combination of endoscopic resection (ER) and radiofrequency ablation (RFA). (a) Antegrade view of BE, (b) a lesion suspicious for early cancer at 2–4 o'clock, (c) view of the esophagus after ER of the lesion in two pieces, (d) histopathological evaluation of the specimens showed a radically resected adenocarcinoma infiltrating the muscularis mucosae (T1m3), (e) same area 6 weeks after the ER showing that the wound has healed completely with scarring, (f) esophagus after primary circumferential RFA, (g) residual islands of Barrett's mucosa 6 weeks after circumferential RFA, visualized with narrow-band imaging, (h) after additional focal RFA of the residual islands of Barrett's mucosa, complete eradication of the Barrett's segment was achieved. *Source: Reproduced with permission of www.endosurgery.eu. Copyright © 2010 Esophageal Research Foundation Amsterdam.*

required that patients undergo at least two high-resolution endoscopies with four-quadrant biopsies every 1–2 cm within 2 months prior to RFA to exclude visible lesions or cancer [8,9,15]. The use of RFA for flat IMC has only been evaluated in retrospective cohort studies [16].

11.2.3 Barrett's Esophagus with Low-Grade Dysplasia

There are several arguments that favor the use of RFA for LGD in BE. First, a confirmed histological diagnosis of LGD in BE represents a significant risk for malignant progression. In one study, patients with a consensus diagnosis of LGD had an 85% cumulative risk of progressing to HGD during follow-up, with an annual incidence of 13.4% per patient per year [17]. In patients with a confirmed histological diagnosis of LGD, RFA has also demonstrated to reduce the risk of malignant progression. In a randomized trial (the SURF trial) that was conducted in 9 European centers with 136 patients with a confirmed histological diagnosis of LGD, patients were assigned to RFA treatment or standard endoscopic surveillance (at 6, 12, 24, and 36 months) [18]. The study was terminated early after an interim analysis showing superiority of ablation over surveillance. RFA significantly reduced the risk of progression to HGD or esophageal adenocarcinoma (1.5% vs 26.5%), and RFA also significantly reduced the risk of progression to esophageal adenocarcinoma alone (1.5% vs 8.8%).

In addition, a cost-effectiveness analysis suggested that RFA is the preferred strategy for LGD, but only if the LGD was confirmed (ie, the diagnosis was agreed on by more than one pathologist) and stable (ie, LGD was seen on biopsies obtained at least 6 months apart) [19].

Given these considerations, many experts believe that the net health benefit of RFA for LGD in BE is favorable, and thus RFA should be available to patients as a primary treatment option, provided that the diagnosis is confirmed by an expert pathologist and that the diagnosis has been confirmed on more than one occasion.

11.2.4 Nondysplastic Barrett's Esophagus

The risk of progression to cancer in patients with nondysplastic BE is small, and no objective markers are available yet to identify patients with an increased risk of developing cancer, although research looking at the risk stratification of nondysplastic BE has shown promising results.

Whether to offer RFA to patients with nondysplastic BE is highly controversial and is influenced by many factors. An argument against RFA in these patients is that the annual risk of malignant progression is low, and many patients with BE are elderly with significant comorbid conditions that limit their life expectancy. Factors that favor treatment include the efficacy and safety profile of RFA and potential cost savings.

For most patients with nondysplastic BE, the net health benefit of RFA may be too low to justify its use. However, RFA could be considered for selected patients (eg, <50 years and a positive family history for esophageal adenocarcinoma or a very long Barrett's segment).

11.3 TECHNICAL ASPECTS

RFA of BE generally starts with a stepwise circumferential ablation procedure, followed by focal ablation for any residual BE (Fig. 11.2). RFA is performed using the Barrx FLEX system, which is comprised of a number of distinct ablation catheters: the Barrx[360] ablation balloon for circumferential RFA and the Barrx[90/60/Ultra] and trough-the-scope device for focal RFA of Barrett's mucosa.

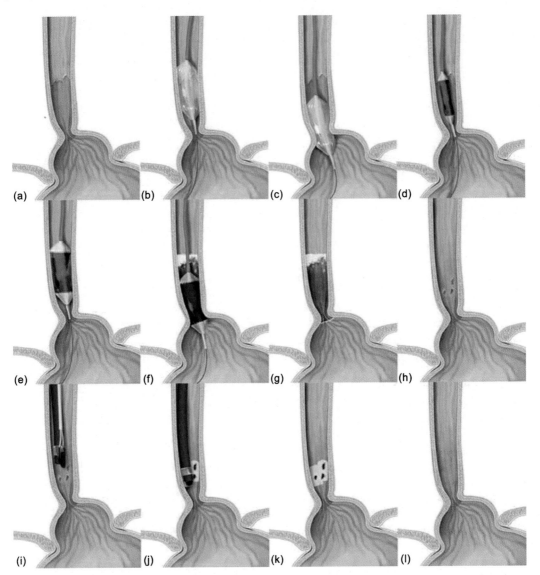

FIGURE 11.2   A schematic illustration of primary circumferential and secondary focal RFA of a Barrett's esophagus. (a) Pretreatment image of a Barrett's segment, (b and c) the esophageal diameter is measured at 1-cm intervals with a sizing balloon placed over a guide wire, (d) introduction of the RFA balloon catheter with the appropriate diameter over the guide wire, (e) the inflated RFA balloon positioned 1 cm above the proximal extent of the Barrett's segment, (f) the RFA balloon repositioned for ablation of the second zone after ablation of the first zone with an overlap of 1 cm, (g) image of the treated Barrett's segment immediately after the RFA ablation with necrosis of the superficial mucosa, (h) image of the healed esophagus 3 months after RFA with three small residual Barrett's islands, (i) introduction of the endoscope with the Barrx[90] catheter for focal ablation, (j) ablation of the third island of Barrett's mucosa, (k) image of the distal esophagus immediately after ablation of the three residual islands of Barrett's mucosa, (l) image of the healed distal esophagus showing complete regeneration with neosquamous mucosa. *Source: Reproduced with permission of www.endosurgery.eu. Copyright* © *2010 Esophageal Research Foundation Amsterdam.*

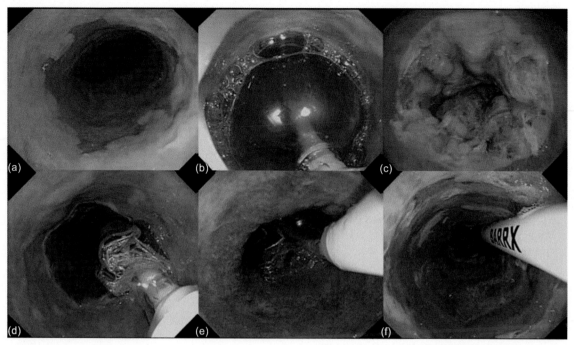

FIGURE 11.3   Endoscopic images of primary circumferential ablation using the Barrx[360] system. (a) Long-segment Barrett's esophagus with high-grade dysplasia, (b) the Barrx[360] catheter is introduced and inflated at the upper end of the Barrett's segment, (c) whitish coagulum resulting from ablation, (d) after ablation of the whole Barrett segment and cleaning of the electrode and ablation zone, the catheter is reintroduced for a second ablation pass, (e) the second ablation pass results in a tan-colored ablation zone, (f) treatment effect after two circumferential ablation passes (standard regimen of $2 \times 12 \, \text{J/cm}^2$ with cleaning). *Source: Reproduced with permission of www.endosurgery.eu. Copyright © 2010 Esophageal Research Foundation Amsterdam.*

11.3.1 Circumferential Ablation

Circumferential RFA with the Barrx[360] catheter involves the inflation of a balloon-based ablation catheter within the esophagus at the site of the BE. The ablation catheter holds a coiled electrode array on its outer surface, through which radiofrequency energy is applied, resulting in ablation of the mucosa (Fig. 11.3). The Barrx[360] catheter uses the Barrx FLEX energy generator.

1. *Landmark determination.* The first step in circumferential ablation is cleaning of the esophageal wall. Initially, this was done with 1% acetylcysteine and flushing with water to remove excessive mucus. A randomized trial has suggested that standard water rinsing through the water jet channel of the endoscope is just as effective [20]. We have therefore abandoned the cleaning with acetylcysteine. Next, the location of the top of the gastric folds and the proximal extent of the BE (including islands) are recorded for reference during the sizing and ablation procedures. A stiff guide wire or metal wire is then introduced, and the endoscope is removed, leaving the guide wire in place.

2. *Esophageal sizing.* Once the guide wire is in place, a sizing catheter is connected to the Barrx FLEX generator, calibrated, and introduced over the guide wire. The sizing

catheter is used to measure the inner esophageal diameter prior to circumferential ablation. It consists of a 165-cm-long shaft with 1-cm markings and a clear, 4-cm-long noncompliant balloon at its distal end. Upon activation via a footswitch, the sizing balloon is inflated to 4.3 psi (0.30 atm) by the generator using an integrated pressure: volume system. Based on the baseline balloon volume and geometry, the mean esophageal inner diameter is calculated along the entire length of the 4-cm-long balloon.

The sizing procedure can be performed as a "blind" procedure, using the 1-cm scale on the catheter shaft for reference. For the first measurement, the distal end of the balloon is placed 6 cm above the proximal extent of the Barrett's mucosa. After the first measurement cycle, the catheter is advanced 1 cm, and the sizing process is repeated. This sequence is reiterated until an increase in measured diameter indicates the transition into a hiatal hernia or the stomach.

3. *Ablation catheter selection.* Based on the esophageal inner diameter measurements, an appropriate Barrx^{360} ablation catheter is selected. The Barrx^{360} ablation catheter consists of a 165-cm-long shaft with a balloon at its distal end that contains a 3-cm-long bipolar electrode. The electrode contains 60 electrode rings that alternate in polarity and completely encircle the balloon (Fig. 11.4). The ablation catheter is available in five outer diameters (18, 22, 25, 28, and 31 mm).

The outer diameter of the ablation balloon should be smaller than the narrowest measured esophageal diameter. In patients who underwent prior ER, the ablation catheter should be selected conservatively (by taking an additional step down), keeping in mind that the sizing balloon calculates a mean inner diameter

Barrx90 ablation catheter

Barrx360 ablation catheter

FIGURE 11.4   The Barrx^{360} and Barrx^{90} ablation cathe-ters. *Source: Reproduced with permission from BARRX Medical Inc. Copyright © 2011. All rights reserved.*

over a length of 4 cm, which might result in an overestimation of the esophageal inner diameter at the site of an ER scar. For example, if the smallest measured diameter is 30 mm, a 28-mm balloon would be appropriate in a patient who had not undergone ER, whereas a 25-mm balloon would be chosen for a patient who had undergone ER [9].

4. *Ablation regimens.* Two different ablation regimens for circumferential ablation are currently in use. The standard ablation regimen consisting of two applications of 12 J/cm^2 with a cleaning phase in between is the most widely used regimen and has been studied extensively. A simplified regimen without a cleaning phase, however, has recently been proven equally effective in a randomized study [20].

a. Standard circumferential ablation. The ablation catheter is introduced over the guide wire, followed by the endoscope, which is advanced alongside the ablation catheter. Under endoscopic visualization, the proximal margin of the electrode is placed 1 cm above the most proximal extent of the BE. The ablation catheter is then inflated to 3 psi. Upon activation, radiofrequency energy is delivered to the electrode (12 J/cm^2). Energy delivery typically lasts less than 1.5 seconds, after which the balloon automatically deflates. Moving distally, the balloon is repositioned, allowing a small amount of overlap with the previous ablation zone (5–10 mm). We treat the entire BE segment in a single session, irrespective of its length [21]. After the entire length of BE has undergone one ablation cycle, the guide wire, ablation catheter, and endoscope are removed. Once outside the patient, the catheter is inflated and any adherent coagulum on the electrode surface is removed using wet gauze. A soft distal attachment cap is then fitted on the tip of the endoscope, and the scope is reintroduced into the patient. The soft extending rim of the cap is used to gently slough off the coagulum from the esophageal wall in the ablation zone. After most of the coagulum has been removed with the cap, forceful spraying of water through a spraying catheter using a high-pressure pistol can be used to wash off residual coagulum. After the cleaning procedure, the entire length of BE is ablated again using the same energy settings. A circumferential ablation treatment using the Barrx^{360} catheter takes approximately 30–40 minutes, depending on the length of the BE.

We would advise using the standard regimen in patients with a complex or tortuous BE (eg, a relative stenosis, narrowing at the ER site). The cleaning step of the standard regimen is a good way to assess the completeness of the first ablation pass and allows for necessary adjustments of the balloon position to treat skipped zones.

b. Simplified circumferential ablation. The ablation catheter is introduced over the guide wire, followed by the endoscope, which is advanced alongside the ablation catheter. Under endoscopic visualization, the proximal margin of the electrode is placed 1 cm above the most proximal extent of the BE. The ablation catheter is then inflated to 3 psi. Upon activation, radiofrequency energy is delivered to the electrode (12 J/cm^2). After the balloon has deflated, it is immediately inflated and activated for a second hit, while keeping the catheter in the same position. After two ablations of 12 J/cm^2, the catheter is moved distally and repositioned, allowing a small amount of overlap with the previous ablation zone (5–10 mm). A randomized trial demonstrated that this simplified regimen omitting the cleaning in between ablations is easier and faster than the standard regimen, but equally safe and effective [20]. The procedure time was reduced to 25 minutes. At 3 months after the circumferential ablation, the percentage of Barrett's surface regression did not differ significantly between those who underwent a simplified ablation and those who underwent a standard ablation. In addition, complete eradication of neoplasia and IM was similar in the two groups (100% and 90%, respectively).

Based on these results, we currently use the simplified protocol in patients with uncomplicated BE (without scarring or stenosis).

11.3.2 Follow-Up After Circumferential Ablation

Twelve weeks after the first circumferential ablation treatment, patients undergo a follow-up endoscopy, and additional therapy is performed. A second circumferential ablation is performed if:

- there is residual circumferential BE measuring 2 cm or more;
- there are multiple islands or tongues of BE.

Focal ablation using the Barrx^{90} catheter is performed in case of:

- residual BE with a circumferential extent less than 2 cm;
- circular treatment of the Z-line (at least once);
- small tongues of Barrett's mucosa;
- scattered islands of Barrett's mucosa.

11.3.3 Focal Ablation

Focal RFA with the Barrx^{90} catheter also uses radiofrequency energy to ablate small areas of BE. For focal ablation, the electric current is delivered through an electrode array attached to the tip of the endoscope (Fig. 11.5). The electrode array is mounted on an articulated platform (Fig. 11.4), allowing the electrode to move front to back and left to right, ensuring optimal tissue contact. It can be attached with a flexible strap to the distal end of any endoscope with a diameter of 8.6–12.8 mm without impairing endoscopic view or function. The electrode array is 20.6 mm long and 13.2 mm wide with an active electrode surface of 20 mm × 13 mm.

The Barrx^{90} catheter uses the Barrx FLEX energy generator. The treatment is performed by following the steps below:

1. *Electrode introduction.* The Barrx^{90} electrode array fits on the tip of the endoscope and is placed at the 12 o'clock position in the endoscopic video image. The device and endoscope are then introduced under visual guidance. When the laryngeal cavity is seen, the tip of the endoscope is deflected slightly downward. The endoscope is gently advanced into the esophagus, passing the leading edge of the catheter behind the arytenoids.

   In about 10% of cases, introducing the electrode array may prove difficult. In those cases, a Zenker's diverticulum should be excluded. Introduction of the device should never be forced due to the risk of perforation. In these cases, we will sometimes blindly pass a biopsy forceps or the spraying catheter into the esophagus to guide the endoscope into the proximal esophagus.

2. *Ablation regimens*

   a. Standard focal ablation. Residual Barrett's epithelium is positioned at the 12 o'clock position in the endoscopic video image, corresponding to the position of the electrode. The electrode is brought into close contact with the mucosa, deflected upward, and activated.

   Without separating the electrode from the esophageal wall, the electrode is immediately activated a second time, resulting in a "double" application of radiofrequency energy at 15 J/cm^2. Using 12 J/cm^2 has been favored in the United States. After all residual Barrett's mucosa has been ablated, the coagulum is carefully pushed off the esophageal wall with the leading edge of the electrode array. The electrode surface is then cleaned outside the patient. Finally, the ablation zone in the esophagus is rinsed with a spraying catheter and pressure pistol, as described above for standard circumferential ablation.

   Using the ablation zones from the first ablation pass as a guide, all ablated areas are treated with a second "double" application of radiofrequency energy at 15 J/cm^2 (resulting in a total of four applications).

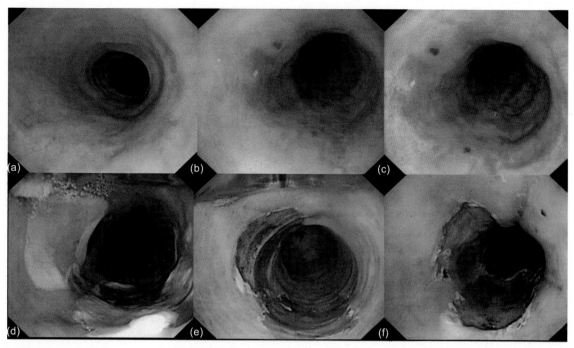

FIGURE 11.5 Endoscopic images of a focal ablation procedure using the Barrx⁹⁰ system. (a) Antegrade view of an esophagus after primary circumferential ablation for long-segment Barrett's esophagus, (b) residual islands of Barrett's mucosa at 6 and 11 o'clock, (c) corresponding image with narrow-band imaging, (d) ablation effect immediately after ablation with the Barrx⁹⁰ system. The distal end of the catheter is visible at the 12 o'clock position in the endoscopic field, (e) endoscopic appearance after the first ablation pass (two applications of 15 J/cm²) and cleaning of the ablation zones, (f) after the second ablation pass, the ablation zones have a tan-colored appearance. *Source: Reproduced with permission of www.endosurgery.eu. Copyright © 2010 Esophageal Research Foundation Amsterdam.*

In addition to treatment of any visually apparent Barrett's mucosa, ablation of the entire Z-line is recommended (even if no clear tongues of Barrett's are observed) to ensure eradication of all IM at the gastroesophageal junction [22].

b. Simplified focal ablation. In a randomized trial, the standard treatment regimen was compared with a simplified regimen consisting of three applications of radiofrequency energy at 15 J/cm², without a cleaning phase in between [23]. This procedure requires only a single introduction of the Barrx⁹⁰ electrode.

In 41 patients, the efficacy of both regimens was compared in pairs of Barrett areas or islands. The simplified regimen of three applications of radiofrequency energy at 15 J/cm² was noninferior to the standard regimen.

Potential indications for this simplified procedure are small residual islands, or in patients in whom introduction of the endoscope and the Barrx⁹⁰ electrode is difficult. However, the triple application has not been evaluated over larger surface areas and may in theory induce stenosis when applied on a larger scale.

11.3.4 New Ablation Devices

There are three new ablation devices available as an alternative to the Barrx[90] catheter, for focal ablation of Barrett's mucosa; however, no studies have yet evaluated the use of these devices in clinical practice. Recommendations for the use of these catheters are therefore based on previous experiences with the Barrx[90] device. "The Barrx[90] Ultra device," with an electrode array of 40 mm long and 13 mm wide, has a 200% larger electrode surface as compared with the regular Barrx[90] device. Patients can be treated with focal ablation using the Barrx[90] Ultra device if there are large tongues of residual BE or if there is short-segment BE. "The Barrx[60] device," with an electrode array of 15 mm long and 10 mm wide, has an active electrode surface area of 60% of the surface area of the regular Barrx[90] device. Patients can be treated with the Barrx[60] device for small islands of Barrett's mucosa in the presence of a stenosis. The latest is the "The channel RFA device," a through-the-scope device that fits through the working channel of a standard gastroscope with a recommended diameter of 2.8 mm or larger. The electrode array is 15.7 mm long and 7.5 mm wide and has approximately the same active electrode surface area as the Barrx[60] device.

11.3.5 Posttreatment Care

After RFA, acid suppressive therapy is important not only to minimize patient discomfort but also to allow the esophagus to heal optimally and regenerate with squamous epithelium. Studies suggest that ongoing gastroesophageal reflux has an adverse effect on treatment outcome [24,25]. All patients should, therefore, receive high-dose proton pump inhibitors as maintenance therapy. In addition, extra acid suppression after each treatment session is advisable. We prescribe all patients esomeprazole 40 mg twice daily, supplemented with ranitidine 300 mg at bedtime, and sucralfate suspension (5 mL of a 200 mg/mL suspension) three times a day for 2 weeks after each ablation session [24,25]. The proton pump inhibitor is continued as maintenance therapy.

After RFA, patients should adhere to a liquid diet for 24 hours. After 24 hours, patients may gradually advance to a soft and then normal diet. Patients may experience symptoms of chest discomfort, sore throat, difficulty or pain with swallowing, and/or nausea, which usually improve daily. If necessary acetaminophen 500–1000 mg up to four times per day may be given as needed. Acetaminophen can be supplemented with diclofenac supplements 50 mg up to twice daily.

11.3.6 Follow-Up Regimen

Three months after the last treatment, the absence of residual Barrett's epithelium should be confirmed by endoscopic inspection. Detailed inspection of the neosquamous mucosa after RFA is important for two reasons: first, to detect small areas of BE that can be additionally treated (Fig. 11.5); second, if random biopsies are obtained and accidentally small residual islands of Barrett's are sampled, this can result in a histological finding of buried Barrett's, causing doubts on the efficacy of the treatment and resulting in a missed opportunity to treat endoscopically visible remnants of Barrett's mucosa. We recommend inspection with high-resolution endoscopy and narrow-band imaging (NBI), or a comparable technique (FICE, i-scan), to carefully inspect the esophagus and neosquamocolumnar junction in antegrade and retroflexed position to rule out the presence of small remnants of Barrett's mucosa. If no visible residual Barrett's mucosa is detected, we obtain biopsies immediately distal (<5 mm) to the neosquamocolumnar junction to evaluate for residual IM. Given the very low incidence of buried Barrett's reported in multiple studies, extensive biopsies from the neosquamous mucosa are, in our opinion, not necessary provided that detailed inspection with high-resolution endoscopy with NBI did not show any columnar mucosa or mucosal irregularities. If residual BE is found, ablation can be repeated every 12 weeks until it has been eradicated both visually and histologically. Most patients will need one circumferential ablation session and one to two focal ablation sessions to eradicate all dysplasia and Barrett's mucosa. We suggest a maximum number of two circumferential and three focal ablation sessions, which should be sufficient in most patients.

The recommended follow-up interval depends on the initial grade of dysplasia:

- For patients with baseline IMC/HGD and complete eradication of dysplasia and IM, we recommend follow-up endoscopy at 3, 9, and 21 months after the last treatment sessions, then annually during the first 5 years. If there is sustained eradication of IM at that stage, surveillance can be stopped or intervals prolonged. Others perform surveillance endoscopies every 3 months for the first year, every 6 months for the second year, and annually thereafter [26].
- For patients with baseline LGD, we recommend follow-up endoscopy at 3, 9, and 21 months after the last treatment session, then annually during the first 3 years. If there is sustained eradication of IM at that stage, surveillance can be stopped or intervals prolonged.

11.4 EFFICACY OF RADIOFREQUENCY ABLATION

The efficacy of RFA has been studied in porcine models, pre-esophagectomy human subjects, and human subjects undergoing surveillance for BE [14,27−35]. Because the focal ablation device became available after the circumferential device, earlier studies focused on outcomes of circumferential RFA alone [27,33,36,37]. More recent trials have included a stepwise approach of circumferential and focal RFA, as well as combining ER with RFA [8,9,14,15,18,21,30−32,34,35,38−42].

Evidence from a number of well-designed studies, including a randomized, sham-controlled trial, suggests that RFA is highly effective at removing all BE at both the endoscopic and histologic levels with a favorable safety profile [14]. Although long-term follow-up studies are still limited, the 5-year follow-up data suggest that eradication of the Barrett's mucosa is maintained in more than 90% of patients [22]. In addition, studies on the properties of the neosquamous mucosa that regenerates after RFA show the absence of the preexisting oncogenetic abnormalities, suggesting a permanent transition to a low-risk epithelium [43,44].

11.4.1 Eradication of Nondysplastic Barrett's Mucosa

Studies have shown that RFA is effective for eradicating nondysplastic BE [30,31,33,34,38]. In a series that included 448 patients with BE with variable degrees of dysplasia, complete remission was achieved in 26% of patients by 1 year, 56% of patients by 2 years, and 71% of patients by 3 years [38]. This study was performed in a community setting. The study, however, lacked a strict ablation protocol, as ablation could be stopped before complete endoscopic and histological eradication of BE was achieved, with a mean of 0.9 cm residual BE after treatment. The incidence of recurrence 2 years after achieving complete remission was 33%, 78% of which were nondysplastic. It should be noted that in this study, complete remission was defined as the absence of IM from both esophageal and gastroesophageal junction biopsies. Similarly, recurrence included isolated IM of the cardia/gastro-esophageal junction, and this accounted for almost half of the reported recurrences.

In the Ablation of Intestinal Metaplasia Trial II (AIM-II) trial, complete eradication of Barrett's mucosa at 12 months was achieved in 48 out of 70 subjects (70%) with nondysplastic BE using only the Barrx system for circumferential ablation [33]. In a follow-up study, the Barrx device was used for additional ablation in patients from the AIM-II trial who had residual BE [30]. At 30-month follow-up, complete remission of BE was found in 97% of patients by intention-to-treat analysis. None of the patients presented with esophageal stenosis and no buried Barrett's glands were found in any of the more than 4000 biopsies obtained during follow-up. Fifty of the patients who had complete remission of BE at 30 months in the AIM-II trial were reevaluated at 5 years [31]. Complete remission of BE was noted in 46 patients (92%), whereas 4 patients (8%) had focal nondysplastic BE. The four patients with nondysplastic BE were all successfully treated with a single session of focal RFA.

11.4.2 Effect on Dysplastic BE

RFA has been shown to decrease the risk of malignant progression in patients with dysplastic BE. RFA eradicates all Barrett's mucosa in 66–100% of patients and eradicates dysplasia in 79–100% [8,9,14,15,32,35,39–45]. In a meta-analysis of 20 studies, complete eradication of dysplastic Barrett's mucosa was achieved in 91% of patients [46]. Similarly, in a systematic review that included 12 studies of RFA for the eradication of HGD or early cancer (EC) in patients with BE, RFA resulted in eradication of HGD or EC in 92% and complete eradication of BE in 88% [7].

For ablation of BE in patients with LGD or HGD, strong evidence that RFA reduces the risk of malignant progression comes from a randomized sham-controlled trial (the AIM-dysplasia trial) that was conducted in 19 US centers and included 127 patients [14]. Patients in the ablation group could receive up to four treatments during the first year. At 1-year follow-up, significantly higher proportions of RFA subjects had achieved complete remission for dysplasia and IM compared with sham-treated patients. In the HGD group, 81% of the patients who received RFA were free of dysplasia, compared with 19% of those who received sham therapy in an intention-to-treat analysis. In the LGD group, the rates were 90% and 23%, respectively. Overall, the rate of progression to higher grades of dysplasia or cancer was significantly lower in the RFA group than in the sham group (4% vs 16%).

Patients who were initially assigned to the sham group were offered RFA after 1 year, and in total, 119 patients were treated with RFA [32]. RFA was repeated at 15 months in patients who failed to achieve complete eradication of IM at 1 year. In total, at the start of the follow-up period, there was complete eradication of dysplasia in 110 patients (92%) and eradication of IM in 108 patients (91%). Two-year follow-up data were available for 106 patients. After 2 years of follow-up, 101 of 106 patients (95%) had complete eradication of dysplasia, including 50 of 54 patients (93%) with HGD and 51 of 52 patients (98%) with LGD. Eradication of IM was seen in 99 of 106 patients (93%), including 48 of 54 patients (89%) with HGD and 51 of 52 patients (98%) with LGD. With no additional therapy, more than 85% of patients with HGD or LGD who were free of dysplasia at the start of follow-up were free of dysplasia after a mean of 3 years. Similarly, more than 75% of patients were free of IM. Cancer developed in two patients (one with LGD and one with HGD) for an annual cancer risk of 0.55%.

A large European multicenter randomized trial, the SURF trial, recently demonstrated that RFA reduces the risk of malignant progression in BE patients with confirmed LGD [18]. A total of 136 patients were enrolled and their baseline histology results were reviewed by an expert pathology panel called "The Dutch Barrett's Advisory Committee" (BAC). The results showed that prophylactic RFA in BE patients with confirmed LGD reduces progression to HGD/EC from 26.5% to 1.5% over 3 years of follow-up. Patients in the ablation group were also less likely to progress to adenocarcinoma alone in comparison with patients in the surveillance arm (1.5% vs 8.8%). In the ablation arm, 93% of the patients were free of dysplasia and 88% of patients were free of IM after treatment. This was maintained during follow-up in 98% and 90% of patients, respectively. In the control arm, 28% of patients were free of dysplasia during follow-up, and 0% of patients were free of IM. All recurrences in the ablation arm were small islands or tongues less than 10 mm and were effectively managed endoscopically. The study was terminated early after an interim analysis showing superiority of ablation over surveillance.

In a cohort study from the Netherlands, 54 patients with HGD or EC were followed for 5 years after RFA [22]. Seventy-two percent of the patients in this study were treated with ER prior to RFA. Follow-up biopsies were obtained in 54 patients and after 5 years, endoscopic ultrasound and ER of neosquamous epithelium were performed. Quality of follow-up was ensured by using a rigorous follow-up protocol, with high-resolution endoscopy scheduled at predefined time points, and by obtaining a large number of samples (both biopsies and ER specimens). Kaplan–Meier analysis demonstrated sustained eradication of dysplasia and IM in more than 90% of patients after 5 years of follow-up.

In a prospective cohort study that included 63 patients with LGD and HGD, after a median follow-up of 24 months, 79% of patients were free of BE and 89% were free of dysplasia [35]. For the subset of 39 patients with LGD, 87% were free of BE and 95% were free of dysplasia, whereas for the subset of 24 patients with HGD, the rates were 67% and 79%, respectively.

A retrospective series examined 112 patients with dysplastic BE who had complete eradication of dysplasia and IM following RFA and were then followed with endoscopic surveillance for a median of 397 days (13.3 months) [41]. Disease recurrence was seen in eight patients (7%) and included progression to cancer in three patients (one treated with ER, one treated with esophagectomy, and one treated palliatively due to metastasis), HGD (one patient), LGD (one patient), and nondysplastic BE (three patients).

11.4.3 Buried Barrett's

Although buried Barrett's are an extremely rare finding after RFA of BE in all studies performed thus far with incidences ranging from 0% to 5% [14,30,32,40,44,47], there is still concern that residual BE could be hidden beneath the neosquamous epithelium following RFA. The possibility of occult malignant progression of the buried glands has been suggested by incidental cases of adenocarcinoma arising underneath neosquamous epithelium after ablation therapy using PDT and APC [48–50]. It has been suggested that presence of buried Barrett's following RFA is underrecognized because the biopsies do not sample the neosquamous epithelium deeply enough [44,51]. A study of 16 patients examined the sampling depth and presence of buried glands in biopsies and ER specimens from neosquamous epithelium after RFA [44]. Immediately after each routine biopsy, a second "keyhole" biopsy was taken from the same biopsy site from the neosquamous epithelium and biopsies were obtained from the untreated squamous epithelium of the proximal esophagus. In addition, a tissue sample from the neosquamous epithelium was obtained using ER. This study showed that biopsies obtained from squamous mucosa following RFA sample the mucosa just as deep as biopsies from untreated squamous mucosa. No buried Barrett's glands were detected in any of the primary, keyhole, or ER specimens. This suggests that RFA may be associated with complete eradication of all Barrett's epithelium and these findings make it highly unlikely that the low rate of buried glands reported after RFA in comparison with other ablative techniques is a reflection of a sampling error specific to RFA.

What needs to be addressed, however, is the risk of a false-positive finding of buried Barrett's. Biopsies from neosquamous epithelium near the neosquamocolumnar junction may lead to sampling of the transition from neosquamous to columnar epithelium, leading to a histologic finding of glandular mucosa beneath the neosquamous epithelium, which may mistakenly be interpreted as buried Barrett's glands. In the case of a residual island of BE that was not treated with RFA, tangential sampling of the island combined with tangential sectioning of the biopsy may also result in an erroneous finding of buried Barrett's [52]. A diagnosis of buried Barrett's glands should therefore only be made if the endoscopist is positive that there were no residual BE islands after detailed inspection with NBI, and if the biopsies were not obtained at the level of the neosquamocolumnar junction [53].

11.4.4 Intestinal Metaplasia of the Cardia

There are limited data available on the natural history of recurrent IM of the cardia/gastroesophageal junction. After RFA treatment, the gastroesophageal junction is frequently biopsied given the known increased risk of recurrences in this area, as endoscopic differentiation between gastric mucosa and IM is nearly impossible [54]. However, the clinical relevance of IM, when detected in this area, is uncertain because focal IM in this area may reflect insufficient treatment, recurrence of disease, or an irrelevant normal finding. Several studies have reflected on the relevance of IM at the cardia/gastroesophageal junction.

A Netherlands cohort study of 54 patients followed patients for 5 years after RFA, obtaining four-quadrant biopsies at every follow-up endoscopy (median 20 biopsies per patient) immediately below the gastroesophageal junction [22]. In 35% of patients, focal IM of the cardia was detected during follow-up. In 89% of these patients, focal IM of the cardia was only diagnosed on a single occasion. In total, 53 of 1143 biopsies contained focal IM. All of the patients with focal IM of the cardia had a normal endoscopic appearance of the neosquamocolumnar junction, and none of these patients developed dysplasia after a median of 61 months of follow-up. None of these patients was retreated endoscopically [22]. In this study, IM of the cardia was mostly observed in a single biopsy, this diagnosis was generally not reproduced during further follow-up, and there was no increased incidence over time. If IM of the cardia reflects residual disease, one would expect to find IM more than once in a single patient. If IM of the cardia results from ongoing reflux after treatment, one would expect an increased incidence over time. It should be noted that all patients in this study received high-dose maintenance therapy with esomeprazole 40 mg twice daily. Studies have shown that IM of the cardia can be detected in biopsies of 25% of the normal population, and this is generally not considered a premalignant condition [55]. The clinical relevance of focal IM of the cardia after RFA is therefore unknown, but these long-term data do not suggest that this is related to residual BE or recurrent disease.

In a United States series that included 448 patients with BE with variable degrees of dysplasia, 17 patients developed recurrence of BE at the gastroesophageal junction only, after having achieved complete remission after RFA [38]. Of these, 72% of patients had IM. Forty-seven percent of patients with IM or dysplasia at the cardia had a normal endoscopic appearance of the neosquamocolumnar junction. Follow-up results after detection of focal IM are not available.

These studies show that the management of focal IM of the gastroesophageal junction varies, as formal guidelines are lacking. The gastroesophageal junction remains an area at risk after endoscopic treatment and should be carefully inspected during follow-up with high-resolution endoscopy combined with advanced imaging techniques, and biopsies should be obtained immediately distal to the junction. Focal IM without dysplasia diagnosed on a single occasion does not appear to warrant treatment, as the majority of patients do not develop dysplasia during follow-up. If dysplasia is detected at the gastroesophageal junction, this can usually be managed endoscopically.

11.4.5 Adverse Events

Adverse events reported with RFA include esophageal strictures, upper gastrointestinal hemorrhage, and chest pain [14,15]. Stricture rates of 0−56% have been described with other endoscopic ablation techniques [56]. Overall, studies of RFA for BE have shown lower rates of stricturing (0−6%) [7,14,15,30,40,57]. The stricturing seen with RFA has generally been associated with either prior ER or a narrow esophagus at baseline due to underlying reflux disease.

A study of 12 patients that compared measurements of esophageal inner diameter, motility, and compliance before RFA treatment and 2 months after the last ablation session showed no significant differences, suggesting that RFA does not impair the functional integrity of the esophagus [57].

In a sham-controlled study of 127 patients, 6% of the RFA cohort experienced a stricture, but all resolved with a mean of 2.6 dilations [14]. There were no perforations or deaths.

In a community-based registry of 429 patients, 9 strictures occurred following 788 procedures (1% of cases, 2% of patients) [40]. All of the strictures resolved after a median of three dilations. There were no cases of bleeding (other than one patient who vomited blood-tinged mucus), perforation, or death.

Performing ER prior to RFA may increase the risk of complications. In a study of 65 patients, there were no complications in the 18 patients who had not previously undergone ER [58]. In the 47 patients who had undergone ER prior to circumferential RFA, mucosal lacerations were observed in patients who underwent RFA with a catheter that exceeded the smallest measured inner esophageal diameter and in patients whose ER involved more than one-third of the esophageal circumference and was greater than 2.5 cm in length. Five cases of esophageal stenosis after RFA occurred, all in patients whose ER involved more than 50% of the esophageal circumference and was longer than 2 cm in length.

Based on these observations, it is advisable to choose the size of the ablation catheters carefully in cases of prior ER and to limit the extent of prior ER to 2 cm in length and 50% of the circumference if possible.

11.5 DIRECTIONS FOR FUTURE RESEARCH

11.5.1 Technical Improvements

When it comes to patient comfort during treatment, the developments in endoscopic technique of RFA treatment have already improved.

Certain vital aspects remain to be studied in the future. First, the introduction of the ablation balloon and the focal device remains uncomfortable even though we have partly succeeded in optimizing the ablation regimen (less introductions of endoscope and devices). Studies on the safety and efficacy of a self-sizing catheter for circumferential ablation are currently underway. This device has the advantage that it is self-sizing, making the sizing step of the procedure unnecessary. In addition, the simplified focal ablation regimen may be improved further by reevaluating energy settings. In Europe, we have always used the focal device at $15 \, J/cm^2$, both for the standard and simplified regimen. Lowering this setting to $12 \, J/cm^2$ (conform the US standard protocol) when using the simplified triple application may reduce the risk of fibrosis and stenosis even further.

11.5.2 Risk Stratification

Two important questions still are how to diagnose LGD objectively and how to risk-stratify patients into low-risk and high-risk categories. The more "purified" a LGD cohort is, the higher the risk of progression and the higher the net health benefit of an intervention such as endoscopic RFA will be. In our view, future basic research focusing on LGD should be directed to two main pillars: optimizing the diagnosis of LGD and searching for biomarkers.

11.5.2.1 Optimizing Diagnosis of LGD

The SURF trial identified clinical predictors for progression such as the number of years since the diagnosis of BE, the number of endoscopies with dysplasia prior to inclusion and circumferential Barrett's length in centimeters. Unfortunately, multivariable analysis could not identify clinical predictors of regression of LGD [18]. A predictive model is currently being developed based on the histological diagnosis

of all patients screened for the SURF trial. This research will aim to identify the relation between spatial and temporal distribution of a diagnosis of LGD and the extent of agreement between expert pathologists for the diagnosis of LGD with the subsequent risk of progression. Combining these histological features with the previously identified clinical predictors, this model will likely aid in the selection of LGD patients for ablation in the future.

11.5.2.2 The Search for Biomarkers

Ideally, histology is eventually replaced by molecular biomarkers that are able to objectively identify patients who carry a low or high risk of neoplastic progression. This selection of patients will greatly improve the cost-effectiveness of endoscopic surveillance, as unnecessary endoscopies can potentially be avoided in the low-risk group, whereas high-risk patients may undergo prophylactic eradication of their BE [18,59].

Research on potential biomarkers for neoplastic progression includes markers of DNA content abnormalities, abnormalities in p53 and p16 tumor suppressor genes, clonal diversity, epigenetic changes, etc. [60,61]. Although potential biomarkers have been identified in retrospective studies, the final step before biomarkers can be implemented in clinical practice has not yet been reached. To achieve this, prospective validation or cancer control studies are required to evaluate the impact of a biomarker test on population disease burden, with primary outcomes including costs and mortality [62].

References

[1] Spechler SJ, Souza RF. Barrett's esophagus. N Engl J Med 2014;371:836—45.
[2] Van Lanschot JJ, Hulscher JB, Buskens CJ, Tilanus HW, ten Kate FJ, Obertop H, et al. Hospital volume and hospital mortality for esophagectomy. Cancer 2001;91:1574—8.

[3] Wouters MW, Wijnhoven BP, Karim-Kos HE, Blaauwgeers HG, Stassen LP, Steup WH, et al. High-volume versus low-volume for esophageal resections for cancer: the essential role of case-mix adjustments based on clinical data. Ann Surg Oncol 2008;15:80−7.

[4] Chang AC, Ji H, Birkmeyer NJ, Orringer MB, Birkmeyer JD. Outcomes after transhiatal and transthoracic esophagectomy for cancer. Ann Thorac Surg 2008;85:424−9.

[5] Sharma P, Falk GW, Weston AP, Reker D, Johnston M, Sampliner RE, et al. Dysplasia and cancer in a large multicenter cohort of patients with Barrett's esophagus. Clin Gastroenterol Hepatol 2006;4:566−72.

[6] Pouw RE, Sharma VK, Bergman JJ, Fleischer DE, et al. Radiofrequency ablation for total Barrett's eradication: a description of the endoscopic technique, its clinical results and future prospects. Endoscopy 2008;40:1033−40.

[7] Chadwick G, Groene O, Markar SR, Hoare J, Cromwell D, Hanna GB, et al. Systematic review comparing radiofrequency ablation and complete endoscopic resection in treating dysplastic Barrett's esophagus: a critical assessment of histologic outcomes and adverse events. Gastrointest Endosc 2014;79:718−31.

[8] Gondrie JJ, Pouw RE, Sondermeijer CMT, Peters FP, Curvers WL, Rosmolen WD, et al. Stepwise circumferential and focal ablation of Barrett's esophagus with high-grade dysplasia: results of the first prospective series of 11 patients. Endoscopy 2008;40:359−69.

[9] Gondrie JJ, Pouw RE, Sondermeijer CMT, Peters FP, Curvers WL, Rosmolen WD, et al. Effective treatment of early Barrett's neoplasia with stepwise circumferential and focal ablation using the HALO system. Endoscopy 2008;40:370−9.

[10] Peters FP, Brakenhoff KPM, Curvers WL, Rosmolen WD, Fockens P, ten Kate FJ, et al. Histologic evaluation of resection specimens obtained at 293 endoscopic resections in Barrett's esophagus. Gastrointest Endosc 2008;67:604−9.

[11] Van Sandick JW, van Lanschot JJ, ten Kate FJ, Offerhaus GJ, Fockens P, Tytgat GN, et al. Pathology of early invasive adenocarcinoma of the esophagus or esophagogastric junction: implications for therapeutic decision making. Cancer 2000;88:2429−37.

[12] Westerterp M, Koppert LB, Buskens CJ, Tilanus HW, ten Kate FJ, Bergman JJ, et al. Outcome of surgical treatment for early adenocarcinoma of the esophagus or gastro-esophageal junction. Virchows Arch 2005;446:497−504.

[13] May A, Gossner L, Pech O, Frtiz A, Günter E, Mayer G, et al. Local endoscopic therapy for intraepithelial high-grade neoplasia and early adenocarcinoma in Barrett's oesophagus: acute-phase and intermediate results of a new treatment approach. Eur J Gastroenterol Hepatol 2002;14:1085−91.

[14] Shaheen NJ, Sharma P, Overholt BF, Wolfsen HC, Sampliner RE, Wang KK, et al. Radiofrequency ablation in Barrett's esophagus with dysplasia. N Engl J Med 2009;360:2277−88.

[15] Pouw RE, Wirths K, Eisendrath P, Sondermeijer CM, ten Kate FJ, Devierere J, et al. Efficacy of radiofrequency ablation combined with endoscopic resection for Barrett's esophagus with early neoplasia. Clin Gastroenterol Hepatol 2010;8:23−9.

[16] Kim HP, Bulsiewicz WJ, Cotton CC, Dellon ES, Spacek MB, Chen X, et al. Focal endoscopic mucosal resection before radiofrequency ablation is equally effective and safe compared with radiofrequency ablation alone for the eradication of Barrett's esophagus with advanced neoplasia. Gastrointest Endosc 2012;76:733−9.

[17] Curvers WL, ten Kate FJ, Krishnadath KK, Visser M, Elzer B, Baak LC, et al. Low-grade dysplasia in Barrett's esophagus: overdiagnosed and underestimated. Am J Gastroenterol 2010;105:1523−30.

[18] Phoa KN, van Vilsteren FGI, Weusten BLAM, Bisschops R, Schoon EJ, Ragunath K, et al. Radiofrequency ablation vs endoscopic surveillance for patients with Barrett esophagus and low-grade dysplasia: a randomized clinical trial. JAMA 2014;311:1209−17.

[19] Hur C, Choi SE, Rubenstein JH, Kong CY, Nishioka NS, Provenzale DT, et al. The cost effectiveness of radiofrequency ablation for Barrett's esophagus. Gastroenterology 2012;143:567−75.

[20] Van Vilsteren FGI, Phoa KN, Alvarez Herrero L, Pouw RE, Sondermeijer CM, van Lijnschoten I, et al. Circumferential balloon-based radiofrequency ablation of Barrett's esophagus with dysplasia can be simplified, yet efficacy maintained, by omitting the cleaning phase. Clin Gastroenterol Hepatol 2013;11:491−8.

[21] Alvarez Herrero L, van Vilsteren FGI, Pouw RE, ten Kate FJ, Visser M, Seldenrijk CA, et al. Endoscopic radiofrequency ablation combined with endoscopic resection for early neoplasia in Barrett's esophagus longer than 10 cm. Gastrointest Endosc 2011;73:682−90.

[22] Phoa KN, Pouw RE, van Vilsteren FGI, Sondermeijer CM, ten Kate FJ, Visser M, et al. Remission of Barrett's esophagus with early neoplasia 5 years after radiofrequency ablation with endoscopic resection: a Netherlands cohort study. Gastroenterology 2013;145:96−104.

[23] Van Vilsteren FGI, Phoa KN, Alvarez Herrero L, Pouw RE, Sondermeijer CM, Visser M, et al. A simplified regimen for focal radiofrequency ablation of Barrett's mucosa: a randomized multicenter trial

comparing two ablation regimens. Gastrointest Endosc 2013;78:30−8.

[24] Krishnan K, Pandolfino JE, Kahrilas PJ, Keefer L, Boris L, Komanduri S, et al. Increased risk for persistent intestinal metaplasia in patients with Barrett's esophagus and uncontrolled reflux exposure before radiofrequency ablation. Gastroenterology 2012;143:576−81.

[25] Akiyama J, Marcus SN, Triadafilopoulos G. Effective intra-esophageal acid control is associated with improved radiofrequency ablation outcomes in Barrett's esophagus. Dig Dis Sci 2012;57:2625−32.

[26] Bedi AO, Kwon RS, Rubenstein JH, Piraka CR, Elta GH, Scheiman JM, et al. A survey of expert follow-up practices after successful endoscopic eradication therapy for Barrett's esophagus with high-grade dysplasia and intramucosal adenocarcinoma. Gastrointest Endosc 2013;78:696−701.

[27] Ganz RA, Utley DS, Stern RA, Jackson J, Batts KP, Termin P, et al. Complete ablation of esophageal epithelium with a balloon-based bipolar electrode: a phased evaluation in the porcine and in the human esophagus. Gastrointest Endosc 2004;60:1002−10.

[28] Dunkin BJ, Martinez J, Bejarano PA, Smith CD, Chang K, Livingstone AS, et al. Thin-layer ablation of human esophageal epithelium using a bipolar radiofrequency balloon device. Surg Endosc Other Interv Tech 2006;20:125−30.

[29] Smith CD, Bejarano PA, Melvin WS, Patti MG, Muthusamy R, Dunkin BJ, et al. Endoscopic ablation of intestinal metaplasia containing high-grade dysplasia in esophagectomy patients using a balloon-based ablation system. Surg Endosc Other Interv Tech 2007;21:560−9.

[30] Fleischer DE, Overholt BF, Sharma VK, Reymunde A, Kimmey MB, Chuttani R, et al. Endoscopic ablation of Barrett's esophagus: a multicenter study with 2.5-year follow-up. Gastrointest Endosc 2008;68:867−76.

[31] Fleischer DE, Overholt BF, Sharma VK, Reymunde A, Kimmey MB, Chuttani R, et al. Endoscopic radiofrequency ablation for Barrett's esophagus: 5-year outcomes from a prospective multicenter trial. Endoscopy 2010;42:781−9.

[32] Shaheen NJ, Overholt BF, Sampliner RE, Wolfsen HC, Wang KK, Fleischer DE, et al. Durability of radiofrequency ablation in Barrett's esophagus with dysplasia. Gastroenterology 2011;141:460−8.

[33] Sharma VK, Wang KK, Overholt BF, Lightdale CJ, Fennerty MB, Dean PJ, et al. Balloon-based, circumferential, endoscopic radiofrequency ablation of Barrett's esophagus: 1-year follow-up of 100 patients. Gastrointest Endosc 2007;65:185−95.

[34] Sharma VK, Kim HJ, Das A, Dean P, DePetris G, Fleischer DE, et al. A prospective pilot trial of ablation of Barrett's esophagus with low-grade dysplasia using stepwise circumferential and focal ablation (HALO system). Endoscopy 2008;380−7.

[35] Sharma VK, Jae Kim H, Das A, Wells CD, Nguyen CC, Fleischer DE, et al. Circumferential and focal ablation of Barrett's esophagus containing dysplasia. Am J Gastroenterol 2009;104:310−17.

[36] Ganz RA, Overholt BF, Sharma VK, Fleischer DE, Shaheen NJ, Lightdale CJ, et al. Circumferential ablation of Barrett's esophagus that contains high-grade dysplasia: a U.S. Multicenter Registry. Gastrointest Endosc 2008;68:35−40.

[37] Avilés A, Reymunde A, Santiago N. Balloon-based electrode for the ablation of non-dysplastic Barrett's esophagus: ablation of intestinal metaplasia (AIM II Trial). Bol Asoc Med P R 2006;98:270−5.

[38] Gupta M, Iyer PG, Lutzke L, Gorospe EC, Abrams JA, Falk GW, et al. Recurrence of esophageal intestinal metaplasia after endoscopic mucosal resection and radiofrequency ablation of Barrett's esophagus: results from a US Multicenter Consortium. Gastroenterology 2013;145:79−86.

[39] Van Vilsteren FGI, Pouw RE, Seewald S, Alvarez Herrero L, Sondermeijer CM, Visser M, et al. Stepwise radical endoscopic resection versus radiofrequency ablation for Barrett's oesophagus with high-grade dysplasia or early cancer: a multicentre randomised trial. Gut 2011;60:765−73.

[40] Lyday WD, Corbett FS, Kuperman DA, Kalvaria I, Mavrelis PG, Shughoury AB, et al. Radiofrequency ablation of Barrett's esophagus: outcomes of 429 patients from a multicenter community practice registry. Endoscopy 2010;42:272−8.

[41] Orman ES, Kim HP, Bulsiewicz WJ, Cotton CC, Dellon ES, Spacek MB, et al. Intestinal metaplasia recurs infrequently in patients successfully treated for Barrett's esophagus with radiofrequency ablation. Am J Gastroenterol 2013;108:187−95.

[42] Bulsiewicz WJ, Kim HP, Dellon ES, Cotton CC, Pasricha S, Madanick RD, et al. Safety and efficacy of endoscopic mucosal therapy with radiofrequency ablation for patients with neoplastic Barrett's esophagus. Clin Gastroenterol Hepatol 2013;11:636−42.

[43] Finkelstein SD, Lyday WD. M1944 The molecular pathology of radiofrequency mucosal ablation of Barrett's esophagus [abstract]. Gastroenterology 2008; 134:A-436.

[44] Pouw RE, Gondrie JJ, Rygiel AM, Sondermeijer CM, ten Kate FJ, Odze RD, et al. Properties of the neosquamous epithelium after radiofrequency ablation of

Barrett's esophagus containing neoplasia. Am J Gastroenterol 2009;104:1366–73.

[45] Van Vilsteren FGI, Alvarez Herrero L, Pouw RE, Schrijnders D, Sondermeijer CM, Bisschops R, et al. Predictive factors for initial treatment response after circumferential radiofrequency ablation for Barrett' s esophagus with early neoplasia: a prospective multicenter study. Endoscopy 2013;45:516–25.

[46] Orman E, Li N, Shaheen N. Efficacy and durability of radiofrequency ablation for Barrett's esophagus: systematic review and meta-analysis. Clin Gastroenterol Hepatol 2013;11:1245–55.

[47] Gray NA, Odze RD, Spechler SJ. Buried metaplasia after endoscopic ablation of Barrett's esophagus: a systematic review. Am J Gastroenterol 2011;106:1899–908.

[48] Mino-Kenudson M, Ban S, Ohana M, Puricelli W, Deshpande V, Shimizu M, et al. Buried dysplasia and early adenocarcinoma arising in Barrett esophagus after porfimer-photodynamic therapy. Am J Surg Pathol 2007;31:403–9.

[49] Van Laethem JL, Peny MO, Salmon I, Cremer M, Deviere J. Intramucosal adenocarcinoma arising under squamous re-epithelialisation of Barrett's oesophagus. Gut 2000;46:574–7.

[50] Titi M, Overhiser A, Ulusarac O, Falk GW, Chak A, Wang K, et al. Development of subsquamous high-grade dysplasia and adenocarcinoma after successful radiofrequency ablation of Barrett's esophagus. Gastroenterology 2012;143:564–6.

[51] Gupta N, Mathur SC, Dumot JA, Singh V, Gaddam S, Wani SB, et al. Adequacy of esophageal squamous mucosa specimens obtained during endoscopy: are standard biopsies sufficient for postablation surveillance in Barrett's esophagus? Gastrointest Endosc 2012;75:11–18.

[52] Pouw RE, Gondrie JJ, Sondermeijer CM, ten Kate FJ, van Gulik TM, Krishnadath KK, et al. Eradication of Barrett esophagus with early neoplasia by radiofrequency ablation, with or without endoscopic resection. J Gastrointest Surg 2008;12:1627–36.

[53] Pouw RE, Visser M, Odze RD, Sondermeijer CM, ten Kate FJ, Weusten BL, et al. Pseudo-buried Barrett's post radiofrequency ablation for Barrett's esophagus, with or without prior endoscopic resection. Endoscopy 2014;46:105–9.

[54] Alvarez Herrero L, Curvers WL, Bisschops, Kara MA, Schoon EJ, ten Kate FJ, et al. Narrow band imaging does not reliably predict residual intestinal metaplasia after radiofrequency ablation at the neo-squamo columnar junction. Endoscopy 2014;46:98–104.

[55] Morales T, Camargo E, Bhattacharyya A, Sampliner RE. Long-term follow-up of intestinal metaplasia of the gastric cardia. Am J Gastroenterol 2000;95:1677–80.

[56] Overholt BF, Lightdale CJ, Wang KK, Canto MI, Burdick S, Haggitt RC, et al. Photodynamic therapy with porfimer sodium for ablation of high-grade dysplasia in Barrett's esophagus: international, partially blinded, randomized phase III trial. Gastrointest Endosc 2005;63:359.

[57] Beaumont H, Gondrie JJ, Mcmahon BP, Pouw RE, Gregersen H, Bergman JJ, et al. Stepwise radiofrequency ablation of Barrett's esophagus preserves esophageal inner diameter, compliance, and motility. Endoscopy 2009;41:2–8.

[58] Pouw RE, Gondrie JJ, Van Vilsteren FG, Sondermeijer CM, Rosmolen WD, Curvers WL, et al. Complications following circumferential radiofrequency energy ablation of Barrett's esophagus containing early neoplasia. Gastrointest Endosc 2008;67: AB145.

[59] Corley DA, Mehtani K, Quesenberry C, Zhao W, de Boer J, Weiss NS. Impact of endoscopic surveillance on mortality from Barrett's esophagus-associated esophageal adenocarcinomas. Gastroenterology 2013; 145:312–19.

[60] Timmer M, Sun G, Gorospe E, Leggett CL, Lutzke L, Krishnadath KK, et al. Predictive biomarkers for Barrett's esophagus: so near and yet so far. Dis Esophagus 2013;26:574–81.

[61] Varghese S, Lao-Sirieix P, Fitzgerald RC. Identification and clinical implementation of biomarkers for Barrett's esophagus. Gastroenterology 2012;142:435–41.

[62] Pepe MS, Etzioni R, Feng Z, Potter JD, Thompson ML, Thornquist M, et al. Phases of biomarker development for early detection of cancer. J Natl Cancer 2001;93: 1054–61.

Cryospray Ablation

John A. Dumot[1,2]

[1]University Hospitals Digestive Health Institute, Case Western Reserve University, Cleveland, OH,
United States [2]University Hospitals Ahuja Medical Center, Beachwood, OH, United States

12.1 INTRODUCTION

Endoscopic cryospray ablation or cryotherapy is a unique method of ablation that offers physicians a versatile choice in the treatment of Barrett's and early esophageal cancer. The concept of endoscopic cryospray ablation began in the late 1990s with the development of two delivery platforms using liquid nitrogen (LN) [1] and carbon dioxide (CO_2). Both methods share some unique characteristics compared to other ablation techniques using radiofrequency energy and bipolar or monopolar electrocautery. The spray application allows the physician to paint the cryogen over the target tissue making it ideal for both flat and uneven surfaces. Similar to a professional painter choosing a spray gun for delivery of their wares, physicians can choose spray cryoablation when the topography of the target lesion is uneven or nodular. Technical challenges in performing cryospray ablation have limited its use to centers specializing in the treatment of Barrett's esophagus (BE) and esophageal cancer. Cryotherapy is currently indicated in a variety of neoplastic stages

including well-established low-grade dysplasia (LGD), high-grade dysplasia (HGD), intramucosal carcinoma (IMC), and palliation of more advanced forms of carcinoma in high-risk patients or for residual disease after definitive traditional oncologic therapy with radiation or chemotherapy. Attempts to use cryospray ablation as a primary palliative ablation therapy for malignant dysphagia have been successful in some patients. Prior to spray cryotherapy, alternative use of cryotherapy with a contact probe has been associated with transmural injury with the risk of perforation in the esophagus [2].

All types of cryotherapy rely on the polarization of water molecules to disrupt cellular membranes leading to cell death when the tissue freezes. Endoscopic spray cryoablation is performed with a catheter delivering liquid nitrogen ($-196°C$) under low pressure (2–4 psi), carbon dioxide ($-70°C$) cooling tissue with the Joules–Thompson principle, or a newer balloon delivery device using nitrous oxide. The familiarity of cryotherapy with patients is due to the widespread application of LN in dermatology, which makes the treatment

D. Pleskow & T. Erim (Eds): Barrett's Esophagus.
DOI: http://dx.doi.org/10.1016/B978-0-12-802511-6.00012-0

appealing to patients. Similar stages of tissue destruction are noted in the dermis and esophagus: initial edema followed by submucosal hemorrhage and blistering and then slough of the epithelium. In an acid suppressed environment, slough of the epithelium prompts healing of the ulcerated lumen with normal squamous or more specifically a "neosquamous" epithelium.

Proper informed consent is required before initiation of any ablation therapy. Important details for consideration include number of treatment sessions, adverse events, as well as success and failure rates. Cryospray ablation is similar to other ablation methods in terms of the number of sessions required to obtain a complete response. On average, three to five sessions are performed 2−3 months apart to allow healing and avoid therapy during periods of ulceration. For more advanced lesions, there may be urgency to induce a remission and procedures can be performed as frequently as every 2−3 weeks.

Preparation for endoscopic cryospray ablation includes a prolonged fast for patients with delayed gastric emptying. Deep sedation is generally recommended by most physicians performing cryospray ablation. Very little data is known about the use of cryospray ablation in patients taking anticoagulation medications, however, it is our practice to not discontinue any anticoagulation medications before the endoscopic ablation session due to the risk of thrombosis. High dose acid suppression with twice daily dosing of proton pump inhibitors is required to promote neosquamous epithelialization after ablation of the unwanted columnar lining (Figs. 12.1 and 12.2).

12.2 LIQUID NITROGEN

The first use of LN cryogen provided proof of concept and early dosimetry data in the swine model using small Yorkshire pigs (20 kg)

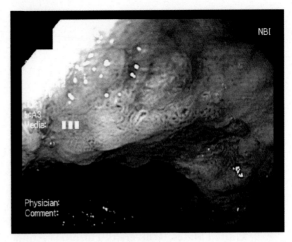

FIGURE 12.1 LN cryotherapy in circumferential nodular high-grade dysplasia in a high-risk patient on anticoagulation for valvular heart disease. *Source: First CSA 4.*

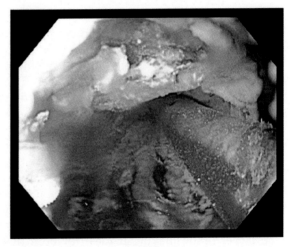

FIGURE 12.2 LN induced freeze with decompression tube in lumen. *Source: First CSA 10.*

[1]. The current commercial device is the second generation with significant improvements over the prototype and first commercial release (CSA Medical, Lutherville, MD). The initial treatment changes observed immediately after LN cryospray were nil to minimal while targeting normal squamous mucosal of swine

esophagus. The treated swine mucosa would slough after a blistering stage identical to the dermatologic applications. Later, immediate treatment effects from cryospray were recognized in BE mucosa with intense reddening and slight hemorrhage from the mucosa, which help guide therapy. In the initial animal model, transmural inflammation was found only with higher doses of 30−60 seconds spray duration.

Initial human use was limited to a relatively small number of referral centers where the technology was refined [3,4−6]. Initial complications related to gaseous distention responded to an improved dual chamber decompression tube with both active suction and passive venting. In an early retrospective cohort of 98 patients from 9 academic and community centers, complete eradication of dysplasia was noted in 87% of patients completing therapy [6]. An early complete response rate of 84% was found in 16 of 19 patients with T1a mucosal cancer receiving only cryospray ablation and 60% in 6 of 10 patients with T1b submucosal cancers (75% overall success in T1 group) [5]. Cryospray ablation demonstrated remarkable ability to provide endoluminal control in several patients with advanced esophageal cancer (T2−T4), which served as the basis for using LN for salvage treatment after traditional oncologic therapy with chemotherapy and radiation. LN cryospray has demonstrated effectiveness in treating residual BE with dysplasia after primary chemoradiation therapy.

The long-term results of cryospray ablation in BE have been reported by both single center and multicenter groups [7]. In the single center, the authors achieved a complete eradication of HGD in 97% with 31 of 32 subjects with a follow-up range of 24−57 months.

12.3 DOSIMETRY

The energy applied to the neoplastic tissue is measured in terms of duration of freeze time

multiplied by the number of freeze cycles. In general, the longer the cryogen is applied, the deeper the penetration or depth of ablation. The duration of the freeze time is measured in seconds and begins at the moment the entire targeted area is frozen. The usual target area is approximately 2 cm × 2 cm but smaller areas are commonly chosen when there are worrisome lesions or areas of the targeted neoplasia that could be difficult to achieve a complete ablation. Overlap of the treatment areas is extremely important to reduce the risk of incomplete response resulting in persistent neoplasia. The resultant tissue changes immediately after cryospray ablation are the appearance of a uniform cherry red color with slight hemorrhage. Identification of these immediate changes is extremely important to be assured that the neoplastic tissue was properly treated. Nodular areas can be given extra cycles to ensure adequate tissue response.

The typical dosimetry for BE with HGD is 20 second cryospray × 2 cycles based initially on expert opinion then confirmed with a small but important human study in patients undergoing esophagectomy. Cryospray ablation demonstrated consistent mucosal destruction and treatment effect well into the submucosa [8]. In this clinical trial of depth of injury, seven male esophageal cancer patients were treated with LN cryospray ablation applied to segments with less advanced pathology to avoid confounding information on the pathologic review of the most advanced lesion. Two dosimetry treatments were compared using either 4 cycles of 10 second duration cryospray or 2 cycles of 20 second cryospray. Deeper injury was noted in the 2 cycles × 20 second cryospray group with 5.4 mm depth compared to 4.0 mm for the 4 cycles × 10 second group. The two dosimetry protocols were found to be similar in terms of maximum depth of necrosis into the submucosal layer. The 20 second × 2 cycles protocol induced a deeper level of inflammation into the adventitia layer more often.

Efforts to reduce the overall treatment session time with single sessions of longer freeze time (30 seconds) did not provide a reproducible treatment effect (unpublished data) confirming the importance of the freeze—thaw—freeze cycle. Higher doses of LN cryospray ablation with 30 second spray repeated with two to three cycles should be considered for unusually thick or nodular lesions and persistent BE after multiple prior ablation sessions.

12.4 LIMITATIONS

Visibility during the ablation procedure can be significantly impaired by the overspray as cryogen comes in contact with the endoscope's lens and is a major disadvantage of LN spray cryotherapy. Although poor visibility is not a problem with every treatment session, the majority are affected with some degree of obscuration of the endoscopic view. The degree of frosting the lens in inherent in the intensely cold temperatures LN reaches during the procedure. Warming the endoscope is the only current solution to a frozen lens and this inevitably prolongs the treatment time. Several methods have been tried to overcome this obstacle including using a clear cap, removing the water bottle, and improving decompression of the esophageal lumen.

Distention during the application of LN requires a dedicated decompression tube, which is placed into the gastric cavity over a spring tip guide wire. Markings allow the proper location of the tube across the esophagogastric junction. Retained food will greatly hinder the function of the decompression tube and result in over distention of the gastric lumen (see Section 12.5). The endoscopy assistant should constantly monitor the patient for abdominal distention as a sign of over distention. Also, any unusual signs of discomfort should trigger cessation of cryospray application until proper decompression can be confirmed.

12.5 ADVERSE EVENTS

Pain is the most common adverse event after cryospray ablation. Many experts believe cryospray ablation induces less pain than other ablative techniques in general. Symptoms usually persist for 1—14 days depending on patient factors and dosimetry. Persistent pain, odynophagia, or dysphagia lasting longer than 7 days should prompt a follow-up visit and consideration for an early repeat endoscopy to assess and treat an early stricture. Patients should be prepared to manage their symptoms with modification of the diet and medications. Routine use of sucralfate suspension 1 g four times a day immediately after the ablation reduces pain significantly and may augment healing. Occasionally opiate medications are required to control symptoms after ablation.

Strictures resulting from cryotherapy occur in approximately 4% of patients. Most respond easily to endoscopic dilation especially when treated early during the healing phase. More severe strictures can result from cryospray ablation used in conjunction with other endoscopic therapy, such as mucosal resection, prior radiation therapy, or with higher than usual doses. Less common adverse events include aspiration pneumonia, lip ulcers from contact with the frozen endoscope (Fig. 12.3).

12.6 CARBON DIOXIDE

The first use of CO_2 was described almost simultaneously with the first report of LN cryospray ablation (Pasricha et al., 1999). CO_2 cryospray ablation (GI Supply, Camp Hill, PA) operates at a higher temperature compared to the LN system. The more modest temperature results in a system that reduces the technical challenges of extended periods of poor visibility and stiffness of the endoscope. On the other hand, the lack of extreme temperature necessitates dosimetry with higher number of

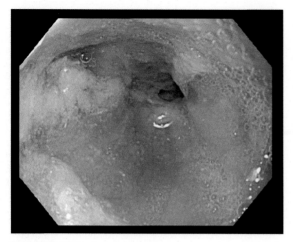

FIGURE 12.3 LN induced stricture at 2 weeks after initial cryotherapy. *Source: Strict 1.*

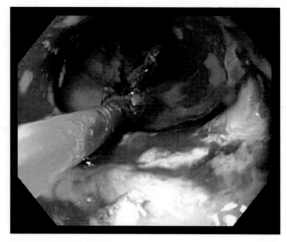

FIGURE 12.4 Stricture treated with balloon dilation. *Source: Strict 4.*

treatment cycles. Other advantages of CO_2 cryospray ablation include a lower cost of the system and catheter cost per case. Refilling the CO_2 system requires a simple exchange of a medical grade CO_2 tank commonly available in most endoscopy units. A second study of 22 patients with BE reported CO_2 to be relatively effective and safe treatment [9]. More data is needed about the overall effectiveness of CO_2 cryospray ablation.

12.7 NITROUS OXIDE

The most recent technology delivers nitrous oxide with a balloon tipped catheter (C2 Therapeutics, Redwood City, CA). The device is self-contained with small cartridges delivering the cryogen through the scope catheter or a parallel delivery device under direct visualization. The first publication was in a swine model to determine the dosimetry of 12 second spray as the maximal effect on the mucosa with higher doses associated with stricture formation [10]. Early results in humans show a treatment effect with both 6 second and 8 second sprays with adverse events including chest pain minor

lacerations of the mucosa [11]. The data from the first human use study is due out this year (Scholvinck et al., in press). Currently, the device is limited to treating focal areas of BE mucosa but more sophisticated catheters are under development to treat circumferential segments (Fig. 12.4).

12.8 SURVEILLANCE

Recurrence of Barrett's mucosa and neoplasia is relatively common and requires lifelong surveillance. Unfortunately, several cases of recurrence of interval cancers may not respond to endoscopic or surgical therapy. Subsquamous or "buried" BE is a form of incomplete response when BE glands are found under neosquamous epithelium. It must be mentioned that subsquamous BE occurs even in treatment naïve patients around the border areas of all squamocolumnar junctions and requires physicians to overlap cryospray ablation with normal appearing squamous epithelium. In the early multicenter cohort of BE HGD, subsquamous BE was found in 3% of patients [6]. More detail on surveillance is

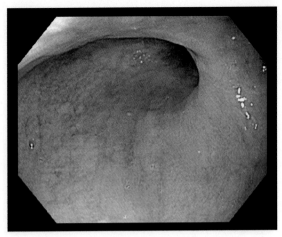

FIGURE 12.5 Follow-up after three sessions of LN cryotherapy for long-segment Barrett's esophagus. *Source: Last cryo.*

provided in Chapter 15: "Posttreatment Surveillance, Risk for Recurrence of Barrett's Esophagus, and Adenocarcinoma After Treatment" (see also Fig. 12.5).

12.9 FUTURE DIRECTIONS

Applying LN cryotherapy to difficult locations like the esophagogastric junction is more feasible with the newer 7Fr catheter. Attempts to treat the esophagogastric junction in retroflexion may increase response rates in this difficult area. Device improvements are directed toward improving visualization throughout the cryogen delivery and effective decompression. The most recent technology developed to deliver spray cryoablation captures the nitrous oxide cryogen in a balloon and a circumferential delivery device is under development.

References

[1] Johnston M, Schoenfeld P, Mysore J. Endoscopic spray cryotherapy for esophageal mucosal ablation. Gastrointest Endosc 1999;50:86–92.

[2] Rodgers BM, Pappelis P. Profound endoesophageal cryotherapy. Cryobiology 1985;22:295–301.

[3] Dumot JA, Vargo JJ, Falk GW, Frey L, Lopez R, Rice TW. An open-label, prospective trial of cryospray ablation for Barrett's esophagus high-grade dysplasia and early esophageal cancer. Gastrointest Endosc 1999;70:635–44.

[4] Johnston MH, Eastone JA, Horwhat JD, Cartledge J, Mathews JS, Foggy JR. Cryoablation of Barrett's esophagus: a pilot study. Gastrointest Endosc 2005;62: 842–8.

[5] Greenwald BD, Dumot JA, Abrams JA, Lightdale CJ, David DS, Nishioka NS, et al. Endoscopic spray cryotherapy for esophageal cancer: safety and efficacy. Gastrointest Endosc 2010;71:686–93.

[6] Shaheen NJ, Greenwald BD, Peery AF, Dumot JA, Nishioka NS, Wolfsen HC, et al. Safety and efficacy of endoscopic spray cryotherapy for Barrett's esophagus with high-grade dysplasia. Gastrointest Endosc 2010; 71:680–5.

[7] Gosain SK, Mercer WS, Twaddell L Uradomo, Greenwald BD, et al. Liquid nitrogen spray cryotherapy in Barrett's esophagus with high-grade dysplasia: long-term results. Gastrointest Endosc 2013;78:260–5.

[8] Ribeiro A, Bejarano P, Livingstone A, Sparling L, Franceschi D, Ardalan B. Depth of injury caused by liquid nitrogen cryospray: study of human patients undergoing planned esophagectomy. Dig Dis Sci. 2014;59:1296–301.

[9] Xue HB, Tan HH, Liu WZ, Chen XY, Feng N, Gao, YJ. A pilot study of endoscopic spray cryotherapy by pressurized carbon dioxide gas for Barrett's esophagus. Endoscopy 2011;43:379–85.

[10] Friedland S, Triadafilopoulos G. A novel device for ablation of abnormal esophageal mucosa (with video). Gastrointest Endosc 2011;74(1):182–8.

[11] Jansen M, Schölvinck DW, Kushima R, Sekine S, Weusten BL, Wang GQ, et al. Is it justified to ablate flat-type esophageal squamous cancer? An analysis of endoscopic submucosal dissection specimens of lesions meeting the selection criteria of radiofrequency studies. Gastrointest Endosc 2014;80(6):995–1002.

Endoscopic Surgical Therapies for Barrett's Esophagus

Brian J. Dunkin[1,2] *and Kathryn Boom*[2]

[1]Section of Endoscopic Surgery, Institute for Academic Medicine, Houston Methodist Hospital, Houston, TX, United States [2]Department of Surgery, Houston Methodist Hospital, Houston, TX, United States

13.1 INTRODUCTION

Earlier chapters in this book describe how ablative therapies have essentially replaced surgery for the management of Barrett's esophagus (BE) with high-grade dysplasia (HGD). However, resectional therapies, performed either with flexible endoscopy or surgery, are still required for managing this disease either as an adjunct to ablation or for definitive therapy. This chapter describes the indications for using resectional therapies in the management of BE, illustrates the available techniques, discusses which option to choose, and explores prevention and management of complications. Surveillance strategies for recurrence following resection of BE are described in a later chapter.

13.2 TECHNIQUES FOR PERFORMING ENDOSCOPIC MUCOSAL RESECTION AND ENDOSCOPIC SUBMUCOSAL DISSECTION

Endoscopic mucosal resection (EMR) and endoscopic submucosal dissection (ESD) are techniques used to resect tissue from within the lumen of the gastrointestinal (GI) tract using a flexible endoscopic platform. Unlike mucosal ablation, which removes the superficial layers of the mucosa only, EMR and ESD resect mucosa through the submucosal layer and down to the muscularis propria (Fig. 13.1). These techniques provide for larger tissue samples and intact pathologic architecture which aid in staging disease, but are also associated with higher risks of perforation, bleeding, and postprocedure stricture.

D. Pleskow & T. Erim (Eds): Barrett's Esophagus.
DOI: http://dx.doi.org/10.1016/B978-0-12-802511-6.00013-2

177

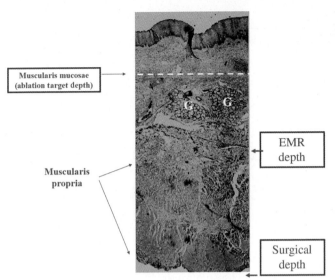

FIGURE 13.1   Histology of ablation and endoscopic mucosal resection/endoscopic submucosal dissection depths of tissue removal.

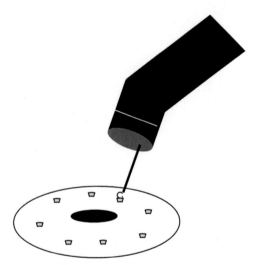

FIGURE 13.2   Marking the periphery of a lesion prior to endoscopic mucosal resection.

13.2.1 Endoscopic Mucosal Resection

There are two main methods for performing EMR: nonsuction ("lift-and-cut") and suction ("suck-and-cut") techniques. Both techniques begin by marking the planned resection margins with brief bursts of electrosurgery using an endoscopic snare with the wire minimally deployed (Fig. 13.2):

"Lift-and-Cut" technique: The lift-and-cut technique is similar to performing a saline lift polypectomy in the colon (Fig. 13.3) [1]. A sclerotherapy needle is used to inject fluid into the submucosal space to elevate the lesion away from the underlying muscularis propria and create a less flat target for resection. The injection is often done with saline, but other solutions have been used to achieve longer maintenance of the bleb including hypertonic saline (3.75% NaCl), 20% dextrose, or sodium hyaluronate [2]. Indigo carmine (0.004%) or methylene blue is often added to the injectate to stain the submucosa and provides a better evaluation of the depth of resection. The submucosal injection can also be used to determine if a lesion is appropriate for endoscopic resection. Lack of elevation during injection indicates adherence to the muscularis propria and is a relative contraindication to proceeding with EMR. After creating the submucosal elevation, the lesion is grasped with a rat tooth forceps that has been passed

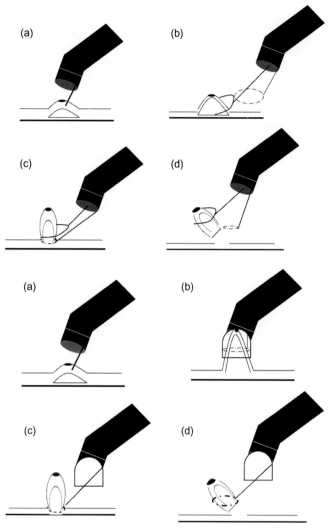

FIGURE 13.3 Lift-and-cut technique for endoscopic mucosal resection. (a) Submucosal injection, (b) passage of grasping forceps through the open polypectomy snare, (c) tightening of the snare at the base of the lesion, and (d) completion of the snare excision.

FIGURE 13.4 Cap and snare technique for endoscopic mucosal resection. (a) Submucosal injection, (b) aspiration of target tissue into cap with premounted snare, (c) tightening of the snare around the base of the lesion, and (d) completion of the snare excision.

through an open polypectomy snare. The forceps lifts the lesion and the snare is pushed down around its base and resection ensues. This "reach-through" technique requires a double lumen endoscope which can be cumbersome to use in the esophagus. As a result, lift-and-cut techniques are used less commonly for esophageal lesions. *"Suck-and-Cut" techniques*: This method aspirates the lesion into a cap attached to the tip of the endoscope and then resects it. The most common methods utilize either a cap only (EMR-C) or a cap with an integrated band ligating device (EMR-L). For EMR-C (Fig. 13.4), a specialized cap is fitted to the end of the endoscope providing a chamber for aspiration of mucosa. The lesion may or may not be elevated with a submucosal injection and a crescent shaped snare is mounted into the distal inner rim of

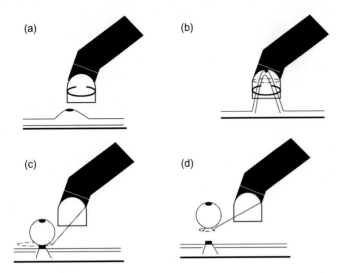

FIGURE 13.5   Band ligation endoscopic mucosal resection. (a) Placement of the cap of the band ligator over the target tissue, (b) aspiration of the target tissue into the cap, (c) placement of the ligation band around the base of the target tissue, and (d) snare excision of the lesion beneath the ligating band.

the cap. The lesion is then aspirated into the cap and the snare tightened around its base. The snared "pseudopolyp" is then excised. The special EMR caps come in various sizes and shapes, the largest of which is 18 mm in diameter.

When performing EMR-L (Fig. 13.5), submucosal injection may not be required. This technique utilizes a variceal band ligator to ligate the mucosal lesion and create a pseudopolyp. The banded tissue is then simply snare excised either above or below the band.

Both EMR-C and EMR-L are limited in their capacity to accomplish en bloc resection. The maximum diameter amenable to one-piece excision is approximately 20 mm. If multiple resections are required, they should be accomplished at the initial setting if possible as submucosal lift may not be attainable once scar tissue has formed.

13.2.2 Endoscopic Submucosal Dissection

ESD allows for larger en bloc resections and begins similar to EMR by marking the periphery

of the lesion with small electrosurgery burns. A margin of at least 5 mm is planned and marks are placed approximately every 2 mm (Fig. 13.6a). A submucosal injection is then accomplished around the periphery of the lesion utilizing a solution such as sodium hyaluronate with Indigo carmine so that the elevation will persist throughout the procedure (Fig. 13.6b). A circumferential incision is then made to isolate the lesion using an ESD electrosurgical knife (Fig. 13.6c and d). There are a number of ESD knifes to choose from (Fig. 13.7):

- *Needle knife*: This knife has a fine tip and small contact area that allows sharp incision. Because of its sharp nature, it can easily cause perforation if not controlled carefully. Mucosal incisions are usually begun with a needle knife, but then a switch is made to a protected tip knife to minimize the risk of perforation.
- *IT knife*: The insulated tip knife is a needle knife with the tip covered by a ceramic ball. This blunt, nonthermal tip reduces the risk of perforation. A second generation IT knife has a conducting surface on the bottom of the ball tip to allow for better tissue division when drawing back on the knife.

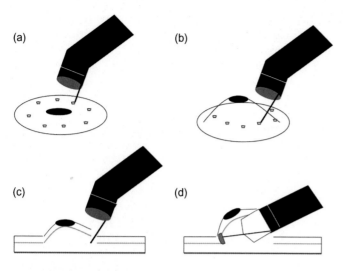

FIGURE 13.6 Technique of endoscopic submucosal dissection (ESD). (a) Marking of the planned resection margin, (b) submucosal injection, (c) incision into the submucosa with an ESD knife, and (d) elevation of the lesion off the muscularis mucosa with dissection into the submucosal plane using an endoscopic cap.

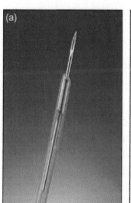    

FIGURE 13.7   (a) Needle knife, (b) insulated tip (IT) knife, (c) hook knife, (d) flex knife, and (e) triangle tip (TT) knife.

- *Hook knife*: This is a needle knife with the distal 1 mm of the tip bent at a right angle. The knife also rotates for optimal positioning.
- *Flex knife*: This knife has a rounded tip made of a twisted wire like a snare. Its length can be adjusted as needed. The shaft of the catheter is flexible with a thickened tip that acts as a tissue stop to minimize the chance of perforation.
- *Triangle tip knife*: Has a triangular conductive tip that facilitates cutting mucosa. This knife was designed to be used for all parts of the ESD procedure.
- *Hybrid knife*: A specialized tool that permits submucosal injection, fulguration, and cutting with one device. This device may permit the endoscopist a more efficient device for ESD.

Once the circumferential incision is complete, additional solution is injected into the submucosa in the center to obtain a more complete lift. One of the ESD knives is then used to excise the lesion in the submucosal plane.

Meticulous hemostasis is critical in order to facilitate visualization, and the lesion is often positioned opposite the ground to utilize gravity to clear the field. A transparent hood is mounted on the end of the endoscope to facilitate visualization and dissection into the submucosal plane.

For BE, EMR and ESD may be used for circumferential resection. EMR is done piecemeal as described earlier. ESD elevates the mucosa off the underlying muscularis propria and then excises a "sleeve" of esophageal mucosa over the desired length.

13.3 PREVENTION AND MANAGEMENT OF COMPLICATIONS FOLLOWING ENDOSCOPIC MUCOSAL RESECTION AND ENDOSCOPIC SUBMUCOSAL DISSECTION

Complications of EMR and ESD include bleeding, perforation, and stricture formation. Immediate bleeding after EMR can occur in up to 10% of patients with perforation rates ranging from 3% to 7% in experienced hands [3]. Stricture formation risk is related to the extent of resection with rates as high as 37% when EMR is performed for >50% of the esophageal circumference. ESD has an immediate bleed rate as high as 5% and a perforation rate of up to 7%. Stricture rate is also related to the extent of resection with rates as high as 70% if >50% of the esophageal circumference is resected [4,5]. Methods to minimize the risk of bleeding and manage it during EMR and ESD include adding epinephrine to the submucosal injectate, clipping the mucosa closed after resection, and coapting exposed blood vessels during the submucosal dissection using bipolar forceps. Perforation can often be managed by closure with endoscopic clips but can even be managed supportively with nasogastric tube decompression and antibiotic administration.

The most common strategy employed to prevent and manage stricture formation after extensive EMR or ESD is the systemic administration of steroids coupled with sequential planned dilations. Stricture rates in patients treated with systemic steroids after ESD are 5% compared to 32% in patients without steroids. The number of required dilations is also significantly lower in steroid treated patients (mean 1.7, range 0−7) compared to no steroids (mean 15.6, range 0−48) [6].

13.4 INDICATIONS FOR ENDOSCOPIC MUCOSAL RESECTION AND ENDOSCOPIC SUBMUCOSAL DISSECTION IN BARRETT'S ESOPHAGUS

EMR and ESD are used to remove tissues from the esophagus for diagnostic and therapeutic purposes. EMR can be used to remove lesions up to 2 cm in size, ESD for larger lesions or areas of patchy disease. ESD is much less commonly deployed than EMR. These techniques may be used as stand-alone treatments for BE or as adjuncts to ablation.

EMR is most commonly used as a one-step definitive therapy for short, noncircumferential segments of dysplastic BE, staging of nodular disease, and treatment of superficial esophageal adenocarcinoma (EAC). For nodular disease, EMR provides complete removal of the lesion to allow for accurate pathologic staging which is necessary to determine the appropriateness of endoscopic therapy. The work flow usually proceeds as follows: during initial endoscopic evaluation of the esophagus, BE is found. A careful evaluation is performed to look for irregularities in the mucosa or nodular lesions (Fig. 13.8). The morphology of the BE is described using the Prague C&M criteria and the Paris classification is used to describe the nodular lesions (see Chapter 5: Diagnosis of

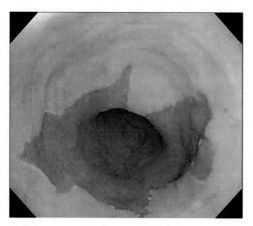

FIGURE 13.8 Nodularity in a field of Barrett's esophagus.

Barrett's Esophagus). Nodular lesions are then excised using EMR and submitted for pathologic evaluation. It is important that the specimens be oriented and pinned appropriately to aid in this evaluation. The remaining BE is then biopsied using the Seattle protocol. If the pathology for the irregular or nodular lesions is benign and there is dysplasia in the surrounding flat BE, then the patient undergoes ablation of the remaining BE. If the lesion is malignant, the depth of invasion and resection margins must be assessed. Inadequate resection margins are potentially amenable to additional EMR. Tumor depth determines the risk of lymph node metastasis. Table 13.1 and Fig. 13.9 illustrate staging of esophageal cancer. T1a lesions with a depth of penetration no deeper than m2 (tumor invades but does not penetrate lamina propria) have a risk of lymph node metastasis less than 6% and are potentially amenable to curative treatment using EMR or ESD. T1b lesions have a significantly increased risk of lymph node metastasis and surgery is usually recommended. Histologic features are also important in this consideration. Absence of lymphovascular invasion and moderate differentiation (G1 and G2) are favorable features for endoscopic resection. Even in T1a

lesions completely excised with EMR or ESD and with favorable histologic features, endoscopic ultrasound (EUS) is performed to assess the presence of lymph node metastasis. Suspicious lymph nodes (hypoechoic, round, in proximity to tumor, greater than 1 cm diameter) are sampled with fine needle aspiration.

13.5 RESULTS OF ENDOSCOPIC MUCOSAL RESECTION AND ENDOSCOPIC SUBMUCOSAL DISSECTION IN BARRETT'S ESOPHAGUS

Both EMR and ESD are effective strategies alone, or in combination with mucosal ablation, for complete eradication of dysplastic BE with or without intramucosal EAC. In expert hands, ESD achieves complete resection of BE at a rate of 84.5% [7]. Interestingly, because Barrett's neoplasia is often characterized by multifocal growth, ESD for early EAC is actually worse than that for EMR using the suck-and-cut method. Curative resection rates of ESD for early EAC are 60–70% with a stricture rate of up to 60% [8]. In contrast, EMR using a suck-and-cut method has a low overall major complication rate of 1.5% with a complete remission rate of 94% [9].

Combining EMR with mucosal ablation results in an initial complete eradication of neoplastic change in 96.6% of patients. However, metachronous lesions develop in 21.5% of patients with 75% occurring within 24 months, thus illustrating the need for continued surveillance endoscopy. Fortunately, the majority of these lesions can be managed endoscopically with a 95.7% long-term complete response rate [10].

When comparing the efficacy of ESD and EMR, in experienced hands ESD achieves a higher en bloc and curative resection rate with lower overall recurrence rates [4]. However, lesions 1.5 cm in diameter or less may be managed with either EMR or ESD with similar

TABLE 13.1   T category Definitions for Esophageal Cancer

T_0	No evidence of primary tumor	
T_{is}, m1	HGD, limited to mucosal layer	
T1, m2	Tumor invades the lamina propria	T1a
T1, m3	Tumor invades into but not through the muscularis mucosa	T1b
T1, sm1	Tumor invades into the upper one-third of the submucosa	T1b
T1, sm2	Tumor invades into the middle one-third of the submucosa	T1b
T1, sm3	Tumor penetrates the deepest one-third of the submucosa	T1b
T2	Tumor invades the muscularis propria	
T3	Tumor invades the adventitia	
T4a	Resectable tumor that invades pleura, pericardium, or diaphragm	
T4b	Unresectable tumor invading aorta, vertebra, trachea, or other structures	

T and N

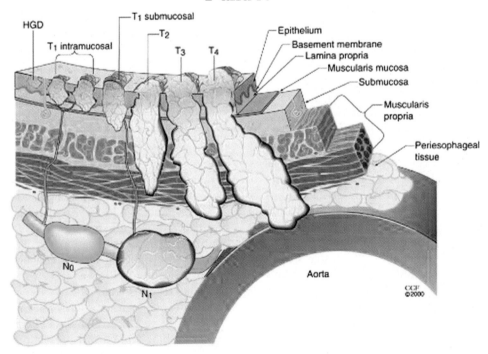

FIGURE 13.9   Depth of tumor invasion and risk of lymph node metastasis. *Source: From Rice WR. Diagnosis and staging of esophageal carcinoma. In: Pearson FG, Cooper JD, Deslauriers J, et al., editors. Esophageal surgery, 2nd ed. New York: Churchill Livingstone; 2002. p. 687.*

results [11]. Because of these findings and the technical difficulty in performing ESD, EMR is the procedure of choice for most endoscopists managing nodular disease in the esophagus.

13.6 INDICATIONS FOR SURGERY IN BARRETT'S ESOPHAGUS

For decades, esophagectomy was the standard of care for BE with HGD or IMC. The rational was based on the finding that up to 37% of esophagectomy specimens in patients undergoing resection for HGD contained invasive cancer [12]. More recent studies have demonstrated a much lower occurrence in resection specimens and endoscopic therapies have essentially replaced surgery for this disease. However, surgery is still indicated in select cases as indicated by the disease processes or patient factors (Table 13.2). These indications include tumor characteristics with significant risk of lymph node metastasis, dysplastic or neoplastic BE that is difficult to eradicate, failed endoscopic eradication, patient unwilling or unable to comply with endoscopic regimen, or an esophagus not worth salvaging.

Esophagectomy is a complex surgery traditionally associated with high morbidity and mortality. Studies reviewing the Nationwide Inpatient Sample in the United States report mortality for esophagectomy of 8.7% [13]. Unfortunately, these data report the outcomes of all esophagectomy cases with no specific outcomes for esophagectomy performed for early neoplasia. When specifically looking at the literature published for esophagectomy performed for BE and HGD, the mortality rate is 0.94% [12].

When operating for early disease with a low potential for lymph node metastasis and high expectations for cure, the surgeon should employ minimally invasive techniques such as vagal sparing or minimally invasive esophagectomy.

13.7 TECHNIQUES OF MINIMALLY INVASIVE ESOPHAGECTOMY

There are essentially three techniques for performing minimally invasive esophagectomy: (1) Ivor—Lewis defined as gastric mobilization via laparoscopy with esophageal and nodal dissection via right-sided thoracoscopy with an intrathoracic anastomosis. (2) Modified McKeown or three-stage esophagectomy defined as gastric mobilization via laparoscopy with esophageal and nodal dissection via right-sided thoracoscopy, and cervical anastomosis via a neck incision. (3) Transhiatal defined as gastric mobilization by laparoscopy with distal esophageal and mediastinal nodal dissection through the diaphragmatic hiatus with proximal esophageal mobilization and cervical anastomosis via a neck incision. All of these procedures can be performed with a vagal-sparing technique. The transhiatal approach is not commonly used with a minimally invasive approach because of the limited access to the mediastinum via a laparoscopic transhiatal pathway and complications of blind mediastinal dissection. The transthoracic approach for both the Ivor—Lewis and three-stage esophagectomy can be done in the lateral decubitus or prone position. Some studies have suggested that the prone position results in less blood loss and better lymph node yield [14].

TABLE 13.2 Relative Indications for Esophagectomy in Patients with HGD or IMC

Depth of Lesion	HGD m1	T1a m2	T1b m3	T1b sm1	T1b sm2	T1b sm3
Risk of lymph node metastasis	0%	0–6%	0–12%	8–32%	12–28%	20–67%

13.8 COMPLICATIONS FROM ESOPHAGECTOMY

Complications can occur in nearly 40% of patients after esophagectomy. These range from anastomotic leak (4−17%), stricture (38−83%), gastroesophageal reflux (68%), and loose bowel movements (44%) [12,15]. Mortality in experienced centers is less than 1%. Despite these findings, 97% of patients are satisfied with their surgical results with a similar number reporting they would undergo the same operation again.

13.9 ESOPHAGECTOMY VERSUS ENDOSCOPIC THERAPIES FOR EARLY ESOPHAGEAL NEOPLASIA

Three retrospective case series have compared surgical and endoscopic treatment of BE with HGD or IMC. The first from the Mayo Clinic published in 2009 compared outcomes in 178 patients with IMC. The majority of patients had endoscopic therapy (132/178) consisting of EMR alone or combined with photodynamic therapy. At nearly 4 years follow-up, 18.2% (24) of the endoscopically treated patients had persistent or recurrent cancer with nine requiring esophagectomy. At over 5 years of follow-up, all resections were curative. The overall mortality between groups was comparable with 17% in the endoscopically treated group and 20% in the surgery group [16].

The second report in 2011 comes from the University of Southern California for a cohort of 101 patients with HGD or IMC. Forty were treated endoscopically with combined EMR and mucosal ablation. Metachronous neoplasia occurred in 20% of endoscopically treated patients with three (7.5%) requiring surgery.

Morbidity was much lower in the endoscopic group (0% vs 39%) and overall and disease-free survivals were equivalent at 94% [17].

A third European trial reported on the experience with 76 patients having EMR and mucosal ablation compared to 38 patients undergoing esophagectomy. Similar to the results from the Southern California group, endoscopic treatment was associated with equivalent cure rates compared to esophagectomy, but low morbidity and no mortality [18].

13.10 FUTURE DEVELOPMENTS IN ENDOLUMENAL AND TRANSLUMENAL SURGERY

The natural evolution of surgery is to become less invasive and more accurate. We have witnessed this in the modern era with the advent of laparoscopic and thoracoscopic surgery. Recent efforts have accelerated the development of intralumenal surgery as evidenced by ESD. An exciting extension of ESD is the introduction of submucosal surgery where an endoscope is tunneled into the submucosal space to work on the muscular wall of the esophagus or to traverse the wall to gain access to the mediastinum. The first clinically applicable approach for this work was the introduction of per oral endoscopic myotomy (POEM) for the treatment of achalasia by Inoue [19]. This innovative extension of ESD techniques taught the world that endoscopic access to the submucosal space could be safely achieved with secure closure afterward. The same access techniques have been used to traverse the wall of the esophagus and harvest lymph nodes in the mediastinum in a laboratory model [20]. The development of intralumenal robotic surgery platforms combined with these exciting techniques may open the door for more advanced endolumenal resectional procedures that will further obviate the need for transthoracic or transabdominal esophageal surgery.

13.11 CONCLUSION

EMR and ESD techniques combined with ablative mucosal therapies have revolutionized how BE with HGD and intramucosal carcinoma are treated. No longer is resective esophageal surgery required for the majority of patients with these pathologies. The avoidance of morbidity associated with esophageal resection, even in experienced centers, is a true advance in modern medicine. Esophageal surgeons must transition to using an endoscopic platform to manage these diseases or be left behind.

References

[1] Tada M, Shimada M, Murakami F, Shimada M, Mizumachi M, Arima K, et al. Development of strip-off biopsy (in Japanese with English abstract). Gastroenterol Endosc 1984;26:833–9.

[2] Fujishiro M, Yahagi N, Kashimura K, Mizushima Y, Oka M, Enomoto S, et al. Comparison of various submucosal injection solutions for maintaining mucosal elevation during endoscopic mucosal resection. Endoscopy 2004;36:579–83.

[3] ASGE Standards of Practice Committee. The role of endoscopy in Barrett's esophagus and other premalignant conditions of the esophagus. Gastrointest Endosc 2012;76(6):1087–94.

[4] Guo HM, Zhang X, Chen M, Huang SL, Zou XP. Endoscopic submucosal dissection vs endoscopic mucosal resection for superficial esophageal cancer. World J Gastroenterol 2014;20(18):5540–7.

[5] Qumeseya B, Panossian AM, Rizk C, Cangemi D, Wolfsen C, Raimondo M, et al. Predictors of esophageal stricture formation post endoscopic mucosal resection. Clin Endosc 2014;47:155–61.

[6] Kobayashi S, Kanai N, Ohki T, Takagi R, Yamaguchi N, Isomoto H, et al. Prevention of esophageal strictures after endoscopic submucosal dissection. World J Gastroenterol 2014;20(41):15098–109.

[7] Tsujii Y, Nishida T, Nishiyama O, Yamamoto K, Kawai N, Yamaguchi S, et al. Clinical outcomes of endoscopic submucosal dissection for superficial esophageal neoplasms: a multicenter retrospective cohort study. Endoscopy 2015;47(9):775–83.

[8] Chevaux JB, Plessevaux H, Jouret-Mourin A, Yeung R, Danse E, Deprez PH, et al. Clinical outcome in patients treated with endoscopic submucosal dissection for superficial Barrett's neoplasia. Endoscopy 2015;47:103–12.

[9] Pech O, May A, Manner H, Behran SA, Pohl J, Wefering M, et al. Long-term efficacy and safety of endoscopic resection for patients with mucosal adenocarcinoma of the esophagus. Gastroenterology 2014;146:652–60.

[10] Pech O, Behrens A, May A, Nachbar L, Gossner L, Rabenstein T, et al. Long-term results and risk factor analysis for recurrence after curative endoscopic therapy in 349 patients with high-grade intraepithelial neoplasia and mucosal adenocarcinoma in Barrett's oesophagus. Gut 2008;57:1200–6.

[11] Yamaguchi Y, Katusmi N, Aoki K, Toki M, Nakamura K, Abe N, et al. Resection area of 15 mm as dividing line for choosing strip biopsy or endoscopic submucosal dissection for mucosal gastric neoplasm. J Clin Gastroenterol 2007;41(5):472–6.

[12] Williams VA, Watson TJ, Herbella FA, Gellersen O, Raymond D, Jones C, et al. Esophagectomy for high grade dysplasia is safe, curative, and results in good alimentary outcome. J Gastrointest Surg 2007;11(12):1589–97.

[13] Connors RC, Reuben BC, Neumayer LA, Bull DA. Comparing outcomes after transthoracic and transhiatal esophagectomy: a 5-year prospective cohort of 17,395 patients. J Am Coll Surg 2007;205(6):735–40.

[14] Markar SR, Wiggins T, Antonowicz S, Zacharakis E, Henna GB. Minimally invasive esophagectomy: lateral decubitus vs. prone positioning; systematic review and pooled analysis. Surg Oncol 2015;24 (3):212–19.

[15] Zhou C, Ma G, Li X, Li J, Yan Y, Liu P, et al. Is minimally invasive esophagectomy effective for preventing anastomotic leakages after esophagectomy for cancer? A systematic review and meta-analysis. World J Surg Onc 2015;13:269–78.

[16] Prasad GA, Wu TT, Wigle DA, Buttar NS, Wangkeesong LM, Danagan KT, et al. Endoscopic and surgical treatment of mucosal (T1a) esophageal adenocarcinoma in Barrett's esophagus. Gastroenterology 2009;137(3):815–23.

[17] Zehetner J, DeMeester SR, Hagen JA, Ayaz S, Augustin F, Lipham JC, et al. Endoscopic resection and ablation versus esophagectomy for high-grade

dysplasia and intramucosal adenocarcinoma. J Thorac Cardiovasc Surg 2011;141(1):39—47.

[18] Pech O, Bollschweiler E, Manner H, Leers J, Ell C, Holscher AH, et al. Comparison between endoscopic and surgical resection of mucosal esophageal adenocarcinoma in Barrett's esophagus at two high-volume centers. Ann Surg 2011;254(1):67—72.

[19] Inoue H, Minami H, Sato Y, Sato Y, Kaga M, Suzuki M, et al. Peroral endoscopic myotomy (POEM) for esophageal achalasia. Endoscopy 2010;42(4):265—71.

[20] Willingham FF, Gee DW, Lauwers GY, Brugge WR, Rattner DW. Natural orifice transesophageal mediastinoscopy and thoracoscopy. Surg Endosc 2008;22 (4):1042—7.

14

Chemoprevention of Barrett's Esophagus and Adenocarcinoma

Ebubekir Şenateş

Department of Gastroenterology, Göztepe Education and Research Hospital,
Istanbul Medeniyet University, Istanbul, Turkey

14.1 INTRODUCTION

Barrett's esophagus (BE) is an acquired condition in which a metaplastic columnar epithelium replaces the stratified squamous epithelium that normally lines the distal esophagus [1,2]. Metaplasia commonly is a consequence of chronic inflammation, and intestinal metaplasia results from chronic reflux esophagitis caused by the gastroesophageal reflux of acid, bile, and other hazardous substances [3].

Approximately 5% of adult population in the United States is affected by BE [4,5]. Esophageal adenocarcinoma (EAC) is strongly associated with BE and it has become increasingly common in developed countries over the last four decades. While the incidences of most malignancies are decreasing, incidence of EAC is increasing. Based on currently available data, the annual incidence of adenocarcinoma secondary to BE is estimated to range from 0.3% to 0.5% and the prognosis of EAC remains extremely poor, with a 5-year survival rates ranging from 15% to 39% [6,7]. The lifetime risk of EAC in patients with BE was estimated as 10- to 125-fold higher than general population [8].

The most important risk factor for development of EAC is BE, and surveillance and early recognition of high-grade dysplasia (HGD) and/or EAC may improve survival [9,10].

Regular endoscopic surveillance is one of the mainstays in the management of BE [11]. The aim of surveillance is to identify patients at a preclinical or asymptomatic early stage of cancer and initiate therapy leading to improved long-term outcomes. However, it is unclear whether only endoscopic surveillance leads to reduced mortality from EAC in patients with BE [12]. Also, endoscopic surveillance of all patients with BE is expensive and more cost-effective methods are needed [13]. Therefore, current strategies for improved management of EAC target identifying patients at high risk for progression to EAC and identifying chemopreventive agents. In light of the poor outcomes associated with EAC, combined with the presence of a readily identifiable precursor lesion, BE represents an attractive target for chemoprevention.

D. Pleskow & T. Erim (Eds): Barrett's Esophagus.
DOI: http://dx.doi.org/10.1016/B978-0-12-802511-6.00014-4

Chemoprevention refers to the use of chemical compounds to prevent the development and progression of dysplasia, as well as blocking the invasion of dysplastic epithelial cells throughout the basement membrane.

In contrast to the traditional therapeutic paradigms in EAC, chemoprevention is intended for generally healthy individuals. Therefore, several factors need to be considered prior to decision making. These factors include degree of dysplasia in BE, existing comorbidities, patient preferences, and local expertise. The risk of developing EAC increases progressively from 0.12% to 0.33% per year in nondysplastic to up to 20% chance of cancer per year in high-grade dysplastic BE [14,15]. Thus, if the patient is relatively healthy with a low risk of progression, it is reasonable to use chemopreventive agents that have been considered safe for use. In this patient population, even a moderate reduction in cancer risk would translate into significant public health benefit as most patients with BE fall under this category. However, patients with preexisting HGD have a relatively higher risk of undergoing neoplastic transformation or a high probability that they may already have developed early EAC. In this patient group, it is important to use more effective chemopreventive treatment even if it carries a higher risk of adverse effects.

In this chapter, we will define suggested targets for chemoprevention of BE and review data about specific chemoprevention agents.

14.2 SUGGESTED TARGETS AND STRATEGIES FOR CHEMOPREVENTION TO PREVENT OR SLOW MALIGNANT TRANSFORMATION OF BARRETT'S ESOPHAGUS

14.2.1 Gastric Acid and Bile Reflux

It is known that in patients with BE complicated with dysplasia and EAC, gastric acid and duodenal bile reflux are significantly more common than nondysplastic BE and GERD [16]. Despite widespread use of acid-suppressing medications, the incidence of EAC is rising. This suggests that refluxed material other than acid might contribute to carcinogenesis [17]. The controlling of reflux symptoms by medical or surgical (repair of lower esophageal sphincter (LES)) treatment do not prevent development of esophageal cancer [18,19]. A probable explanation for this lack of efficacy is that despite medical treatment, bile reflux into the esophagus persists in approximately one out of three patients. Furthermore, reflux recurs within years of surgical fundoplication in a significant proportion of patients treated with fundoplication [18,19]. More recently, in an ex vivo study [20], it was suggested that the pattern of gastric acid reflux may be an important determinant factor in the neoplastic progression of BE. In this study, it was shown that pulsed acid exposure increased cell proliferation but continuous acid exposure decreased cell proliferation [20].

In animal studies, it has been shown that the reflux of bilious intestinal juice alone is sufficient to cause esophageal cancer in rats [21–23]. Studies in humans have shown that patients with BE have significantly more esophageal exposure to bile and significantly higher esophageal luminal concentrations of bile salts than patients who have GERD without BE [24,25].

In a study conducted on patients with BE [17], it was shown that deoxycholic acid (DCA) causes DNA damage and induces phosphorylation of proteins in the nuclear factor kappa B (NF-κB) signaling pathway in Barrett's epithelial cells in vitro and in vivo. Authors also have reported that DCA-induced DNA damage and NF-κB pathway activation are mediated by the generation of reactive oxygen species (ROS) and reactive nitrogen species (RNS). In addition, it was demonstrated that DCA-mediated activation of the NF-κB pathway allows

Barrett's epithelial cells to resist apoptosis in the setting of DNA injury, events which might contribute to neoplastic progression in BE.

Animal studies have shown that chronic gastric acid and/or bile exposure can lead to dysplasia and EAC. The mechanisms by which reflux causes chronic esophageal damage and induce carcinogenesis in BE are suggested as follows.

14.2.1.1 Chronic Injury and Inflammation

In animal model studies, it has been reported that normal cell volume regulatory mechanisms may be inhibited by gastric acid-induced mucosal injury [26]. Reflux of duodenogastric contents (including bile acids) has been shown to cause esophageal damage in synergy with gastric acid in animal studies [27,28]. These findings have also been confirmed in human studies [29]. Reflux of bile parallels gastric acid and increases with the severity of GERD (BB). However, bile salts can cause esophageal damage in a wide range of pH. Glycine-conjugated bile salts cause damage in pH between 4 and 6, taurine-conjugated bile salts cause damage when the pH is lower than 4, while, unconjugated bile salts cause damage in neutral or alkaline pH states [16,24]. Long-term use of PPIs lead to deconjugation of bile salts by bacterial colonization of the proximal intestine [30]. In the alkaline pH state associated with PPI use, unconjugated bile salts cause chronic low-grade inflammation and induce carcinogenesis in BE [31].

14.2.1.2 Arachidonic Acid Pathway

Arachidonic acid pathway is a central regulator of inflammatory response. Gastric acid and duodenal bile acids may contribute to carcinogenesis in BE through activation of this pathway. Low pH and bile acids induce cyclooxygenase-2 (COX-2), which is a central enzyme of the arachidonic acid pathway both in the human BE ex vivo culture model and the EAC cell lines [32,33]. Expression of COX-2

increases concomitantly with neoplastic progression in BE and this increase supports an association between the arachidonic acid pathway and the development of EAC [32–35]. The COX-2 enzyme catalyzes conversion of arachidonic acid to different prostaglandins such as prostaglandin E2 (PGE2). PGE2 induces proliferation of Barret's epithelial cells and its inhibition may slow their growth [36]. PGE2-induced cell proliferation leads to accumulation of replicative errors in premalignant Barrett's cells. Furthermore, it inhibits tumor surveillance through natural killer cell activity of PGE2. Chronic induction of PGE2 might facilitate accumulation of abnormal cells that have genomic instability and thus inhibition of components of arachidonic acid pathway probably will inhibit carcinogenesis [37].

14.2.1.3 Oxidative Stress

Bile acids can stimulate both esophageal squamous and Barrett's epithelial cells to produce substances which can induce esophageal inflammation (ie, interleukin 8 (IL-8) and COX-2) and bile acids also can lead to oxidative stress and DNA damage in these cells [38].

Epidemiological, animal, and clinical studies suggest that BE develops as a result of injury induced by two major components of refluxate: gastric acid and bile acids. Bile acids in combination with gastric acid induce oxidative stress, DNA damage, and alterations in cell signaling. Moreover, bile acids may induce the expression of proteins associated with a phenotypic switch from normal squamous to intestinal phenotype such as Klf-4, villin, and CDX2 [39].

Gastric acid and bile salt injury induce ROS in BE [40]. Moreover, decreased levels of glutation and vitamin C in the epithelium show that the antioxidant defense mechanisms are diminished in patients with BE [40]. Low levels of antioxidant enzymes like glutathione-S-transferase and glutathione have been demonstrated in biopsies obtained from patients with BE and

EAC. This change in oxidative tissue state promotes mutagenesis to induce neoplasia [41].

DNA damage incurred by free radicals causes mutations in key cell survival regulatory genes and induces carcinogenesis [40]. In a normal individual, mutated cells do not further proliferate because they are forced to cell cycle arrest or apoptose by p53 [42]. However, bile salts can inhibit this process via proteasome-mediated degradation of p53 [43]. In conclusion, genetically abnormal cells accumulate and carcinogenesis progresses.

14.2.2 Obesity, Diet, and Lifestyle

In the last three decades, obesity has reached epidemic proportions in Western countries [44]. Obese patients have a significantly increased risk of developing cancer [45]. Approximately 14% of cancer-related deaths in men and 20% in women are partially attributed to obesity [46] and the relative risk for EAC is increased in parallel to increased body mass index (BMI) (relative risk: 1.52 for each 5 kg/m^2 increase in BMI) [47]. In a multicenter population-based case–control study, it was found that compared with subjects in the lowest 10% of usual BMI (<21.70 for men and <20.18 for women), risk of EAC increased fivefold among those in the highest decile (>29.54 for men and >31.25 for women) [48]. There is a closer association between the risk of EAC and how the fat is distributed in the body, and the distribution of fat in the body is more important for EAC rather than only BMI [49,50]. Risk of developing of EAC is higher in patients with increased waist to hip ratio, which is a surrogate measure of truncal obesity [49]. In CT scan studies, it was reported that the excess visceral fat, compared to subcutaneous fat, more likely predict EAC risk [51,52]. In diet studies, close association was reported between increased risk of EAC and higher consumption of dietary cholesterol and protein

along with a decreased consumption of fiber, vitamins C, B6, E, folate, and beta-carotene [48,53–55].

Several mechanisms have been proposed through which obesity can facilitate carcinogenesis in BE [31].

14.2.2.1 Disruption of Antireflux Mechanisms

It has been suggested that by increasing abdominal or intraabdominal adiposity (representing visceral fat and other fats within the abdominal cavity), intraabdominal pressure increases in parallel to visceral obesity and this increase changes the relationship between gastroesophageal junction and diaphragmatic antireflux mechanisms. Studies conducted in obese patients reported that LES pressure is significantly lower compared to controls [56,57] and obesity is associated with significant drop in LES pressure [58]. Obese patients have also increased incidence of asymptomatic reflux [59]. When the LES decreases below 10 mm Hg, this leads to reflux which causing injury, BE, and EAC [60,61].

14.2.2.2 Altered Composition of Refluxate and Mucosal Response to Reflux Injury

In a systematic review, it was reported that the concentration of carcinogenic bile acid in esophageal aspirate was higher in obese patients [38]. In obesity, mucosal response to this reflux injury is changed along with the increased frequency of reflux episodes and alteration in composition of reflux [31].

Obesity causes a systemic and local proinflammatory condition where adipocytes release high concentration of inflammatory cytokines like tumor necrosis factor alpha (TNFα), IL6, IL1B, IL10, and CRP [62]. Such cytokines lead to a proinflammatory state which has been suggested as an important association between obesity and different cancers including EAC [62].

14.2.2.3 *Endocrine Disorders*

Obesity is characterized by insulin resistance, high circulating levels of insulin, down regulation of adipokines like adiponectin and ghrelin as well as increased leptins [63−66]. High-energy intake and excessive animal fat consumption and low fiber intake as well as physical inactivity contribute to insulin resistance and resulting hyperinsulinemia [52,65,67,68]. Induction of the insulin/insulin-like growth factor (IGF) pathway in obese patients has been suggested in recent studies [67−69]. Activation of IGF-1 receptors by insulin stimulates cellular proliferation and inhibits apoptosis via the oncogenic PI3K−AKT−mTOR−S6K1 signaling cascade which facilitates carcinogenesis in BE [60,63]. Along with insulin/IGF, leptin also stimulates the proliferation of EAC cells and inhibits their apoptosis through PGE2-mediated activation of EGFR and activation of c-Jun NH2-terminal kinase [51]. Adipokine signaling is also downregulated in metabolic syndrome and EAC has been correlated with the expression of adiponectin receptors [70].

14.2.3 Hypergastrinemia

Gastrin has trophic effects on gastric acid secretion and has been implicated in the regulation of cell survival as well as differentiation in the gastrointestinal tract [71−74]. Gastrin-induced signaling by increasing COX-2 expression probably promotes carcinogenesis in BE [74]. Compared to squamous epithelium of esophagus, Barrett's mucosa expresses more CCK2 receptors and gastrin increases Barrett's epithelial cell survival by inducing proliferation and inactivating proapoptotic factors to promote carcinogenesis [73,74].

Gastrin induces proliferation in Barrett's metaplasia via prevention of apoptosis [75], activation of the cholecystokinin-2 receptor [73], and upregulation of COX-2 expression [74].

14.2.4 *Helicobacter pylori* Infection

Infection of the stomach with *Helicobacter pylori* causes chronic inflammation which can result in intestinal metaplasia and cancer in the stomach. However, *H. pylori* does not infect Barrett's esophagus and there is no positive association between *H. pylori* infection and GERD. In contrast, studies have suggested that *H. pylori* infection may protect the esophagus from GERD and its complications like BE, perhaps by causing a chronic gastritis that interferes with acid production. *H. pylori* strains which express cytotoxin-associated gene A (cagA) appear to be important for damage to the stomach, and especially protective for the esophagus. However, a recent study that was controlled for potential confounders such as demographic factors, lifestyle factors, BMI, and smoking did not find a negative association between cagA positive *H. pylori* strains and EAC as had been expected [76]. Despite the theoretical protective benefit, studies evaluating the association between BE and *H. pylori* infection have shown contradictory results [76−80]. Although in vitro studies have shown that cagA expressing *H. pylori* strains leads to apoptosis in esophageal cancer cell lines [81], the clinical relevance of this finding remains unclear because *H. pylori* rarely colonize BE.

Therefore, protective role of *H. pylori* from EAC in BE remains uncertain. In EAC, the epidemiological evidence regarding the protective role of *H. pylori* is controversial.

H. pylori is also defined by World Health Organization (WHO) as a type I carcinogen for gastric cancer and therefore is not an optimal strategy for chemoprevention.

14.2.5 Molecular Alterations

The epigenetic alterations associated with the development of BE have been extensively studied. Caudal homeobox genes 1 (CDX1) and 2 (CDX2) are involved and are critical points in this process [82]. Nuclear factor

TABLE 14.1   A Partial List of Altered Genes/Pathways in BE and EAC

Barret's Esophagus	Esophageal Adenocarcinoma
CDX1, CDX2	
NF-κB targets	NF-κB targets
Genes in the mTOR pathway	Genes in the mTOR pathway
BMP4	
RB pathway genes	
P53	P53
Ras pathway	Ras pathway
VEGF-related genes	VEGF-related genes
Cell cycle-related pathway genes	Cell cycle-related pathway genes
STAT-3	STAT-3
Sox 9	Sox 9
MMPs	MMPs
Wnt signaling	Wnt signaling
c-MET pathway	c-MET pathway

kappa B (NF-κB), fibroblast growth factor (FGF), bone morphogenetic protein 4 (Bmp4), and hedgehog and wnt pathways are related to CDX gene regulation [83−87]. Moreover, p53 suppression which is a determinant of squamous phenotype may also be an important factor [88]. There are multiple genes/pathways alterations in the BE and EAC development process. A partial list of altered genes/pathways in BE and EAC are shown in Table 14.1.

14.3 CHEMOPREVENTION OF ADENOCARCINOMA ASSOCIATED WITH BARRETT'S ESOPHAGUS

In light of the poor outcomes associated with EAC, combined with the presence of a readily identifiable precursor lesion, BE represents an attractive target for chemoprevention. Because the absolute risk of EAC is very low even in patients with BE [89−91], a viable chemoprevention strategy would have to be safe, inexpensive, and effective.

14.3.1 Proton Pump Inhibitors

GERD is a known primary risk factor for BE and EAC [92]. PPIs, since the introduction of omeprazole in 1988, have been largely used for treatment of acid-related disorders, including BE. PPIs decrease secretion of gastric acid by inhibiting H^+/K^+ ATPase of parietal cells and alleviate symptoms associated with acid reflux. Gastric acid suppression with PPI therapy plays a pivotal role in the management of symptoms in persons with chronic GERD and BE. Several epidemiologic studies suggest that acid suppression with PPIs has chemopreventive effects in patients with BE [93−96]. In a meta-analysis, it was reported that PPI use in patients with BE was associated with a 71% reduced risk of progression to HGD or EAC [97]. Although, clinical guidelines do not recommend gastric acid suppression as a means of cancer risk reduction for patients with BE, in daily clinical practice, PPIs have become de facto chemopreventive agents [98]. Currently, almost all patients with BE under surveillance are prescribed PPIs [89,99,100].

Historically, PPIs have been considered safe medications. However, observational data suggest that chronic PPI use is associated with increased risks of bone fractures and of *Clostridium difficile* infection [101−105]. Based on these data, the Food and Drug Administration has issued warnings regarding long-term use of PPIs and bone fracture and use of PPIs and *C. difficile* infection [106,107]. However, given the observational nature of the data, physicians continue to prescribe PPIs to almost all patients with BE.

Although data obtained from clinical studies are lacking, epidemiologic studies suggest that PPI use in patients with BE has chemopreventive effects [94–96,108].

In a study, reported by El-Serag et al. [95], the development of dysplasia was compared in patients with BE treated with or without PPI or histamine 2 receptor antagonist (H2RA) over a 20-year time period. They found that the cumulative incidence of dysplasia was significantly lower among patients who received PPI after BE diagnosis than in those who received no therapy or H2RA. Furthermore, among those on PPIs, a longer duration of PPI use was associated with less frequent occurrence of dysplasia. In another study [108], Hillman et al. examined whether PPI therapy influences the incidence and progression of dysplasia in patients with BE. They found that ongoing PPI therapy appeared beneficial in the prevention of dysplasia and adenocarcinoma in patients with BE and suggested that all patients with this condition, even those with no esophagitis or symptoms, should be encouraged to continue long-term PPI therapy.

Many patients with BE have chronic reflux symptoms; PPIs have obvious therapeutic value in this group and therefore have benefits beyond potential EAC risk reduction. However, in two studies aimed at determining BE prevalence, approximately one half of patients with BE did not report a history of regular reflux symptoms [4,5]. In a cost-effectiveness study, it was found that chemoprevention with PPIs in patients with BE without reflux is cost effective [109].

An important concern regarding long-term PPI use is that gastric acid suppression treatment with PPIs may lead to the bacterial overgrowth and increased reflux of toxic, unconjugated bile acids. This may result in increased cell DNA damage, mutations, and consequently to BE and EAC development [39]. However, this issue remained uncertain and yet to be confirmed with epidemiological and randomized controlled trials. On the other hand, there is no significant association between EAC and the use of antisecretory agents per se [39].

In a retrospective observational study [110] conducted on 344 patients with documented BE, authors performed COX regression analysis in order to examine the association between prescriptions for PPI, NSAID/aspirin or statins and the risk of developing esophageal dysplasia or EAC during follow-up (from 1982 to 2005). They found that after diagnosis of BE, 67.2% of the patients were prescribed PPI for a mean duration of 5.1 years. It was found that after BE diagnosis, PPI therapy was associated with a reduced risk of HGD or EAC and this association persisted after adjustment for gender, age, and the length of BE at time of diagnosis. They conclude that in patients with BE, PPI therapy reduces the risk of neoplasms. A nested case–control observational study in a cohort of patients with BE identified in a cohort of 11,823 patients with BE using PPI treatment might reduce the risk of developing EAC [99].

There are other additional observational studies which have demonstrated an association between PPI use and the reduced incidence of dysplasia in BE [95,108,111].

14.3.2 Aspirin/NSAIDs

The mechanism of EAC prevention is possibly related to inhibition of COX-2 production. Increased levels of COX-2 in esophageal epithelial cells have been observed in BE and noted to increase with disease progression from BE to EAC [112]. In preclinical studies, COX-2 inhibitors inhibited the growth of BE cells, potentially through suppression of basic FGF [36]. Another study confirmed that the end product of COX-2 conversion (prostaglandin E2) is reduced in patients with BE without HGD when using esomeprazole combined with higher doses of aspirin [113].

Epidemiologic data and animal studies suggest that aspirin and other NSAIDs which inhibit COX may protect against the development of BE [114] or, in patients with established BE, the development of cancer [115,116] in the chemoprevention of EAC. COX-2 inhibitors are a new class of NSAIDs that inhibit prostaglandin synthesis by selectively blocking the COX-2 enzyme. The COX-2 enzyme has been reported to be overexpressed in premalignant and malignant states, including BE and EAC [112]. In BE-associated EAC cell lines, inhibition of COX-2 resulted in antiproliferative and proapoptotic effects [117]. In a recent case−control study conducted on 434 patients with BE, patients who used aspirin were less likely to be diagnosed with BE compared with matched patients who did not use aspirin [114].

In a meta-analysis of published cohort studies that investigated the effect of aspirin on the development of EAC in patients with BE, it was estimated that there is an inverse association between aspirin use and EAC (OR 0.64, 95% CI 0.52−0.79) [116]. A similar risk reduction was observed also for NSAIDs. With NSAID and/or aspirin use, reduced risk of EAC in patients with BE was shown in several observational studies and in a meta-analysis [99,110,118,119]. Such observations provided the rational base for trials which have evaluated NSAIDs (also including COX-2 inhibitors) in EAC prevention in patients with BE [120,121]. However, in a study investigating effect of celecoxib given for 48 weeks, no benefit was found regarding the progression of BE to dysplasia or cancer [122].

Although, NSAIDs are effective in preventing the progression of BE to EAC, it is not clear that the high cost and cardiovascular risks of the COX-2 selective NSAIDs will be justified for routine clinical use.

Aspirin is a cheap and nonselective NSAID that can prevent both cardiovascular and neoplastic complications, may be a useful chemopreventive agent if its protective effects can be demonstrated to outweigh its risks of gastrointestinal complications [123].

A phase III randomized study of aspirin and esomeprazole chemoprevention in BE (the ASPECT trial) is being conducted in the United Kingdom, however, the results are not expected to be available until after 2017.

It has been shown that aspirin with and without endoscopic surveillance is a cost-effective strategy [124]. Choi et al. [125] reported that chemoprevention with aspirin could be a cost-effective strategy when added to endoscopic surveillance for nondysplastic BE.

14.3.3 Ursodeoxycholic Acid

Reflux of bile acid into the esophagus induces esophagitis, inflammation-stimulated hyperplasia, BE, and EAC. Caudal-type homeobox 2 via NF-κB induced by bile acid is an important factor in the development of BE and EAC. Hydrophobic bile acids like DCA that cause oxidative DNA injury and activate NF-κB in Barrett's metaplasia might contribute to carcinogenesis in BE [31].

A study conducted on Wistar rats [126], which underwent a duodenoesophageal reflux procedure, assessed whether UDCA may protect against the esophageal inflammation−metaplasia−carcinoma sequence by decreasing the overall proportion of the toxic bile acids. The rats were divided into two groups: one group was given commercial chow (control group) and the other was given experimental chow containing UDCA (UDCA group). After 40 weeks, the rats were sacrificed and their bile and esophagi were examined. They found that, in the UDCA group, the esophagitis was milder and the incidence of BE was significantly lower than in the control group, and EAC was not observed. Expression intensity of Cdx2 and NF-κB was found greater than in the UDCA group. Authors concluded that UDCA may be a chemopreventive agent against EAC by varying the bile acid composition.

In another study reported by Peng et al. [127], authors obtained biopsies of BE from 21 patients before and after esophageal perfusion with DCA at baseline and after 8 weeks of oral UDCA treatment. They found that UDCA increases expression of antioxidants that prevent toxic bile acids from causing DNA damage and NF-κB activation in Barrett's metaplasia. In cells, they found that DCA-induced DNA damage and NF-κB activation were prevented by 24-hour pretreatment with UDCA, but not by mixing UDCA with DCA. They drew a conclusion that elucidation of the molecular pathway for UDCA protection provides rationale for clinical trials on UDCA for chemoprevention in BE.

14.3.4 Statins

Statins have antineoplastic effects through several pathways (both HMG-CoA reductase dependent and HMG-CoA reductase independent). The primary mechanism of action of statins on cholesterol reduction is by competitive inhibition of HMG-CoA reductase. This prevents posttranslational prenylation of the Ras/Rho superfamily, which is an important mediator of cell growth, differentiation, and survival [126]. Due to obesity-associated lipid derangements, statins have also been studied for their potential role in chemoprevention in BE.

Statins, so-called 3-hydroxy-3-methylglutaryl coenzyme A (HMG-CoA) reductase inhibitors, are used for primary and secondary prevention of cardiovascular diseases. Recent observational data suggest that statins have protective effects against the development of cancers [128,129] and may decrease the risk of cancers [130–132]. In vitro and animal studies have shown that in addition to cholesterol reduction, statins have antiproliferative, proapoptotic, antiangiogenic, and immunomodulatory effects, which prevent cancer development and growth [126]. This effect has also been shown in EC cell lines

[46,133–135]. Some recent observational studies have shown that statins may be associated with a lower risk of EC, particularly in patients with BE [136–138] whereas others have shown no beneficial effect [139,140]. Studies of human-derived EAC cell lines suggest that statins may offer exciting potential as chemopreventive agents in EAC, especially in BE [141].

It was shown that statins inhibit proliferation and induce apoptosis in EAC cells via inhibition of Ras farnesylation and inhibition of the ERK and Akt signaling pathways; thus, they may have some potential as chemopreventative agents in EAC, however, human studies are lacking [142].

In a recent retrospective observational study [110], Nguyen et al. examined data from 344 patients diagnosed with BE (mean age: 61 years, 90% Caucasian, 94% male) investigating the association between prescriptions for PPI, NSAIDs/aspirin or statins and the risk of developing esophageal dysplasia or adenocarcinoma during follow-up (from 1982 to 2005). After BE diagnosis, 25.3% were prescribed statins for a mean duration of 2.8 years. They found that while PPI treatment reduces the risk of neoplasms and NSAIDs/aspirin appear to reduce cancer risk in patients with BE, statin use was not significantly associated with lowering the risk of neoplasia in patients with BE.

However, in a recent systematic review with meta-analysis of existing randomized controlled trials and observational studies which investigated the association between statins and risk of development of EAC or progression of dysplasia in patients with BE [62], it was revealed that use of statins was associated with a statistically significant 43% reduction in development of EAC and/or HGD (unadjusted OR, 0.57; 95% CI, 0.44–0.75). This effect remained after adjustment for potential confounders, including use of NSAIDs/aspirin and length and dysplasia status of BE (adjusted OR, 0.59; 95% CI, 0.45–0.78). Authors found that the number needed to treat with statins to prevent one case of EAC per year in patients with BE was a dismal 389.

However, in a more recent population-based nested case–control study [143] within a BE cohort from two primary care settings (located in the United Kingdom and Netherlands), authors aimed to estimate the risk of EAC among patients with BE exposed to NSAIDs, statins, and PPIs. In this study, drug use was assessed from BE diagnosis until study matching date. Controls were matched on age, sex, year of BO diagnosis and database. Authors found that statins did not significantly reduce the risk of HGD and EAC among patients with BE (OR for statin use >3 years, 0.5; 95% CI, 0.1–1.7). Although not statistically significant, authors found that there was a dose–duration–response seen for statins, with lower OR for longer duration of use compared with nonuse of statins. Higher doses of statins showed lower estimates for EAC and HGD. In a more recent study reported by Agrawal et al. in patients with BE and EAC during 20-year period, statin use was found to be protective against EAC in patients with BE ($p = 0.001$) [144].

Finally, in patients with multiple risk factors for EAC, statins may have a chemoprotective effect on dysplasia progression and development of EAC. However, there are no animal models that effectively mimic human disease in terms of risk factors, molecular pathogenesis, and response to interventions [145]. Therefore, chemopreventive potential of statins in phase I/II trials is warranted.

14.3.5 Metformin

As outlined earlier, obesity is associated with neoplasia through several mechanisms, most likely via insulin-mediated cell pathways that affect cell proliferation. Metformin, a commonly used antidiabetic drug shown to be protective agent against different types of cancers, has been proposed to protect against obesity-associated cancers by decreasing serum insulin. However, there are few studies and limited data whether metformin is effective on chemoprevention of BE and EAC.

In a study which investigated patients with BE and EAC between 1992 and 2012, metformin use was neither associated with an increased nor a decreased risk of esophageal cancer [144]. In another study [146] that assessed the effect of metformin on phosphorylated S6 kinase (pS6K1), which is a biomarker of insulin pathway activation, 74 patients with BE (mean age 59 years, 58 men) were randomly assigned to two groups and given metformin daily ($n = 38$) or placebo ($n = 36$) for 12 weeks. Biopsy specimens were collected at baseline and at week 12 by upper endoscopy. The primary end point was percent changes in median levels of pS6K1 between subjects given metformin versus placebo. While it was found that metformin was associated with a significant reduction in serum levels of insulin as well as in homeostatic model assessments of insulin resistance (HOMA-IR), the percent changes in median level of pS6K1 did not differ significantly between two groups.

14.4 FUTURE DIRECTIONS

The use of aspirin and esomeprazole for prevention of EAC in patients with BE has shown significant positive results in preclinical animal model-based studies and is currently being tested in the phase III ASPECT trial. Although there are promising results in preclinical studies, the clinical benefit of combination therapy with UDCA acid and aspirin needs to be confirmed in clinical trials in order to recommend their use as potential chemopreventive agents in clinical practice. Further studies need to be conducted on understanding the molecular mechanisms involved in BE oncogenesis in order to employ effective strategies to prevent or slow down neoplastic progression.

As PPIs suppress acid reflux-related procarcinogenic signaling, combining of PPIs and gastrin-dependent signaling inhibitors appears to be a novel strategy in order to prevent the development of EAC. Expression of serotonin (5-HT4) receptors increases significantly from the controls to patients with GERD and to those with BE. Preliminary data suggest that the serotonin 5-HT4 receptor involvement could be important in development of GERD process, especially at early stages, by inducing cell proliferation via increased activity of both EGF and COX-2 pathway. It is possible 5-HT4 selective antagonists may play a role in the future of GERD treatment and chemoprevention of BE [39]. However, further preclinical and clinical studies in order to support this hypothesis are needed.

Acknowledgment

Special thanks to Yasar Colak, MD, for his valuable contributions and supervision while writing this chapter.

References

[1] Spechler SJ, Fitzgerald RC, Prasad GA, Wang KK. History, molecular mechanisms, and endoscopic treatment of Barrett's esophagus. Gastroenterology 2010;138:854.

[2] Spechler SJ, Souza RF. Barrett's esophagus. N Engl J Med 2014;371:836.

[3] Spechler SJ. Barrett's esophagus: pathogenesis and malignant transformation. <http://www.uptodate.com.lproxy.yeditepe.edu.tr/contents/barretts-esophagus-pathogenesis-and-malignant-transformation?source=search_result&search=barrett&selectedTitle=4%7E69> [accessed 21.05.15].

[4] Ronkainen J, Aro P, Storskrubb T, Johansson SE, Lind T, Bolling-Sternevald E, et al. Prevalence of Barrett's esophagus in the general population: an endoscopic study. Gastroenterology 2005;129:1825–31.

[5] Rex DK, Cummings OW, Shaw M, Cumings MD, Wong RK, Vasudeva RS, et al. Screening for Barrett's esophagus in colonoscopy patients with and without heartburn. Gastroenterology 2003;125:1670–7.

[6] Jemal A, Siegel R, Xu J, Ward E. Cancer statistics, 2010. CA Cancer J Clin 2010;60:277–300.

[7] Refaely Y, Krasna MJ. Multimodality therapy for esophageal cancer. Surg Clin North Am 2002;82 (4):729–46.

[8] Solaymani-Dodaran M, Logan RF, West J, Card T, Coupland C. Risk of oesophageal cancer in Barrett's oesophagus and gastro-oesophageal reflux. Gut 2004; 53:1070–4.

[9] APCSC Collaboration. The burden of overweight and obesity in the Asia-Pacific region. Obes Rev 2007;8:191–6.

[10] Calle EE, Thun MJ, Petrelli JM, Rodriguez C, Heath Jr CW. Body-mass index and mortality in a prospective cohort of U.S. adults. N Engl J Med 1999;341: 1097–105.

[11] Wang KK, Sampliner RE. Updated guidelines 2008 for the diagnosis, surveillance and therapy of Barrett's esophagus. Am J Gastroenterol 2008;103:788–97.

[12] Corley DA, Mehtani K, Quesenberry C, Zhao W, de Boer J, Weiss NS. Impact of endoscopic surveillance on mortality from Barrett's esophagus-associated esophageal adenocarcinomas. Gastroenterology 2013; 145:312 e1–19 e1.

[13] Kopelman PG. Obesity as a medical problem. Nature 2000;404:635–43.

[14] Gupta M, Iyer PG, Lutzke L, Gorospe EC, Abrams JA, Falk GW, et al. Recurrence of esophageal intestinal metaplasia after endoscopic mucosal resection and radiofrequency ablation of Barrett's esophagus: results from a US multicenter consortium. Gastroenterology 2013;145(1):79–86.

[15] Prasad GA, Wang KK, Halling KC, Buttar NS, Wongkeesong LM, Zinsmeister AR, et al. Utility of biomarkers in prediction of response to ablative therapy in Barrett's esophagus. Gastroenterology 2008;135 (2):370–9.

[16] Vaezi MF, Richter JE. Synergism of acid and duodenogastroesophageal reflux in complicated Barrett's esophagus. Surgery 1995;117(6):699–704.

[17] Huo X, Juergens S, Zhang X, Rezaei D, Yu C, Strauch ED, et al. Deoxycholic acid causes DNA damage while inducing apoptotic resistance through NF-κB activation in benign Barrett's epithelial cells. Am J Physiol Gastrointest Liver Physiol 2011;301(2):G278–86.

[18] Spechler SJ, Lee E, Ahnen D, Goyal RK, Hirano I, Ramirez F, et al. Long-term outcome of medical and surgical therapies for gastroesophageal reflux disease: follow-up of a randomized controlled trial. J Am Med Assoc 2001;285(18):2331–8.

[19] Caldwell MT, Lawlor P, Byrne PJ, Walsh TN, Hennessy TP. Ambulatory oesophageal bile reflux monitoring in Barrett's oesophagus. Br J Surg 1995;82(5):657–60.

[20] Fitzgerald RC, Omary MB, Triadafilopoulos G. Dynamic effects of acid on Barrett's esophagus.

An ex vivo proliferation and differentiation model. J Clin Invest 1996;98:2120.

[21] Buttar NS, Wang KK, Leontovich O, Westcott JY, Pacifico RJ, Anderson MA, et al. Chemoprevention of esophageal adenocarcinoma by COX-2 inhibitors in an animal model of Barrett's esophagus. Gastroenterology 2002;122:1101−12.

[22] Fein M, Peters JH, Chandrasoma P, Ireland AP, Oberg S, Ritter MP, et al. Duodenoesophageal reflux induces esophageal adenocarcinoma without exogenous carcinogen. J Gastrointest Surg 1998;2:260−8.

[23] Sarosi G, Brown G, Jaiswal K, Feagins LA, Lee E, Crook TW, et al. Bone marrow progenitor cells contribute to esophageal regeneration and metaplasia in a rat model of Barrett's esophagus. Dis Esophagus 2008;21:43−50.

[24] Nehra D, Howell P, Williams CP, Pye JK, Beynon J. Toxic bile acids in gastro-oesophageal reflux disease: influence of gastric acidity. Gut 1999;44:598−602.

[25] Vaezi MF, Richter JE. Role of acid and duodenogastroesophageal reflux in gastroesophageal reflux disease. Gastroenterology 1996;111:1192−9.

[26] Snow JC, Goidstein JL, Schmidt LN, Lisitza P, Layden TJ. Rabbit esophageal cells show regulatory volume decrease: ionic basis and effect of pH. Gastroenterology 1993;105:102−10.

[27] Salo J, Kivilaaksc E. Role of luminal H in the pathogenesis of experimental esophagitis. Surgery 1982;92:61−8.

[28] Gillison EW, DeCastro VAM, Nyhus LM, Kusakari K, Bombeck CT. The significance of bile in reflux esophagitis. Surg Gynecol Obstet 1972;134:419−24.

[29] Sears RJ, Champion G, Richter JE. Characteristics of partial gastrectomy (PG) patients with esophageal symptoms of duodenogastric reflux. Am J Gastroenterol 1995;90:211−15.

[30] Theisen J, Nehra D, Citron D, Johansson J, Hagen JA, Crookes PF, et al. Suppression of gastric acid secretion in patients with gastroesophageal reflux disease results in gastric bacterial overgrowth and deconjugation of bile acids. J Gastrointest Surg 2000;4 (1):50−4.

[31] Baruah A, Buttar NS. Chemoprevention in Barrett's oesophagus. Best Pract Res Clin Gastroenterol 2015;29 (1):151−65.

[32] Kaur BS, Ouatu-Lascar R, Omary MB, Triadafilopoulos G. Bile salts induce or blunt cell proliferation in Barrett's esophagus in an acid-dependent fashion. Am J Physiol Gastrointest Liver Physiol 2000;278(6):G1000−9.

[33] Zhang F, Altorki NK, Wu YC, Soslow RA, Subbaramaiah K, Dannenberg AJ. Duodenal reflux induces cyclooxygenase-2 in the esophageal mucosa of

rats: evidence for involvement of bile acids. Gastroenterology 2001;121(6):1391−9.

[34] Yen CJ, Izzo JG, Lee DF, Guha S, Wei Y, Wu TT, et al. Bile acid exposure up-regulates tuberous sclerosis complex 1/mammalian target of rapamycin pathway in Barrett's-associated esophageal adenocarcinoma. Cancer Res 2008;68(8):2632−40.

[35] Souza RF, Shewmake K, Terada LS, Spechler SJ. Acid exposure activates the mitogen-activated protein kinase pathways in Barrett's esophagus. Gastroenterology 2002;122(2):299−307.

[36] Buttar NS, Wang KK, Anderson MA, Dierkhising RA, Pacifico RJ, Krishnadath KK, et al. The effect of selective cyclooxygenase-2 inhibition in Barrett's esophagus epithelium: an in vitro study. J Natl Cancer Inst 2002;94(6):422−9.

[37] Nishigaki Y, Ohnishi H, Moriwaki H, Muto Y. Ursodeoxycholic acid corrects defective natural killer activity by inhibiting prostaglandin E2 production in primary biliary cirrhosis. Dig Dis Sci 1996;41 (7):1487−93.

[38] McQuaid KR, Laine L, Fennerty MB, Souza R, Spechler SJ. Systematic review: the role of bile acids in the pathogenesis of gastro-oesophageal reflux disease and related neoplasia. Aliment Pharmacol Ther 2011;34(2):146−65.

[39] Triadafilopoulos G, Taddei A, Bechi P, Freschi G, Ringressi MN, Degli'Innocenti DR, et al. Barrett's esophagus: proton pump inhibitors and chemoprevention I. Ann N Y Acad Sci 2011;1232:93−113.

[40] Wild CP, Hardie LJ. Reflux, Barrett's oesophagus and adenocarcinoma: burning questions. Nat Rev Cancer 2003;3(9):676−84.

[41] Lee OJ, Schneider-Stock R, McChesney PA, Kuester D, Roessner A, Vieth M, et al. Hypermethylation and loss of expression of glutathione peroxidase-3 in Barrett's tumorigenesis. Neoplasia 2005;7(9):854−61.

[42] Barrett MT, Sanchez CA, Prevo LJ, Wong DJ, Galipeau PC, Paulson TG, et al. Evolution of neoplastic cell lineages in Barrett oesophagus. Nat Genet 1999;22 (1):106−9.

[43] Qiao D, Gaitonde SV, Qi W, Martinez JD. Deoxycholic acid suppresses p53 by stimulating proteasome-mediated p53 protein degradation. Carcinogenesis 2001;22(6):957−64.

[44] Hedley AA, Ogden CL, Johnson CL, Carroll MD, Curtin LR, Flegal KM. Prevalence of overweight and obesity among US children, adolescents, and adults, 1999−2002. JAMA 2004;291(23):2847−50.

[45] Wolk A, Gridley G, Svensson M, Nyren O, McLaughlin JK, Fraumeni JF, et al. A prospective study of obesity and cancer risk (Sweden). Cancer Causes Control 2001;12(1):13−21.

[46] Calle EE, Rodriguez C, Walker-Thurmond K, Thun MJ. Overweight, obesity, and mortality from cancer in a prospectively studied cohort of U.S. adults. N Engl J Med 2003;348(17):1625–38.

[47] Renehan AG, Tyson M, Egger M, Heller RF, Zwahlen M. Body mass index and incidence of cancer: a systematic review and meta-analysis of prospective observational studies. Lancet 2008;371(9612):569–78.

[48] Chow WH, Blot WJ, Vaughan TL, Risch HA, Gammon MD, Stanford JL, et al. Body mass index and risk of adenocarcinomas of the esophagus and gastric cardia. J Natl Cancer Inst 1998;90(2):150–5.

[49] Steffen A, Schulze MB, Pischon T, Dietrich T, Molina E, Chirlaque MD, et al. Anthropometry and esophageal cancer risk in the European prospective investigation into cancer and nutrition. Cancer Epidemiol Biomarkers Prev 2009;18(7):2079–89.

[50] Nelsen EM, Kirihara Y, Takahashi N, Shi Q, Lewis JT, Namasivayam V, et al. Distribution of body fat and its influence on esophageal inflammation and dysplasia in patients with Barrett's esophagus. Clin Gastroenterol Hepatol 2012;10(7):728–34. quiz e61–2.

[51] El-Serag HB, Hashmi A, Garcia J, Richardson P, Alsarraj A, Fitzgerald S, et al. Visceral abdominal obesity measured by CT scan is associated with an increased risk of Barrett's oesophagus: a case–control study. Gut 2014;63(2):220–9.

[52] Beddy P, Howard J, McMahon C, Knox M, de Blacam C, Ravi N, et al. Association of visceral adiposity with oesophageal and junctional adenocarcinomas. Br J Surg 2010;97(7):1028–34.

[53] Engel LS, Chow WH, Vaughan TL, Gammon MD, Risch HA, Stanford JL, et al. Population attributable risks of esophageal and gastric cancers. J Natl Cancer Inst 2003;95(18):1404–13.

[54] Lagergren J, Bergström R, Nyrén O. Association between body mass and adenocarcinoma of the esophagus and gastric cardia. Ann Intern Med 1999;130 (11):883–90.

[55] Calle EE, Kaaks R. Overweight, obesity and cancer: epidemiological evidence and proposed mechanisms. Nat Rev Cancer 2004;4(8):579–91.

[56] Iovino P, Angrisani L, Tremolaterra F, Nirchio E, Ciannella M, Borrelli V, et al. Abnormal esophageal acid exposure is common in morbidly obese patients and improves after successful Lap-Band system implantation. Surg Endosc 2002;16:1631–5.

[57] Suter M, Dorta G, Giusti V, Calmes JM. Gatroesophageal reflux and esophageal motility disorders in morbidly obese patients. Obes Surg 2004;14:959–66.

[58] Küper MA, Kramer KM, Kirschniak A, Zdichavsky M, Schneider JH, Stüker D, et al. Dysfunction of the lower esophageal sphincter and dysmotility of the tubular esophagus in morbidly obese patients. Obes Surg 2009;19(8):1143–9.

[59] Ortiz V, Ponce M, Fernández A, Martínez B, Ponce JL, Garrigues V, et al. Value of heartburn for diagnosing gastroesophageal reflux disease in severely obese patients. Obesity (Silver Spring) 2006;14(4):696–700.

[60] Lagergren J. Influence of obesity on the risk of esophageal disorders. Nat Rev Gastroenterol Hepatol 2011;8 (6):340–7.

[61] Koppman JS, Poggi L, Szomstein S, Ukleja A, Botoman A, Rosenthal R. Esophageal motility disorders in the morbidly obese population. Surg Endosc 2007;21 (5):761–4.

[62] Ryan AM, Duong M, Healy L, Ryan SA, Parekh N, Reynolds JV, et al. Obesity, metabolic syndrome and esophageal adenocarcinoma: epidemiology, etiology and new targets. Cancer Epidemiol 2011;35 (4):309–19.

[63] Kendall BJ, Macdonald GA, Hayward NK, Prins JB, O'Brien S, Whiteman DC. The risk of Barrett's esophagus associated with abdominal obesity in males and females. Int J Cancer 2013;132(9):2192–9.

[64] Lee SJ, Jung MK, Kim SK, Jang BI, Lee SH, Kim KO, et al. Clinical characteristics of gastroesophageal reflux disease with esophageal injury in korean: focusing on risk factors. Korean J Gastroenterol 2011;57(5):281–7. [Abstract only].

[65] El-Serag HB, Kvapil P, Hacken-Bitar J, Kramer JR. Abdominal obesity and the risk of Barrett's esophagus. Am J Gastroenterol 2005;100(10):2151–6.

[66] Rubenstein JH, Morgenstern H, Appelman H, Scheiman J, Schoenfeld P, McMahon Jr LF, et al. Prediction of Barrett's esophagus among men. Am J Gastroenterol 2013;108(3):353–62.

[67] Otterstatter MC, Brierley JD, De P, Ellison LF, Macintyre M, Marrett LD, et al. Esophageal cancer in Canada: trends according to morphology and anatomical location. Can J Gastroenterol 2012;26(10):723–7.

[68] Turati F, Tramacere I, La Vecchia C, Negri E. A meta-analysis of body mass index and esophageal and gastric cardia adenocarcinoma. Ann Oncol 2013;24 (3):609–17.

[69] O'Doherty MG, Freedman ND, Hollenbeck AR, Schatzkin A, Abnet CC. A prospective cohort study of obesity and risk of oesophageal and gastric adenocarcinoma in the NIH-AARP diet and health study. Gut 2012;61(9):1261–8.

[70] Larsson SC, Wolk A. Obesity and colon and rectal cancer risk: a meta-analysis of prospective studies. Am J Clin Nutr 2007;86(3):556–65.

[71] Ohsawa T, Hirata W, Higichi S. Effects of three H2-receptor antagonists (cimetidine, famotidine,

ranitidine) on serum gastrin level. Int J Clin Pharmacol Res 2002;22(2):29–35.

[72] Iwao T, Toyonaga A, Kuboyama S, Tanikawa K. Effects of omeprazole and lansoprazole on fasting and postprandial serum gastrin and serum pepsinogen A and C. Hepatogastroenterology 1995;42(5):677–82.

[73] Haigh CR, Attwood SE, Thompson DG, Jankowski JA, Kirton CM, Pritchard DM, et al. Gastrin induces proliferation in Barrett's metaplasia through activation of the CCK2 receptor. Gastroenterology 2003;124(3):615–25.

[74] Abdalla SI, Lao-Sirieix P, Novelli MR, Lovat LB, Sanderson IR, Fitzgerald RC. Gastrininduced cyclooxygenase-2 expression in Barrett's carcinogenesis. Clin Cancer Res 2004;10:4784–92.

[75] Harris JC, Clarke PA, Awan A, Jankowski J, Watson SA. An antiapoptotic role for gastrin and the gastrin/CCK-2 receptor in Barrett's esophagus. Cancer Res 2004;64:1915–19.

[76] Wang C, Yuan Y, Hunt RH. Helicobacter pylori infection and Barrett's esophagus: a systematic review and meta-analysis. Am J Gastroenterol 2009;104:492.

[77] Wu AH, Crabtree JE, Bernstein L, Hawtin P, Cockburn M, Tseng CC, et al. Role of Helicobacter pylori CagAş strains and risk of adenocarcinoma of the stomach and esophagus. Int J Cancer 2003;103(6):815–21.

[78] Rokkas T, Pistiolas D, Sechopoulos P, Robotis I, Margantinis G. Relationship between Helicobacter pylori infection and esophageal neoplasia: a meta-analysis. Clin Gastroenterol Hepatol 2007;5:1413.

[79] Ye W, Held M, Lagergren J, Engstrand L, Blot WJ, McLaughlin JK, et al. Helicobacter pylori infection and gastric atrophy: risk of adenocarcinoma and squamous-cell carcinoma of the esophagus and adenocarcinoma of the gastric cardia. J Natl Cancer Inst 2004;96(5):388–96.

[80] Chow WH, Blaser MJ, Blot WJ, Gammon MD, Vaughan TL, Risch HA, et al. An inverse relation between cagA + strains of Helicobacter pylori infection and risk of esophageal and gastric cardia adenocarcinoma. Cancer Res 1998;58(4):588–90.

[81] Jones AD, Bacon KD, Jobe BA, Sheppard BC, Deveney CW, Rutten MJ. Helicobacter pylori induces apoptosis in Barrett's-derived esophageal adenocarcinoma cells. J Gastrointest Surg 2003;7(1):68–76.

[82] Souza RF, Krishnan K, Spechler SJ. Acid, bile, and CDX: the ABCs of making Barrett's metaplasia. Am J Physiol Gastrointest Liver Physiol 2008;295:G211–18.

[83] Pahl HL. Activators and target genes of Rel/NFkappaB transcription factors. Oncogene 1999; 18:6853–66.

[84] Keenan ID, Sharrard RM, Isaacs HV. FGF signal transduction and the regulation of Cdx gene expression. Dev Biol 2006;299:478–88.

[85] van Baal JW, Milano F, Rygiel AM, Bergman JJ, Rosmolen WD, van Deventer SJ, et al. A comparative analysis by SAGE of gene expression profiles of Barrett esophagus, normal squamous esophagus, and gastric cardia. Gastroenterology 2005;129: 1274–81.

[86] Ioannides AS, Henderson DJ, Spitz L, Copp AJ. Role of sonic hedgehog in the development of the trachea and oesophagus. J Pediatr Surg 2003;38:29–36.

[87] Shimizu T, Bae YK, Muraoka O, Hibi M. Interaction of Wnt and caudal-related genes in zebra Wsh posterior body formation. Dev Biol 2005;279:125–41.

[88] Enginger PC, Mayer RJ. Esophageal cancer. N Engl J Med 2003;349:2241–52.

[89] Wani S, Falk G, Hall M, Gaddam S, Wang A, Gupta N, et al. Patients with nondysplastic Barrett's esophagus have low risks for developing dysplasia or esophageal adenocarcinoma. Clin Gastroenterol Hepatol 2011;9:220–7 quiz e26.

[90] Hvid-Jensen F, Pedersen L, Drewes AM, Sørensen HT, Funch-Jensen P. Incidence of adenocarcinoma among patients with Barrett's esophagus. N Engl J Med 2011;365:1375–83.

[91] Bhat S, Coleman HG, Yousef F, Johnston BT, McManus DT, Gavin AT, et al. Risk of malignant progression in Barrett's esophagus patients: results from a large population-based study. J Natl Cancer Inst 2011;103:1049–57.

[92] Lagergren J, Bergstrom R, Lindgren A, Nyrén O. Symptomatic gastroesophageal reflux as a risk factor for esophageal adenocarcinoma. N Engl J Med 1999;340:825–31.

[93] Jung KW, Talley NJ, Romero Y, Katzka DA, Schleck CD, Zinsmeister AR, et al. Epidemiology and natural history of intestinal metaplasia of the gastroesophageal junction and Barrett's esophagus: a population-based study. Am J Gas-troenterol 2011;106:1447–55 quiz 1456.

[94] Hillman LC, Chiragakis L, Shadbolt B, Kaye GL, Clarke AC. Effect of proton pump inhibitors on markers of risk for high-grade dysplasia and oesophageal cancer in Barrett's oesophagus. Aliment Pharmacol Ther 2008;27:321–6.

[95] El-Serag HB, Aguirre TV, Davis S, Kuebeler M, Bhattacharyya A, Sampliner RE. Proton pump inhibitors are associated with reduced incidence of dysplasia in Barrett's esophagus. Am J Gastroenterol 2004;99:1877–83.

[96] de Jonge PJ, Steyerberg EW, Kuipers EJ, Honkoop P, Wolters LM, Kerkhof M, et al. Risk factors for the development of esophageal adenocarcinoma in Barrett's esophagus. Am J Gastroenterol 2006; 101:1421–9.

[97] Singh S, Garg SK, Singh PP, Iyer PG, El-Serag HB. Acid-suppressive medications and risk of oesophageal adenocarcinoma in patients with Barrett's oesophagus: a systematic review and meta-analysis. Gut 2014;63(8):1229–37.

[98] Spechler SJ. Barrett esophagus and risk of esophageal cancer: a clinical review. JAMA 2013;310:627–36.

[99] Nguyen DM, Richardson P, El-Serag HB. Medications (NSAIDs, statins, proton pump inhibitors) and the risk of esophageal ade-nocarcinoma in patients with Barrett's esophagus. Gastroenterology 2010;138:2260–6.

[100] Laroui H, Dalmasso G, Nguyen HT, Yan Y, Sitaraman SV, Merlin D. Drug-loaded nanoparticles targeted to the colon with polysaccharide hydrogel reduce colitis in a mouse model. Gastroenterology 2010;138: 843e1-2–853e1-2.

[101] Yang YX, Lewis JD, Epstein S, Metz DC. Long-term proton pump inhibitor therapy and risk of hip fracture. JAMA 2006;296:2947–53.

[102] Dial S, Delaney JA, Barkun AN, Suissa S. Use of gastric acid-suppressive agents and the risk of community-acquired *Clostridium difficile*-associated disease. JAMA 2005;294:2989–95.

[103] Gray SL, LaCroix AZ, Larson J, Robbins J, Cauley JA, Manson JE, et al. Proton pump inhibitor use, hip fracture, and change in bone mineral density in postmeno-pausal women: results from the Women's Health Initiative. Arch Intern Med 2010;170:765–71.

[104] Corley DA, Kubo A, Zhao W, Quesenberry C. Proton pump inhibitors and histamine-2 receptor antagonists are associated with hip fractures among at-risk patients. Gastroenterology 2010;139:93–101.

[105] Loo VG, Bourgault AM, Poirier L, Lamothe F, Michaud S, Turgeon N, et al. Host and pathogen factors for *Clostridium difficile* infection and colonization. N Engl J Med 2011;365:1693–703.

[106] Food and Drug Administration. U.S. Department of Health and Human Services. <http://www.fda.gov/drugs/drugsafety/postmarketdrugsafetyinformation-forpatientsandproviders/ucm213206>.

[107] Food and Drug Administration. U.S. Department of Health and Human Services. <http://www.fda.gov/drugs/drugsafety/ucm290510.htm>.

[108] Hillman LC, Chiragakis L, Shadbolt B, Kaye GL, Clarke AC. Proton-pump inhibitor therapy and the development of dysplasia in patients with Barrett's oesophagus. Med J Aust 2004;180:387–91.

[109] Sharaiha RZ, Freedberg DE, Abrams JA, Wang YC. Cost-effectiveness of chemoprevention with proton pump inhibitors in Barrett's esophagus. Dig Dis Sci 2014;59(6):1222–30.

[110] Nguyen DM, El-Serag HB, Henderson L, Stein D, Bhattacharyya A, Sampliner RE. Medication usage and the risk of neoplasia in patients with Barrett's esophagus. Clin Gastroenterol Hepatol 2009;7 (12):1299–304.

[111] Cooper BT, Chapman W, Neumann CS, Gearty JC. Continuous treatment of Barrett's oesophagus patients with proton pump inhibitors up to 13 years: observations on regression and cancer incidence. Aliment Pharmacol Ther 2006;23(6):727–33.

[112] Wilson KT, Fu S, Ramanujam KS, Meltzer SJ. Increased expression of inducible nitric oxide synthase and cyclooxygenase-2 in Barrett's esophagus and associated adenocarcinomas. Cancer Res 1998;58:2929–34.

[113] Falk GW, Buttar NS, Foster NR, Ziegler KL, Demars CJ, Romero Y, et al. A combination of esomeprazole and aspirin reduces tissue concentrations of prostaglandin E(2) in patients with Barrett's esophagus. Gastroenterology 2012;143 917–26.e1.

[114] Omer ZB, Ananthakrishnan AN, Nattinger KJ, Cole EB, Lin JJ, Kong CY, et al. Aspirin protects against Barrett's esophagus in a multivariate logistic regression analysis. Clin Gastroenterol Hepatol 2012;10:722.

[115] Corley DA, Kerlikowske K, Verma R, Buffler P. Protective association of aspirin/NSAIDs and esophageal cancer: a systematic review and meta-analysis. Gastroenterology 2003;124:47.

[116] Abnet CC, Freedman ND, Kamangar F, Leitzmann MF, Hollenbeck AR, Schatzkin A. Non-steroidal anti-inflammatory drugs and risk of gastric and oesophageal adenocarcinomas: results from a cohort study and a meta-analysis. Br J Cancer 2009;100:551.

[117] Souza RF, Shewmake K, Beer DG, Cryer B, Spechler SJ. Selective inhibition of cyclooxygenase-2 suppresses growth and induces apoptosis in human esophageal adenocarcinoma cells. Cancer Res 2000;60:5767.

[118] Kastelein F, Spaander MC, Biermann K, Steyerberg EW, Kuipers EJ, Bruno MJ, et al. Nonsteroidal anti-inflammatory drugs and statins have chemopreventative effects in patients with Barrett's esophagus. Gastroenterology 2011;141:2000.

[119] Zhang S, Zhang XQ, Ding XW, Yang RK, Huang SL, Kastelein F, et al. Cyclooxygenase inhibitors use is associated with reduced risk of esophageal adenocarcinoma in patients with Barrett's esophagus: a meta-analysis. Br J Cancer 2014;110:2378.

[120] Heath EI, Canto MI, Wu TT, Piantadosi S, Hawk E, Unalp A, et al. Chemoprevention for Barrett's esophagus trial. Design and outcome measures. Dis Esophagus 2003;16:177.

[121] Souza RF, Spechler SJ. Barrett's esophagus: chemo-prevention. Gastrointest Endosc Clin N Am 2003;13:419.

[122] Heath EI, Canto MI, Piantadosi S, Montgomery E, Weinstein WM, Herman JG, et al. Secondary chemo-prevention of Barrett's esophagus with celecoxib: results of a randomized trial. J Natl Cancer Inst 2007;99:545.

[123] Bennett C, Vakil N, Bergman J, Harrison R, Odze R, Vieth M, et al. Consensus statements for management of Barrett's dysplasia and early-stage esophageal adenocarcinoma, based on a Delphi process. Gastroenterology 2012;143:336.

[124] Hur C, Nishioka NS, Gazelle GS. Cost-effectiveness of aspirin chemoprevention for Barrett's esophagus. J Natl Cancer Inst 2004;96:316−25.

[125] Choi SE, Perzan KE, Tramontano AC, Kong CY, Hur C. Statins and aspirin for chemoprevention in Barrett's esophagus: results of a cost-effectiveness analysis. Cancer Prev Res (Phila) 2014;7:341−50.

[126] Ojima E, Fujimura T, Oyama K, Tsukada T, Kinoshita J, Miyashita T, et al. Chemoprevention of esophageal adenocarcinoma in a rat model by ursodeoxycholic acid. Clin Exp Med 2014;18.

[127] Peng S, Huo X, Rezaei D, Zhang Q, Zhang X, Yu C, et al. In Barrett's esophagus patients and Barrett's cell lines, ursodeoxycholic acid increases antioxidant expression and prevents DNA damage by bile acids. Am J Physiol Gastrointest Liver Physiol 2014;307(2): G129−39.

[128] Blais L, Desgagne A, Lelorier J. 3-Hydroxy-3-methylglutaryl coenzyme A reductase inhibitors and the risk of cancer: a nested case−control study. Arch Intern Med 2000;160(15):2363−8.

[129] Karp I, Behlouli H, Lelorier J, Pilote L. Statins and cancer risk. Am J Med 2008;121(4):302−9.

[130] Manson JE, Willett WC, Stampfer MJ, Colditz GA, Hunter DJ, Hankinson SE, et al. Body weight and mortality among women. N Engl J Med 1995;333:677−85.

[131] Stevens J, Cai J, Pamuk ER, Williamson DF, Thun MJ, Wood JL. The effect of age on the association between body-mass index and mortality. N Engl J Med 1998;338:1−7.

[132] Bianchini F, Kaaks R, Vainio H. Overweight, obesity, and cancer risk. Lancet Oncol 2002;3:565−74.

[133] Bergstrom A, Pisani P, Tenet V, Wolk A, Adami HO. Overweight as an avoidable cause of cancer in Europe. Int J Cancer 2001;91:421−30.

[134] Parkin DM, Bray F, Ferlay J, Pisani P. Global cancer statistics, 2002. CA Cancer J Clin 2005;55:74−108.

[135] Vizcaino AP, Moreno V, Lambert R, Parkin DM. Time trends incidence of both major histologic types of esophageal carcinomas in selected countries, 1973−1995. Int J Cancer 2002;99:860−8.

[136] Pohl H, Welch HG. The role of overdiagnosis and reclassification in the marked increase of esophageal adenocarcinoma incidence. J Natl Cancer Inst 2005;97:142−6.

[137] Brown LM, Devesa SS, Chow WH. Incidence of ade-nocarcinoma of the esophagus among white Americans by sex, stage, and age. J Natl Cancer Inst 2008;100:1184−7.

[138] Bird-Lieberman EL, Fitzgerald RC. Early diagnosis of oesophageal cancer. Br J Cancer 2009;101:1−6.

[139] Edelstein ZR, Farrow DC, Bronner MP, Rosen SN, Vaughan TL. Central adiposity and risk of Barrett's esophagus. Gastroenterology 2007;133:403−11.

[140] Brown LM, Swanson CA, Gridley G, Swanson GM, Schoenberg JB, Greenberg RS, et al. Adenocarcinoma of the esophagus: role of obesity and diet. J Natl Cancer Inst 1995;87:104−9.

[141] Whiteman DC, Sadeghi S, Pandeya N, Smithers BM, Gotley DC, Bain CJ, et al. Combined effects of obe-sity, acid reflux and smoking on the risk of adenocar-cinomas of the oesophagus. Gut 2008;57:173−80.

[142] Ogunwobi OO, Beales IL. Statins inhibit proliferation and induce apoptosis in Barrett's esophageal adeno-carcinoma cells. Am J Gastroenterol 2008;103 (4):825−37.

[143] Masclee GM, Coloma PM, Spaander MC, Kuipers EJ, Sturkenboom MC. NSAIDs, statins, low-dose aspirin and PPIs, and the risk of oesophageal adenocarci-noma among patients with Barrett's oesophagus: a populationbased case−control study. BMJ Open 2015;5:e006640.

[144] Agrawal S, Patel P, Agrawal A, Makhijani N, Markert R, Deidrich W. Metformin use and the risk of esophageal cancer in Barrett esophagus. South Med J 2014;107(12):774−9.

[145] Siewert JR, Stein HJ. Classification of adenocarci-noma of the oesophagogastric junction. Br J Surg 1998;85:1457−9.

[146] Chak A, Buttar NS, Foster NR, Seisler DK, Marcon NE, Schoen R, et al. Metformin does not reduce mar-kers of cell proliferation in esophageal tissues of patients with Barrett's esophagus. Clin Gastroenterol Hepatol 2015;13(4) 665−72.e1−4.

Posttreatment Surveillance, Risk for Recurrence of Barrett's Esophagus, and Adenocarcinoma After Treatment

George Triadafilopoulos

Stanford University School of Medicine, Stanford, CA, United States

15.1 INTRODUCTION

Endoscopic eradication therapy (EET), either using endoscopic resection (ER) or radiofrequency ablation (RFA), cryoablation, or a combination of resection and ablation, aims not only to completely eliminate Barrett's esophagus (BE) dysplasia and early neoplasia but also a complete eradication of intestinal metaplasia (CEIM) [1]. EET is mostly applied in patients with high-grade dysplasia (HGD) and intramucosal cancer (IMC) and increasingly in patients with BE and low-grade dysplasia (LGD). In general, a successful EET is based on the ability to accurately diagnose the disease burden and extent, remove the responsible endoscopically recognizable lesion or lesions, successfully treat surrounding BE epithelium, and effectively survey for recurrent intestinal metaplasia (IM), dysplasia, or malignancy after the initial effective therapy.

Most studies on EET of BE with advanced histopathology have shown a real, albeit small, risk of cancer progression. In a recent US multicenter study, 2- and 3-year follow-ups found that 5 of 119 patients (4.2%) progressed with two patients progressing to cancer. In this study, the rate of esophageal adenocarcinoma (EAC) was 1 per 181 patient-years (0.55% per patient-years) with no cancer-related morbidity or mortality, while the annual rate of any neoplastic progression was 1 per 73 patient-years (1.37% per patient-years). Therefore, lifelong intensive surveillance is mandatory in patients treated with EET, but the exact surveillance intervals have not been well established [2]. Hence, although EET reduces the risk of progression to invasive cancer, it does not completely eliminate it. Broad and detailed discussion of all available alternative therapies, including esophagectomy, as well as a commitment to lifelong treatment of gastroesophageal acid reflux disease (GERD) and endoscopic surveillance are essential and they are best managed in multidisciplinary and experienced referral centers.

D. Pleskow & T. Erim (Eds): Barrett's Esophagus.
DOI: http://dx.doi.org/10.1016/B978-0-12-802511-6.00015-6

In a similar fashion, esophagectomy aims at complete removal of early neoplasia and the entire BE length. The premise of these approaches is the complete elimination of the risk of EAC occurrence and mortality during the lifetime of the patient (Fig. 15.1). Several studies suggest an enhanced patient survival if early esophageal neoplasia is detected by endoscopic surveillance as compared with symptomatic detection. However, the rates and predictors of BE recurrence after CEIM have not been well defined.

The purpose of this chapter is to review the existing evidence and propose management strategies that would be applicable to the everyday care of patients with BE after they have received endoscopic or surgical eradication. It is important to note that CEIM is a key initial outcome that requires, at present, validation by endoscopic and histologic assessments (Fig. 15.2). Advanced imaging methods, such as narrow band imaging (NBI; Olympus), Fuji intelligent color enhancement (FICE), or i-scan (Pentax), confocal laser endomicroscopy, endocytoscopy, and optical coherence tomography (OCT), are increasingly utilized in the assessment of patients with BE and they are poised to be the methods of choice in the determination of CEIM [3,4].

A key issue is the durability of the treatment effect and the longer term outcomes of therapy. Because durability of treatment effect is a determinant of the cost effectiveness of the intervention and because subjects with recurrent BE after EET are at continued risk for developing EAC, it is vital to know whether the neo-squamous epithelium present after EET is durable. Hence, the posttreatment surveillance and the risk for recurrence of BE and EAC after EET or surgery start with the determination of CEIM, usually 3 months postintervention. In the context of this posttreatment endoscopy, it is important not only to carefully assess the neo-squamous epithelium by white light and advanced imaging, but also to obtain careful random and directed biopsies to evaluate for overt or buried metaplasia or dysplasia [5]. Compared to conventional random biopsies, the use of Wide Area Transepithelial Sampling with 3D computer-assisted analysis (WATS-3D) promises to enhance the yield for intestinal metaplasia and dysplasia by up to 60% [6]. In addition, acid suppressive therapy, using proton pump inhibitors (PPIs) and preferably adjusted using ambulatory pH monitoring, is mandatory. Alternatively, antireflux surgery may be required to control esophageal pH and symptoms of GERD, if any.

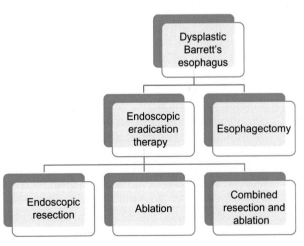

FIGURE 15.1 Options in the management of dysplastic Barrett's esophagus. These options need to be discussed in detail with the patient and a long-term commitment to therapy and surveillance is established in conjunction with aggressive proton pump inhibition therapy or antireflux surgery to control gastroesophageal acid reflux disease.

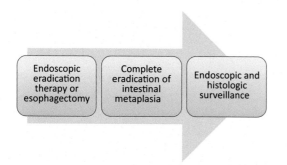

FIGURE 15.2   Key steps in the management of dysplastic Barrett's esophagus. After endoscopic or surgical therapy, confirmation of complete eradication of intestinal metaplasia (CEIM) is essential prior to embarking in long-term surveillance. If CEIM is not achieved, further endoscopic eradication is required.

15.2 SURVEILLANCE AND RISK FOR RECURRENCE AFTER ABLATION

15.2.1 Radiofrequency Ablation

Recent evidence supports the use of RFA as an effective and safe therapeutic modality for patients with dysplastic BE, but in all prospective cohort studies, repeated RFA treatments were needed during follow-up surveillance (Fig. 15.3).

The US RFA registry is a nationwide registry of BE patients who have been receiving RFA and was aimed to determine rates and factors that would predict recurrence of IM. In a recent study of that cohort, 85% of patients achieved CEIM. After an average of 2.4 years, IM recurred in 20% and was nondysplastic or indefinite for dysplasia in 86%. The average length of recurrent BE was 0.6 cm. They also noted that the yearly recurrence rate was worse with higher pretreatment histology, increasing age, BE length, and non-Caucasian race [7].

Another single-center retrospective study evaluated the recurrence and progression rates of patients who had completed RFA for dysplastic BE or intramucosal carcinoma (IMC). The cohort consisted of 231 patients who had

achieved CEIM and underwent subsequent surveillance for a median observation time of 397 days. BE recurred at a rate of 5.2% per year, while disease progression occurred in 1.9% per year. The authors found no clinical characteristics to be associated with disease recurrence [8]. These studies suggest that patients who are undergoing RFA for dysplastic BE should be retained in endoscopic surveillance after CEIM (Fig. 15.4).

15.2.2 Cryoablation

Cryoablation is another endoscopic technique that applies cold nitrogen or carbon dioxide gas to freeze the BE surface. In a retrospective, nonrandomized study of 60 patients with dysplastic BE, 97% had complete eradication of HGD, 87% had complete eradication of all dysplasia, and 57% had CEIM. Buried metaplasia was found in 3%. However, the study had short follow-up of 10.5 months [9]. In another smaller study, 32 patients with BE and HGD were treated with cryoablation. At 2 years of follow-up, 100% had complete eradication of HGD and 84% had CEIM. HGD recurred in 19%; those patients were treated with repeat cryotherapy or argon plasma coagulation and were HGD free on follow-up ranging 24–57 months. One patient progressed to adenocarcinoma, but after cryotherapy downgraded to HGD [10].

15.3 SURVEILLANCE AND RISK FOR RECURRENCE AFTER ENDOSCOPIC RESECTION

ER of BE, either by endoscopic mucosal resection (EMR) or endoscopic submucosal dissection (ESD), allows for removal of visible lesions and accurate histopathological staging. If the lesion is confined to the mucosa and the resection margins are clear, the intervention is considered curative because of the very low risk of lymph node involvement. In contrast, if

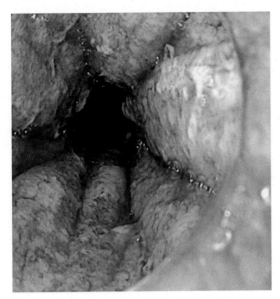 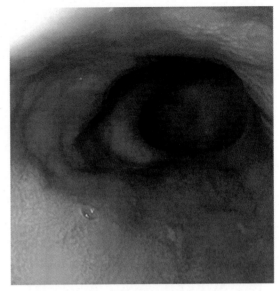

FIGURE 15.3 *Left*: Endoscopic appearance of Barrett's esophagus (BE) immediately after radiofrequency ablation. *Right*: Complete eradication of intestinal metaplasia confirmed histologically 2 months after ablation. There is no visible BE and random biopsies showed no metaplasia.

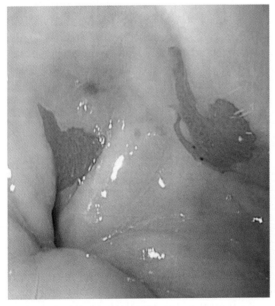 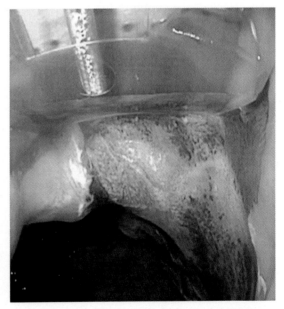

FIGURE 15.4 *Left*: Islands of endoscopically visible Barrett's esophagus (BE) seen 2 months after radiofrequency ablation of long segment BE. Histology showed residual metaplasia. *Right*: Radiofrequency ablation of these islands of metaplasia.

submucosal invasion is found, patients should be referred for esophagectomy because of high risk for metastasis [11]. Although not proven in a randomized prospective fashion in patients with early BE neoplasia, ESD is generally perceived as providing larger specimens than EMR and hence higher *en bloc* curative resection rates potentially reducing the likelihood of recurrence [12].

15.3.1 Endoscopic Resection

ER is the treatment of choice in HGD and early mucosal cancer in BE because the risk for lymph node and distant metastasis is almost absent. Stepwise radical endoscopic resection (SRER) of the complete BE has the advantage of complete removal of all metaplasia and histological correlation. Since the resection involves the complete thickness of the mucosa and the upper part of the submucosa, SRER may be associated with less, if any, buried metaplasia and a more sustained treatment response during follow-up [13]. Large cohort series from several groups have confirmed the long-term safety and efficacy of this approach. In most studies, patients were closely followed every 3 months up to 2 years and annually thereafter. Such follow-up consisted of upper endoscopy with biopsies of all visible lesions, biopsies from the resection margins, and four-quadrant biopsies every 1–2 cm. Endoscopic ultrasound and CT were also used every 6 months for 2 years. Complete remission was assumed if the malignant lesion was completely resected and a follow-up examination was negative (Fig. 15.5). Aggressive control of the esophageal pH with PPI was also sustained and, in many centers, adjusted by pH monitoring [14].

15.3.2 Submucosal Dissection

Clinical experience with ESD is limited to Asian studies with small number of patients.

A recent, single-center, retrospective study from Japan evaluated 23 patients (21 men, 2 women; mean age, 63 years) with 26 superficial Barrett's adenocarcinomas. The endoscopic en bloc resection rate was 100% and the pathological en bloc resection rate was 85% (22/26). Bleeding after ESD occurred in 1 case (4%), and it was controlled using electrocoagulation. No perforation occurred. Four cases (15%) developed esophageal strictures that were treated by balloon dilation. In cases of complete resection, surveillance endoscopy was performed 12 months after ESD and once every 12 months thereafter. In cases of incomplete resection, endoscopic examination was performed 3, 6, and 12 months after the procedure during the first year, and every 12 months thereafter. A tumor detected in close proximity to the scar resulting from ESD was regarded as a local recurrent tumor, whereas one or more primary tumors detected more than 1 cm from the ESD scar were regarded as metachronous tumors. CT scanning was performed annually to detect any lymph node or other organ metastasis. There were no recurrent tumors and no residual tumor was noted in the nine cases where surgical resection was carried out after ESD [15].

In a Belgian retrospective analysis of 75 patients with BE, ESD was performed for visible lesions that were multiple, larger than 15 mm, poorly lifting, or suspected to have submucosal infiltration. In this series, the en bloc resection rate was 90% and curative resection rates of carcinoma and HGD or carcinoma were 85% (47/55) and 64% (42/66), respectively. In this study, endoscopic follow-up was scheduled at 2 weeks for all patients to check for postoperative stricture and perform dilation as needed. Sometimes, self-expandable stent placement was performed. When necessary, subsequent treatments aimed at removing residual BE were administered during follow-up endoscopic procedures at 6, 12, and 18 months, then yearly if residual BE was still

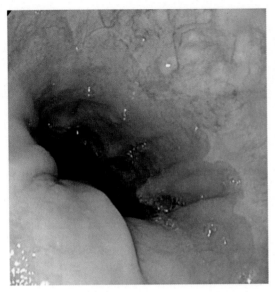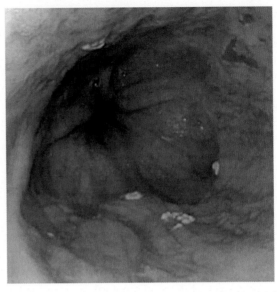

FIGURE 15.5   *Left*: Endoscopic closeup view of the distal esophagus and esophagogastric junction 4 months after endoscopic resection of Barrett's esophagus containing high-grade dysplasia. No visible metaplasia is seen. *Right*: Endoscopic view of the distal esophagus 1 year later reveals islands of intestinal metaplasia.

present. Enhanced imaging using high-resolution video endoscopy combined with narrow band imaging and acetic acid chromoendoscopy were used. Targeted biopsies from both the columnar epithelium and random areas just underneath the neo-squamo-columnar junction were also performed. Medial follow-up was 20 months and complete remission of neoplasia and intestinal metaplasia was found in 92% (54/59) and 73% (43/59) of patients, respectively [16].

15.4 SURVEILLANCE AND RISK FOR RECURRENCE AFTER COMBINATION OF RESECTION AND ABLATION

Combined mucosal resection and RFA is an established treatment for dysplastic BE and is currently the preferred strategy by most groups around the world. Most resection procedures are performed using the multiband mucosectomy method with or without saline injection lift. For suspicious lesions that are greater than 3 cm in diameter, most groups use the cap technique after saline injection lift. In order to aggressively control acid reflux and optimize the efficacy of ablation, patients are maintained on twice daily doses of PPIs. Patients then return for repeat EET every 3–6 months until all esophageal glandular mucosa had been successfully ablated and replaced with neo-squamous epithelium. Surveillance endoscopy is performed every 6–12 months thereafter to detect and treat any recurrent BE using additional ER or ablation [17].

Several studies have assessed the incidence and factors that predict the recurrence of IM after successful initial therapy. In one study of 592 patients with BE treated at 3 tertiary referral centers, 55% of patients underwent EMR before RFA. Before RFA, 71% of patients had HGD or EAC, 15% had LGD, and 14% had nondysplastic BE. Of patients treated, 448 (76%) were assessed after RFA. At 24 months, the incidence of recurrence was 33%; 22% of all recurrences observed were dysplastic BE. There were no demographic or endoscopic factors associated with recurrence. Most recurrences were nondysplastic and manageable using endoscopy [18].

A Dutch study followed patients who received RFA for BE containing high-grade intraepithelial neoplasia and/or early stage cancer for 5 years to determine the durability of treatment response. The authors followed 54 patients with BE (2–12 cm) first underwent focal ER in case of visible lesions (n = 40, 72%), followed by serial RFA every 3 months. Patients underwent high-resolution endoscopy with NBI at 6 and 12 months after treatment and then annually. After 5 years, Kaplan–Meier analysis showed sustained complete remission of neoplasia and intestinal metaplasia in 90% of patients; neoplasia recurred in 3 patients and was managed endoscopically. Focal IM in the cardia was found in 19 of 54 patients (35%), in 53 of 1143 gastric cardia biopsies (4.6%). Buried glands were detected in 3 of 3543 neosquamous epithelium biopsies (0.08%, from 3 patients) [19].

A UK registry examined 335 patients with BE and neoplasia (72% with HGD, 24% with IMC, 4% with LGD) who were treated with RFA and EMR of nodules if necessary. Dysplasia was cleared in 81% of patients, and BE in 62% at 12 months. Shorter BE segments responded better to ablation. EMR before RFA did not provide any benefit with respect to clearance of dysplasia. Invasive cancer developed in 3% of patients by 12 months and disease progression was noted 5.1% after a median follow-up time of 19 months. Symptomatic strictures developed in 9% of patients and were treated by endoscopic dilation. Overall, 19 months after therapy, 94% of patients remained clear of dysplasia [20].

15.5 SURVEILLANCE AND RISK FOR RECURRENCE AFTER ESOPHAGECTOMY

Patients who undergo esophagectomy and reconstruction with a gastric conduit are prone to reflux of both acid and duodenal contents that compromise their quality of life [21]. As a consequence of the disruption of the normal anatomical antireflux mechanisms, most patients after esophagectomy report reflux symptoms and exhibit pathological levels of reflux in the esophageal remnant [22]. The lower esophageal sphincter, angle of His, and diaphragmatic sling are all resected while the position of the gastric tube between the positive pressure environment of the abdomen and the negative pressure of the chest promotes reflux. Many surgeons perform routine pyloroplasty in order to facilitate gastric emptying but this may promote duodenal (bile) reflux [23].

There is clear evidence that BE occurs in a significant proportion of such patients raising concerns about the long-term fate of the esophageal remnant following surgery. Although data on the time to develop postoperative BE is inconsistent, the risk of malignant progression appears to be small. Since surgical or endoscopic resection or ablation of malignancy occurring within postesophagectomy BE is possible, regular endoscopic surveillance is indicated in postesophagectomy patients [24] (Fig. 15.6). Although the number of patients with post-esophagectomy gastric tube cancers is expected to increase because of their improved survival, the number of patients that will benefit is extremely small [25].

15.6 SUMMARY AND FUTURE DIRECTIONS

Because of the possibilities for biopsy sampling errors and persistence of buried

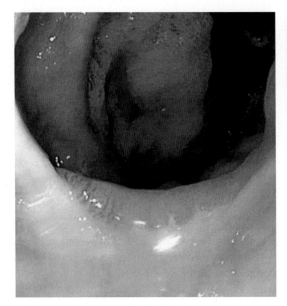 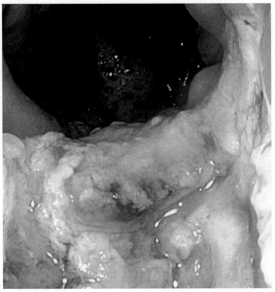

FIGURE 15.6   *Left*: Endoscopically visible Barrett's esophagus at the level of esophagogastrostomy 6 months after an Ivor—Lewis esophagectomy for esophageal adenocarcinoma. Histology revealed high-grade dysplasia. *Right*: Endoscopic appearance immediately after radiofrequency ablation using the Barrx Channel RFA Endoscopic Catheter.

metaplasia, patients receiving EET or esophagectomy for dysplastic BE should be kept in surveillance until definitive long-term, prospective data become available. Currently, it is unclear if EET can maintain lifetime remission of dysplasia and metaplasia in all patients who have achieving CEIM. It is hoped that in the future, biomarkers will facilitate risk stratification of these patients [26].

References

[1] Subramanian CR, Triadafilopoulos G. Endoscopic treatments for dysplastic Barrett's esophagus: resection, ablation, what else? World J Surg 2015;39(3):597—605.

[2] Shaheen NJ, Overholt BF, Sampliner RE, Wolfsen HC, Wang KK, Fleischer DE, et al. Durability of radiofrequency ablation in Barrett's esophagus with dysplasia. Gastroenterology 2011;141(2):460—8.

[3] Goda K, Kato T, Tajiri H. Endoscopic diagnosis of early Barrett's neoplasia: perspectives for advanced endoscopic technology. Dig Endosc. 2014;26(3):311—21.

[4] Boerwinkel DF, Swager A, Curvers WL, Bergman JJ. The clinical consequences of advanced imaging techniques in Barrett's esophagus. Gastroenterology 2014;146(3) 622—9.e42.

[5] Pouw RE, Visser M, Odze RD, Sondermeijer CM, ten Kate FJ, Weusten BL, et al. Pseudo-buried Barrett's post radiofrequency ablation for Barrett's esophagus, with or without prior endoscopic resection. Endoscopy 2014;46(2):105—9.

[6] Kataria R, Thomas R, Smith MS. Wide area transepithelial sampling (Wats3D) detects Barrett's metaplasia missed by forceps biopsies after ablation of short and long segment disease. Gastroenterology 2013;144(5) Suppl. 1, p. S-691

[7] Pasricha S, Bulsiewicz WJ, Hathorn KE, Komanduri S, Muthusamy VR, Rothstein RI, et al. Durability and predictors of successful radiofrequency ablation for Barrett's esophagus. Clin Gastroenterol Hepatol 2014;12 (11) 1840—7.e1.

[8] Orman ES, Kim HP, Bulsiewicz WJ, Cotton CC, Dellon ES, Spacek MB, et al. Intestinal metaplasia recurs infrequently in patients successfully treated for Barrett's esophagus with radiofrequency ablation. Am J Gastroenterol 2013;108(2):187—95. quiz 196.

[9] Shaheen NJ, Greenwald BD, Peery AF, Dumot JA, Nishioka NS, Wolfsen HC, et al. Safety and efficacy of endoscopic spray cryotherapy for Barrett's esophagus with high-grade dysplasia. Gastrointest Endosc 2010;71 (4):680—5.

[10] Gosain S, Mercer K, Twaddell WS, Uradomo L, Greenwald BD. Liquid nitrogen spray cryotherapy in Barrett's esophagus with high-grade dysplasia: long-term results. Gastrointest Endosc 2013;78(2):260−5.

[11] Chennat J, Konda VJ, Ross AS, de Tejada AH, Noffsinger A, Hart J, et al. Complete Barrett's eradication endoscopic mucosal resection: an effective treatment modality for high-grade dysplasia and intramucosal carcinoma—an American single-center experience. Am J Gastroenterol 2009;104(11):2684−92.

[12] Cao Y, Liao C, Tan A, Gao Y, Mo Z, Gao F. Meta-analysis of endoscopic submucosal dissection versus endoscopic mucosal resection for tumors of the gastrointestinal tract. Endoscopy 2009;41(9):751−7.

[13] Peters FP, Kara MA, Rosmolen WD, ten Kate FJ, Krishnadath KK, van Lanschot JJ, et al. Stepwise radical endoscopic resection is effective for complete removal of Barrett's esophagus with early neoplasia: a prospective study. Am J Gastroenterol 2006;101 (7):1449−57.

[14] Ell C, May A, Gossner L, Pech O, Günter E, Mayer G, et al. Endoscopic mucosal resection of early cancer and high-grade dysplasia in Barrett's esophagus. Gastroenterology 2000;118(4):670−7.

[15] Kagemoto K, Oka S, Tanaka S, Miwata T, Urabe Y, Sanomura Y, et al. Clinical outcomes of endoscopic submucosal dissection for superficial Barrett's adenocarcinoma. Gastrointest Endosc 2014;80(2):239−45.

[16] Chevaux JB, Piessevaux H, Jouret-Mourin A, Yeung R, Danse E, Deprez PH. Clinical outcome in patients treated with endoscopic submucosal dissection for superficial Barrett's neoplasia. Endoscopy 2015;47(2):103−12.

[17] Qumseya BJ, Panossian AM, Rizk C, Cangemi DJ, Wolfsen C, Raimondo M, et al. Survival in esophageal high-grade dysplasia/adenocarcinoma post endoscopic resection. Dig Liver Dis 2013;45(12):1028−33.

[18] Gupta M, Iyer PG, Lutzke L, Gorospe EC, Abrams JA, Falk GW, et al. Recurrence of esophageal intestinal metaplasia after endoscopic mucosal resection and radiofrequency ablation of Barrett's esophagus: results from a US Multicenter Consortium. Gastroenterology 2013;145(1) 79−86.e1.

[19] Phoa KN, Pouw RE, van Vilsteren FG, Sondermeijer CM, Ten Kate FJ, Visser M, et al. Remission of Barrett's esophagus with early neoplasia 5 years after radiofrequency ablation with endoscopic resection: a Netherlands cohort study. Gastroenterology 2013;145 (1):96−104.

[20] Haidry RJ, Dunn JM, Butt MA, Burnell MG, Gupta A, Green S, et al. Radiofrequency ablation and endoscopic mucosal resection for dysplastic barrett's esophagus and early esophageal adenocarcinoma: outcomes of the UK National Halo RFA Registry. Gastroenterology 2013;145(1):87−95.

[21] Dunn LJ, Shenfine J, Griffin SM. Columnar metaplasia in the esophageal remnant after esophagectomy: a systematic review. Dis Esophagus 2015;28(1):32−41.

[22] Aly A, Jamieson G. Reflux after oesophagectomy. Br J Surg 2004;91:137−41.

[23] D'Journo X, Martin J, Ferraro P, Duranceau A. The esophageal remnant after gastric interposition. Dis Esophagus 2008;21:377−88.

[24] da Rocha J, Ribeiro U, Sallum R, Szachnowicz S, Cecconello I. Barrett's esophagus (BE) and carcinoma in the esophageal stump (ES) after esophagectomy with gastric pull-up in achalasia patients: a study based on 10 years follow-up. Ann Surg Oncol 2008;15:2903−9.

[25] Hanif F, Kerr J, Going JJ, Fullarton G. Gastric neo-adenocarcinoma arising in a gastric tube after Ivor Lewis oesophagectomy for oesophageal adenocarcinoma. Scott Med J 2015; pii: 0036933015570520

[26] Levine DM, Ek WE, Zhang R, Liu X, Onstad L, Sather C, et al. A genome-wide association study identifies new susceptibility loci for esophageal adenocarcinoma and Barrett's esophagus. Nat Genet 2013;45(12):1487−93.

Index

Note: Page numbers followed by "*f*" and "*t*" refer to figures and tables, respectively.

Printed in the United States
By Bookmasters